CLINICAL ANATOMY

199

m.

LED.

Due 11 a.m.

To my wife and late parents

Clinical Anatomy

A revision and applied anatomy for clinical students

HAROLD ELLIS

CBE, MA, MCh, FRCS, FRCOG, FACS (Hon)
Clinical Anatomist, United Medical and Dental Schools
of Guy's and St Thomas's Hospitals;
Emeritus Professor of Surgery, Charing Cross
and Westminster Medical School, London;
Formerly Examiner in Anatomy, Primary FRCS (Eng)

NINTH EDITION

Blackwell
Science

© 1960, 1962, 1966, 1969, 1971, 1977,
1983, 1992, 1997 by
Blackwell Science Ltd
Editorial Offices:
Osney Mead, Oxford OX2 0EL
25 John Street, London WC1N 2BL
23 Ainslie Place, Edinburgh EH3 6AJ
350 Main Street, Malden
  MA 02148-5018, USA
54 University Street, Carlton
  Victoria 3053, Australia

Other Editorial Offices:
Arnette Blackwell SA
  224, Boulevard Saint Germain
  75007 Paris, France

Blackwell Wissenschafts-Verlag GmbH
  Kurfürstendamm 57
  10707 Berlin, Germany

  Zehetnergasse 6
  A-1140 Wien
  Austria

First published 1960
Second edition 1962
Reprinted 1963
Third edition 1966
Fourth edition 1969
Fifth edition 1971
Reprinted 1972, 1973, 1974
Sixth edition 1977
Reprinted 1978, 1980
Seventh edition 1983
Revised reprint 1986
Eighth edition 1992
Ninth edition 1997

Greek edition 1969

Set by Excel Typesetters, Hong Kong
Printed and bound in Italy
by G. Canale & SpA, Turin

DISTRIBUTORS

Marston Book Services Ltd
PO Box 269
Abingdon
Oxon OX14 4YN
(*Orders*: Tel: 01235 465500
          Fax: 01235 465555)

USA
Blackwell Science, Inc.
Commerce Place
350 Main Street
Malden, MA 02148-5018
(*Orders*: Tel: 800 759 6102
          617 388 8250
          Fax: 617 388 8255)

Canada
Copp Clark Professional
200 Adelaide St West, 3rd Floor
Toronto, Ontario M5H 1W7
(*Orders*: Tel: 416 597 1616
          800 815 9417
          Fax: 416 597 1617)

Australia
Blackwell Science Pty Ltd
54 University Street
Carlton, Victoria 3053
(*Orders*: Tel: 3 9347 0300
          Fax: 3 9347 5001)

A catalogue record for this title
is available from the British Library

ISBN 0-86542-882-4 (BSL)
    0-86542-694-5 (IE)

Library of Congress
Cataloging-in-publication Data
Ellis, Harold, 1926–
    Clinical anatomy: a revision and
applied anatomy for clinical students/
Harold Ellis.—9th ed.
      p.   cm.
    Includes bibliographical references
and index.
    ISBN 0-86542-882-4
    1. Human anatomy. I. Title.
    [DNLM: 1. Anatomy.   QS 4 E47c
1997]
QM23.2.E43   1997
611—dc20
DNLM/DLC
for Library of Congress   96-33109
                                     CIP

Contents

Preface

Experience of teaching clinical students at five medical schools and of examining them in fifteen cities and in eight countries has convinced me that there is still an unfortunate hiatus between the anatomy which the student learns in the pre-clinical years and that which is later encountered in the wards and operating theatres.

This book attempts to counter this situation. It does so by highlighting those features of anatomy which are of clinical importance, in radiology, pathology, medicine and midwifery as well as in surgery. It presents the facts which students might reasonably be expected to carry with them during their years on the wards, through their final examinations and into their postgraduate years; it is designed for the clinical student.

Anatomy is a vast subject and, therefore, in order to achieve this goal, I have deliberately carried out a rigorous selection of material so as to cover only those of its thousands of facts which I consider form the necessary anatomical scaffolding for the clinician. Wherever possible practical applications are indicated throughout the text — they cannot, within the limitations of a book of this size, be exhaustive, but I hope that they will act as signposts to the student and indicate how many clinical phenomena can be understood and remembered on simple anatomical grounds.

In this ninth edition a complete revision of the text has been carried out. Many pages have been re-drafted, new figures have been added and many other illustrations modified. Representative computerized axial tomography and magnetic resonance imaging films have been included, since these techniques have given increased impetus to the clinical importance of topographical anatomy.

The continued success of this volume, now in its thirty-seventh year of publication, owes much to the helpful comments which the author has received from readers all over the world. Every suggestion is given the most careful consideration in an attempt to keep the material abreast of the needs of today's medical students.

Harold Ellis

Acknowledgements

I wish to thank the many students who have sent suggestions to me, many of which have been incorporated into this new edition.

Professor Martin Berry of the Department of Anatomy at UMDS reviewed the section on the CNS and Mr Roger Parker FRCS gave much helpful advice. Dr Sheila Rankin, of the Department of Radiology at Guy's Hospital, provided the new scans while the new and revised illustrations were skilfully produced by Jane Fallows. To Mrs Katherine Ellis go my grateful thanks for invaluable secretarial assistance.

I am grateful to the following authors for permission to reproduce illustrations:

The late Lord Brock for Figs 20 and 21 (from *Lung Abscess*);

Professor R. G. Harrison for Figs 12, 32 and 69 (from *A Textbook of Human Embryology*); and

Professor Dame Sheila Sherlock for Fig. 71 (from *Diseases of the Liver and Biliary System*).

Finally, I wish to express my debt to the staff of Blackwell Science for their continued and unfailing help.

Harold Ellis

Part 1
The Thorax

Surface anatomy and surface markings

The experienced clinician spends much of his working life relating the surface anatomy of his patients to their deep structures (Fig. 1).

The following bony prominences can usually be palpated in the living subject (corresponding vertebral levels are given in brackets):
- superior angle of the scapula (T2);
- upper border of the manubrium sterni, the suprasternal notch (T2/3);
- spine of the scapula (T3);
- sternal angle (of Louis) — the transverse ridge at the manubriosternal junction (T4/5);
- inferior angle of scapula (T8);
- xiphisternal joint (T9);
- lowest part of costal margin—10th rib (L3).

Note from Fig. 1 that the manubrium corresponds to the 3rd and 4th thoracic vertebrae and overlies the aortic arch, and that the sternum corresponds to the 5th to 8th vertebrae and neatly overlies the heart.

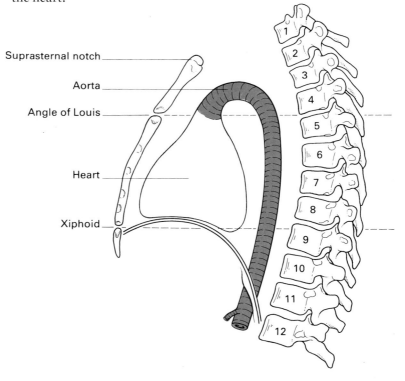

Fig. 1 Lateral view of the thorax—its surface markings and vertebral levels. (Note that the angle of Louis (T4/5) demarcates the superior mediastinum, the upper margin of the heart and the beginning and end of the aortic arch.)

3

Since the 1st and 12th ribs are difficult to feel, the ribs should be enumerated from the 2nd costal cartilage, which articulates with the sternum at the angle of Louis.

The spinous processes of all the thoracic vertebrae can be palpated in the midline posteriorly, but it should be remembered that the first spinous process that can be felt is that of C7 (the vertebra prominens).

The position of the *nipple* varies considerably in the female, but in the male it usually lies in the 4th intercostal space about 4 in (10 cm) from the midline. The *apex beat*, which marks the lowest and outermost point at which the cardiac impulse can be palpated, is normally in the 5th intercostal space 3.5 in (9 cm) from the midline (just below and medial to the nipple).

The *trachea* is palpable in the suprasternal notch midway between the heads of the two clavicles.

Surface markings of the more important thoracic contents (Figs 2–4)

The trachea

The trachea commences in the neck at the level of the lower border of the cricoid cartilage (C6) and runs vertically downwards to end at the level of the sternal angle of Louis (T4/5), just to the right of the midline, by dividing to form the right and left main bronchi. In the erect position and in full inspiration the level of bifurcation is at T6.

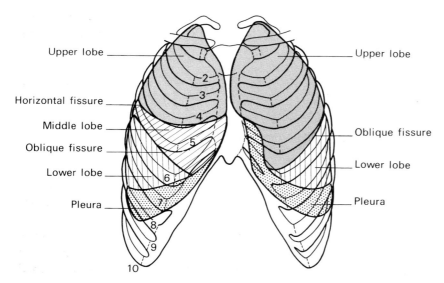

Fig. 2 The surface markings of the lungs and pleura—anterior view.

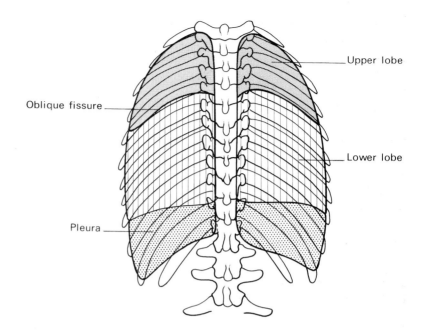

Upper lobe

Oblique fissure

Lower lobe

Pleura

Fig. 3 The surface markings of the lungs and pleura—posterior view.

The pleura

The *cervical pleura* can be marked out on the surface by a curved line drawn from the sternoclavicular joint to the junction of the medial and middle thirds of the clavicle; the apex of the pleura is about 1 in (2.5 cm) above the clavicle. This fact is easily explained by the oblique slope of the first rib. It is important because the pleura can be wounded (with consequent pneumothorax) by a stab wound—and this includes the surgeon's knife and the anaesthetist's needle—above the clavicle.

The lines of pleural reflexion pass from behind the sternoclavicular joint on each side to meet in the midline at the 2nd costal cartilage (the angle of Louis). The right pleural edge then passes vertically downwards to the 6th costal cartilage and then crosses:

- the 8th rib in the midclavicular line;
- the 10th rib in the midaxillary line;
- the 12th rib at the lateral border of the erector spinae.

On the left side the pleural edge arches laterally at the 4th costal cartilage and descends lateral to the border of the sternum; apart from this, its relationships are those of the right side.

The pleura actually descends just below the 12th rib margin at its medial extremity—or even below the edge of the 11th rib if the 12th is unusually short; obviously in this situation the pleura may be opened accidentally in making a loin incision to expose the kidney, perform an adrenalectomy or to drain a subphrenic abscess.

The lungs

The surface projection of the lung is somewhat less extensive than that of the parietal pleura as outlined above, and in addition it varies quite considerably with the phase of respiration. The *apex* of the lung closely follows the line of the cervical pleura and the surface marking of the *anterior border of the right lung* corresponds to that of the right mediastinal pleura. On the left side, however, the *anterior border* has a distinct notch (the *cardiac notch*) which passes behind the 5th and 6th costal cartilages. The *lower border* of the lung has an excursion of as much as 2–3 in (5–8 cm) in the extremes of respiration, but in the neutral position (midway between inspiration and expiration) it lies along a line which crosses the 6th rib in the midclavicular line, the 8th rib in the midaxillary line, and reaches the 10th rib adjacent to the vertebral column posteriorly.

The *oblique fissure*, which divides the lung into upper and lower lobes, is indicated on the surface by a line drawn obliquely downwards and outwards from 1 in (2.5 cm) lateral to the spine of the 5th thoracic vertebra to the 6th costal cartilage about 1.5 in (4 cm) from the midline. This can be represented approximately by abducting the shoulder to its full extent; the line of the oblique fissure then corresponds to the position of the medial border of the scapula.

The surface markings of the *transverse fissure* (separating the middle and upper lobes of the right lung) is a line drawn horizontally along the 4th costal cartilage and meeting the oblique fissure where the latter crosses the 5th rib.

The heart

The outline of the heart can be represented on the surface by the irregular quadrangle bounded by the following four points (Fig. 4):
1   the 2nd *left* costal cartilage 0.5 in (12 mm) from the edge of the sternum;
2   the 3rd *right* costal cartilage 0.5 in (12 mm) from the sternal edge;
3   the 6th *right* costal cartilage 0.5 in (12 mm) from the sternum;
4   the 5th *left* intercostal space 3.5 in (9 cm) from the midline (corresponding to the apex beat).

The *left border* of the heart (indicated by the curved line joining points 1 and 4) is formed almost entirely by the left ventricle, the *lower border* (the horizontal line joining points 3 and 4) corresponds to the right ventricle and the apical part of the left ventricle; the *right*

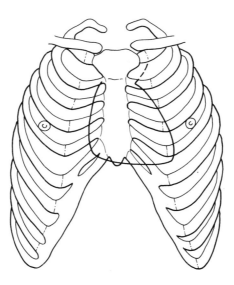

Fig. 4 The surface markings of the heart (see text).

border (marked by the line joining points 2 and 3) is formed by the right atrium.

A good guide to the size and position of your own heart is given by placing your clenched right fist palmar surface down immediately inferior to the manubriosternal junction. Note that the heart is about the size of the subject's fist, lies behind the body of the sternum (therefore anterior to thoracic vertebrae 5–8), and bulges over to the left side.

The surface markings of the vessels of the thoracic wall are of importance if these structures are to be avoided in performing aspiration of the chest. The *internal thoracic (internal mammary) vessels* run vertically downwards behind the costal cartilages half an inch from the lateral border of the sternum. The intercostal vessels lie immediately below their corresponding ribs (the vein above the artery) so that it is safe to pass a needle immediately *above* a rib, dangerous to pass it immediately *below* (see Fig. 8).

The thoracic cage

The thoracic cage is formed by the vertebral column behind, the ribs and intercostal spaces on either side and the sternum and costal cartilages in front. Above, it communicates through the 'thoracic inlet' with the root of the neck; below, it is separated from the abdominal cavity by the diaphragm (Fig. 1).

Above the diaphragm

The thoracic vertebrae

See 'vertebral column' in Part 5, page 352.

The ribs

The greater part of the thoracic cage is formed by the twelve pairs of ribs. Of these, the first seven are connected anteriorly by way of their costal cartilages to the sternum, the cartilages of the 8th, 9th and 10th articulate each with the cartilage of the rib above ('false ribs') and the last two ribs are free anteriorly ('floating ribs').

Each typical rib (Fig. 5) has a *head* bearing two articular facets, for articulation with the numerically corresponding vertebra and the vertebra above, a stout *neck*, which gives attachment to the costo-transverse ligaments, a *tubercle* with a rough non-articular portion and a smooth facet, for articulation with the transverse process of the corresponding vertebra, and a long shaft flattened from side to side and divided into two parts by the 'angle' of the rib. The angle demarcates the lateral limit of attachment of the erector spinae muscle.

The following are the significant features of the 'atypical' ribs.

1st Rib (Fig. 6). This is flattened from above downwards. It is not only the flattest but also the shortest and most curvaceous of all the ribs. It has a prominent *tubercle* on the inner border of its upper surface for the insertion of scalenus anterior. In front of this tubercle, the subclavian vein crosses the rib; behind the tubercle is the *subclavian groove* where the subclavian artery and lowest trunk of the brachial plexus lie in relation to the bone. It is here that the anaesthetist can infiltrate the plexus with local anaesthetic.

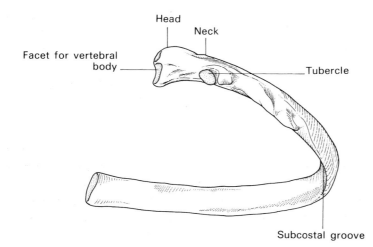

Fig. 5 A typical rib.

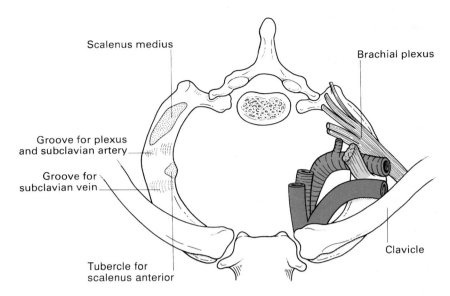

Scalenus medius

Brachial plexus

Groove for plexus
and subclavian artery

Groove for
subclavian vein

Clavicle

Fig. 6 Structures
crossing the first rib.

Tubercle for
scalenus anterior

Crossing the neck of the first rib from the medial to the lateral side are the sympathetic trunk, the superior intercostal artery and the large branch of the first thoracic nerve to the brachial plexus.

The *2nd rib* is much less curved than the 1st and about twice as long.

The *10th rib* has only one articular facet on the head.

The *11th* and *12th ribs* are short, have no tubercles and only a single facet on the head. The 11th rib has a slight angle and a shallow subcostal groove; the 12th has neither of these features.

Clinical features

Rib fractures

The chest wall of the child is highly elastic and therefore fractures of the rib in children are rare. In adults, the ribs may be fractured by direct violence or indirectly by crushing injuries; in the latter the rib tends to give way at its weakest part in the region of its angle. Not unnaturally, the upper two ribs, which are protected by the clavicle, and the lower two ribs, which are unattached and therefore swing free, are the least commonly injured.

In a severe crush injury to the chest several ribs may fracture in front and behind so that a whole segment of the thoracic cage becomes torn free ('stove-in chest'). With each inspiration this loose flap sucks in, with each expiration it blows out, thus undergoing paradoxical respiratory movement. The associated swinging movements of the mediastinum produce severe shock and this injury calls for urgent treatment by insertion of a chest drain with underwater seal, followed by endotracheal intubation, or tracheostomy, combined with positive pressure respiration.

Coarctation of the aorta

In coarctation of the aorta, the intercostal arteries derived from the aorta receive blood from the superior intercostals (from the costocervical trunk of the subclavian artery) and from the arteries anastomosing around the scapula. Together with the communication between the internal thoracic and inferior epigastric arteries, they provide the principal collaterals between the aorta above and below the block. In consequence, the intercostal arteries undergo dilatation and tortuosity and erode the lower borders of the corresponding ribs to give the characteristic irregular *notching of the ribs*, which is very useful in the radiographic confirmation of this lesion.

Cervical rib

A cervical rib (Fig. 7) occurs in 0.5% of subjects and is bilateral in half of these. It is attached to the transverse process of the 7th cervical vertebra and articulates with the 1st (thoracic) rib or, if short, has a free distal extremity. Pressure of such a rib on the lowest trunk of

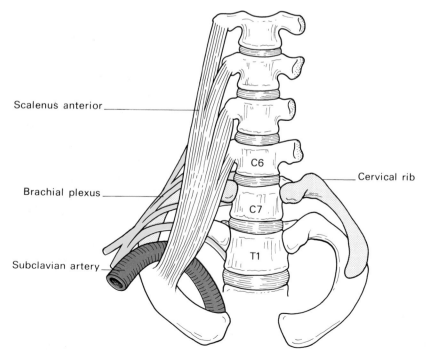

Scalenus anterior

Brachial plexus

Subclavian artery

C6

C7

T1

Cervical rib

Fig. 7 Bilateral cervical ribs. On the right side the brachial plexus is shown arching over the rib and stretching its lowest trunk.

the brachial plexus arching over it may produce paraesthesiae along the ulnar border of the forearm and wasting of the small muscles of the hand (T1). Less commonly vascular changes, even gangrene, may be caused by pressure of the rib on the overlying subclavian artery. This results in post-stenotic dilatation of the vessel distal to the rib in which a thrombus forms from which emboli are thrown off.

The costal cartilages

These bars of hyaline cartilage serve to connect the upper seven ribs directly to the side of the sternum and the 8th, 9th and 10th ribs to the cartilage immediately above. The cartilages of the 11th and 12th ribs merely join the tapered extremities of these ribs and end in the abdominal musculature.

Clinical features

1   The cartilage adds considerable resilience to the thoracic cage and protects the sternum and ribs from more frequent fracture.
2   In old age (and sometimes also in young adults) the costal cartilages undergo progressive ossification; they then become radio-opaque and may give rise to some confusion when examining a chest radiograph of an elderly patient.

The sternum

This dagger-shaped bone, which forms the anterior part of the thoracic cage, consists of three parts. The *manubrium* is roughly triangular in outline and provides articulation for the clavicles and for the first and upper part of the 2nd costal cartilages on either side. It is situated opposite the 3rd and 4th vertebrae. Opposite the disc between T4 and T5 it articulates at an oblique angle at the manubriosternal joint (the angle of Louis), with the *body of the sternum* (placed opposite T5 to T8). This is composed of four parts or 'sternebrae' which fuse between puberty and 25 years of age. Its lateral border is notched to receive part of the 2nd and the 3rd to the 7th costal cartilage. The *xiphoid process* is the smallest part of the sternum and usually remains cartilaginous well into adult life. The cartilaginous manubriosternal joint and that between the xiphoid and the body of the sternum may also become ossified after the age of 30.

Clinical features

1   The attachment of the elastic costal cartilages largely protects the

sternum from injury, but indirect violence accompanying fracture dislocation of the thoracic spine may be associated with a sternal fracture. Direct violence to the sternum may lead to displacement of the
relatively mobile body of the sternum backwards from the relatively
fixed manubrium.

2   In a sternal puncture a wide-bore needle is pushed through the
thin layer of cortical bone covering the sternum into the highly vascular spongy bone beneath, and a specimen of bone marrow aspirated
with a syringe.

3   In operations on the thymus gland, and occasionally for a retrosternal goitre, it is necessary to split the manubrium in the midline
in order to gain access to the superior mediastinum. A complete vertical split of the whole sternum is one of the standard approaches to the
heart and great vessels used in modern cardiac surgery.

The intercostal spaces

There are slight variations between the different intercostal spaces,
but typically each space contains three muscles, comparable to those
of the abdominal wall, and an associated neurovascular bundle (Fig.
8). The muscles are:

1   the *external intercostal*, the fibres of which pass downwards and
forwards from the rib above to the rib below and reach from the vertebrae behind to the costochondral junction in front, where muscle is
replaced by the *anterior intercostal membrane*;

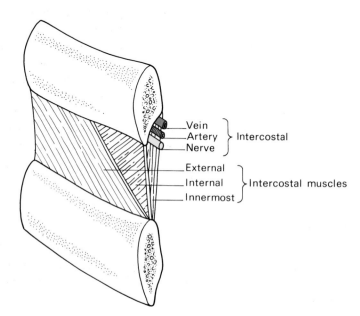

Fig. 8 The relationship of
an intercostal space.
(Note that a needle passed
into the chest immediately
above a rib will avoid the
neurovascular bundle.)

2   the *internal intercostal*, which runs downwards and backwards from the sternum to the angles of the ribs where it becomes the *posterior intercostal membrane*;

3   the *innermost intercostal*, which is only incompletely separated from the internal intercostal muscle by the neurovascular bundle.

The fibres of this sheet cross more than one intercostal space and it may be incomplete. Anteriorly it has a more distinct portion which is fan-like in shape, termed the *transversus thoracis* (or sternocostalis), which spreads upwards from the posterior aspect of the lower sternum to insert onto the inner surfaces of the second to the sixth costal cartilages.

Just as in the abdomen, the nerves and vessels of the thoracic wall lie between the middle and innermost layers of muscles. This neurovascular bundle consists, from above downwards, of vein, artery and nerve, the vein lying in a groove on the undersurface of the corresponding rib (remember—v,a,n).

The vessels comprise the posterior and anterior intercostals.

The *posterior intercostal arteries* of the lower nine spaces are branches of the thoracic aorta, while the first two are derived from the superior intercostal branch of the costocervical trunk, the only branch of the second part of the subclavian artery. Each runs forward in the subcostal groove to anastomose with the anterior intercostal artery. Each has a number of branches to adjacent muscles, to the skin and to the spinal cord. The corresponding veins are mostly tributaries of the azygos and hemiazygos veins. The first posterior intercostal vein drains into the brachiocephalic or vertebral vein. On the left, the 2nd and 3rd veins often join to form a superior intercostal vein, which crosses the aortic arch to drain into the left brachiocephalic vein.

The *anterior intercostal arteries* are branches of the internal thoracic artery (1st–6th space) or of its musculophrenic branch (7th–9th spaces). The lowest two spaces have only posterior arteries. Perforating branches pierce the upper five or six intercostal spaces; those of the 2nd–4th spaces are large in the female and supply the breast.

The *intercostal nerves* are the anterior primary rami of the thoracic nerves, each of which gives off a collateral muscular branch and lateral and anterior cutaneous branches for the innervation of the thoracic and abdominal walls (Fig. 9).

Clinical features

1   Local irritation of the intercostal nerves by such conditions as Pott's disease of the thoracic vertebrae (tuberculosis) may give rise to pain which is referred to the front of the chest or abdomen in the region of the peripheral termination of the nerves.

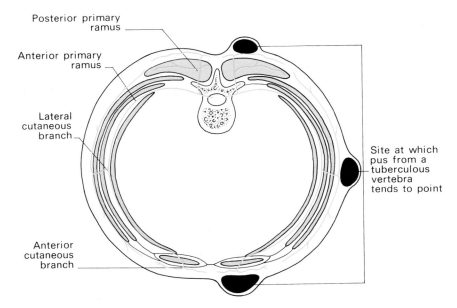

Fig. 9 Diagram of a typical spinal nerve and its body-wall relationships. On the left side the sites of eruption of a tuberculous cold abscess tracking forwards from a diseased vertebra are shown—these occur at the points of emergence of the cutaneous branches.

2 Local anaesthesia of an intercostal space is easily produced by infiltration around the intercostal nerve trunk and its collateral branch—a procedure known as *intercostal nerve block*.

3 In a conventional posterolateral *thoracotomy* (e.g. for a pulmonary lobectomy) an incision is made along the line of the 5th or 6th rib; the periosteum over a segment of the rib is elevated, thus protecting the neurovascular bundle, and the rib is excised. Access to the lung or mediastinum is then gained though the intercostal space, which can be opened out considerably owing to the elasticity of the thoracic cage.

4 Pus from the region of the vertebral column tends to track around the thorax along the course of the neurovascular bundle and to 'point' to the three sites of exit of the cutaneous branches of the intercostal nerves, which are lateral to erector spinae, in the midaxillary line and just lateral to the sternum (Fig. 9).

The diaphragm

The diaphragm is the dome-shaped septum dividing the thoracic from the abdominal cavity. It comprises two portions: a peripheral muscular part which arises from the margins of the thoracic outlet and a centrally placed aponeurosis (Fig. 10).

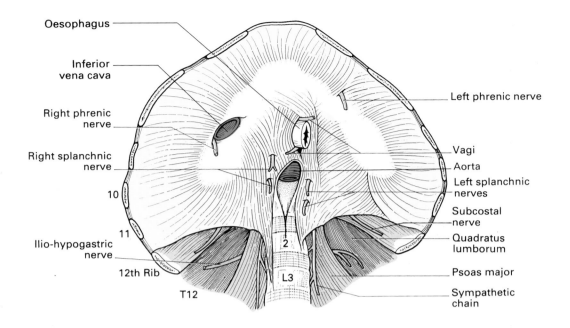

Oesophagus

Inferior vena cava

Right phrenic nerve

Right splanchnic nerve

10

11

Ilio-hypogastric nerve

12th Rib

T12

Left phrenic nerve

Vagi

Aorta

Left splanchnic nerves

Subcostal nerve

Quadratus lumborum

Psoas major

Sympathetic chain

Fig. 10 The diaphragm—inferior aspect. The three major orifices, from above downwards, transmit the inferior vena cava, oesophagus and aorta.

The muscular fibres are arranged in three parts.

1   A *vertebral part* from the crura and from the arcuate ligaments. The *right crus* arises from the front of the bodies of the upper three lumbar vertebrae and intervertebral discs; the *left crus* is only attached to the first two vertebrae. The arcuate *ligaments* are a series of fibrous arches, the *medial* being a thickening of the fascia covering psoas major and the *lateral* of fascia overlying quadratus lumborum. The fibrous medial borders of the two crura form a *median arcuate ligament* over the front of the aorta.

2   A *costal part* is attached to the inner aspect of the lower six ribs and costal cartilages.

3   A *sternal portion* consists of two small slips from the deep surface of the xiphisternum.

The central tendon, into which the muscular fibres are inserted, is trefoil in shape and is partially fused with the undersurface of the pericardium.

The diaphragm receives its entire motor supply from the phrenic nerve (C3, 4, 5) whose long course from the neck follows the embryological migration of the muscle of the diaphragm from the cervical region (see below). Injury or operative division of this nerve results in paralysis and elevation of the corresponding half of the diaphragm.

Radiographically, paralysis of the diaphragm is recognized by its elevation and paradoxical movement; instead of descending on inspiration it is forced upwards by pressure from the abdominal viscera.

The sensory nerve fibres from the central part of the diaphragm also run in the phrenic nerve, hence irritation of the diaphragmatic pleura (in pleurisy) or of the peritoneum on the undersurface of the diaphragm by subphrenic collections of pus or blood produces referred pain in the corresponding cutaneous area, the shoulder-tip.

The peripheral part of the diaphragm, including the crura, receives sensory fibres from the lower intercostal nerves.

Openings in the diaphragm

The three main openings in the diaphragm (Figs 10 and 11) are:
1   the *aortic* (at the level of T12) which transmits the abdominal aorta, the thoracic duct and often the azygos vein;
2   the *oesophageal* (T10) which is situated between the muscular fibres of the right crus of the diaphragm and transmits, in addition to the oesophagus, branches of the left gastric artery and vein and the two vagi;
3   *the opening for the inferior vena cava* (T8) is placed in the central tendon and also transmits the right phrenic nerve.

In addition to these structures, the greater and lesser splanchnic

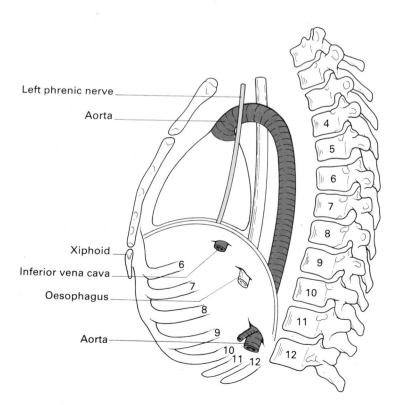

Fig. 11 Schematic lateral view of the diaphragm to show the levels at which it is pierced by major structures.

nerves pierce the crura and the sympathetic chain passes behind the diaphragm deep to the medial arcuate ligament.

The development of the diaphragm and the anatomy of diaphragmatic herniae

The diaphragm is formed (Fig. 12) by fusion in the embryo of:
1   the septum transversum (forming the central tendon);
2   the dorsal oesophageal mesentery;
3   a peripheral rim derived from the body wall;
4   the pleuroperitoneal membranes, which close the primitive communication between the pleural and peritoneal cavities.

The septum transversum is the mesoderm which, in early development, lies in front of the head end of the embryo. With the folding off of the head, this mesodermal mass is carried ventrally and caudally, to lie in its definitive position at the anterior part of the diaphragm. During this migration, the cervical myotomes and nerves contribute muscle and nerve supply respectively, thus accounting for the long course of the phrenic nerve (C3, 4, and 5) from the neck to the diaphragm.

With such a complex embryological story, one may be surprised to know that congenital abnormalities of the diaphragm are unusual.

However, a number of defects may occur, giving rise to a variety of congenital herniae through the diaphragm. These may be:

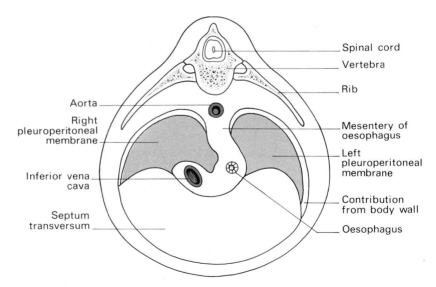

Fig. 12 The development of the diaphragm. This drawing shows the four elements contributing to the diaphragm—(1) the septum transversum, (2) the dorsal mesentery of the oesophagus, (3) the body wall and (4) the pleuroperitoneal membrane.

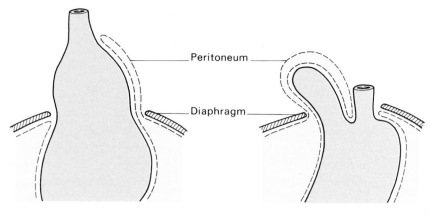

Peritoneum

Diaphragm

(a) 'Sliding hernia'   (b)'Rolling hernia'

Fig. 13 (a) A sliding
hiatus hernia. (b) A
rolling hiatus hernia.

1   through the foramen of Morgagni; anteriorly between the xiphoid
and costal origins;
2   through the foramen of Bochdalek—the pleuroperitoneal canal—
lying posteriorly;
3   through a deficiency of the whole central tendon (occasionally
such a hernia may be traumatic in origin);
4   through a congenitally large oesophageal hiatus.
    Far more common are the *acquired* hiatus herniae (subdivided
into sliding and rolling herniae). These are found in patients usually
of middle age where weakening and widening of the oesophageal
hiatus has occurred (Fig. 13).
    In the *sliding herniae* the upper stomach and lower oesophagus
slide upwards into the chest through the lax hiatus when the patient
lies down or bends over; the competence of the cardia is often disturbed and peptic juice can therefore regurgitate into the gullet. This
may be followed by oesophagitis with consequent heartburn, bleeding and, eventually, stricture formation.
    In the *rolling hernia* (which is far less common) the cardia remains
in its normal position and the cardio-oesophageal junction is intact,
but the fundus of the stomach rolls up through the hiatus in front of
the oesophagus, hence the alternative term of para-oesophageal
hernia. In such a case there may be epigastric discomfort, flatulence
and even dysphagia, but *no* regurgitation because the cardiac mechanism is undisturbed.

The movements of respiration

During inspiration the movements of the chest wall and diaphragm result in an increase in all diameters of the thorax. This, in turn, brings about an increase in the negative intrapleural pressure and an expansion of the lung tissue. Conversely, in expiration the relaxation of the respiratory muscles and the elastic recoil of the lung reduce the thoracic capacity and force air out of the lungs.

In quiet *inspiration* the first rib remains relatively fixed, but contraction of the external intercostals elevates and, at the same time, everts the succeeding ribs. In the case of the 2nd–7th ribs this principally increases the anteroposterior diameter of the thorax (by the forward thrust of the sternum). The corresponding movement of the lower ribs raises the costal margin and leads mainly to an increase in the transverse diameter of the thorax. The depth of the thorax is increased by the contraction of the diaphragm which draws down its central tendon. Normal quiet *expiration* is aided by the tone of the abdominal musculature which, acting through the contained viscera, forces the diaphragm upwards.

In deep and in forced inspiration additional muscles are called into play (e.g. scalenus anterior, sternocleidomastoid, serratus anterior and pectoralis major) to increase further the capacity of the thorax. Similarly, in deep expiration, forced contraction of the abdominal muscles aids the normal expulsive factors described above.

The pleurae

The two pleural cavities are totally separate from each other (Fig. 2). Each *pleura* consists of two layers: a *visceral layer* intimately related to the surface of the lung, and a *parietal layer* lining the inner aspect of the chest wall, the upper surface of the diaphragm and the sides of the pericardium and mediastinum. The two layers are continuous in front and behind the root of the lung, but below this the pleura hangs down in a loose fold, the *pulmonary ligament*, which forms a 'dead-space' for distension of the pulmonary veins. The surface markings of the pleura and lungs have already been described in the section on surface anatomy.

Notice that the lungs do not occupy all the available space in the pleural cavity even in forced inspiration.

Clinical features

1 Normally the two pleural layers are in close apposition and the space between them is only a potential one. It may, however, fill with air (pneumothorax), blood (haemothorax) or pus (empyema).

2 Fluid can be drained from the pleural cavity by inserting a wide-

bore needle through an intercostal space (usually the 7th posteriorly). The needle is passed along the superior border of the lower rib, thus avoiding the intercostal nerves and vessels (Fig. 8). Below the 7th intercostal space there is danger of penetrating the diaphragm.

3   Since the parietal pleura is segmentally innervated by the inter-costal nerves, inflammation of the pleura results in pain referred to the cutaneous distribution of these nerves (i.e. to the thoracic wall or, in the case of the lower nerves, to the anterior abdominal wall).

The lower respiratory tract

The trachea (Figs 14 and 15)

The trachea is about 4.5 in (11.5 cm) in length and nearly 1 in (2.5 cm) in diameter. It commences at the lower border of the cricoid cartilage (C6) and terminates by bifurcating at the level of the sternal angle of Louis (T4/5) to form the right and left main bronchi. (In the living subject, the level of bifurcation varies slightly with the phase of respi-ration; in deep inspiration is descends to T6 and in expiration it rises to T4.)

Relations

Lying partly in the neck and partly in the thorax, its relations are:

Thyroid isthmus

Brachiocephalic artery

Left brachiocephalic vein

Superior vena cava

Aortic arch

Right pulmonary artery

Fig. 14 The trachea and its anterior relationships.

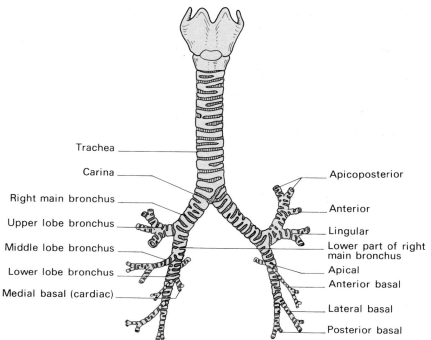

Trachea

Carina

Right main bronchus

Upper lobe bronchus

Middle lobe bronchus

Lower lobe bronchus

Medial basal (cardiac)

Apicoposterior

Anterior

Lingular

Lower part of right main bronchus

Apical

Anterior basal

Lateral basal

Posterior basal

Fig. 15 The trachea and main bronchi viewed from the front.

Cervical

• anteriorly — the isthmus of thyroid gland, inferior thyroid veins, sternohyoid and sternothyroid muscles;
• laterally — the lobes of thyroid gland and the common carotid artery;
• posteriorly — the oesophagus with the recurrent laryngeal nerve lying in the groove between oesophagus and trachea (Fig. 16).

Thoracic

In the superior mediastinum its relations are:
• anteriorly — commencement of the brachiocephalic (innominate) artery and left carotid artery, both arising from the arch of the aorta, the left brachiocephalic (innominate) vein, and the thymus;
• posteriorly — oesophagus and left recurrent laryngeal nerve;
• to the left — arch of the aorta, left common carotid and left subclavian arteries, left recurrent laryngeal nerve and pleura;
• to the right — vagus, azygos vein and pleura (Fig. 17).

Structure

The patency of the trachea is maintained by a series of 15–20 U-

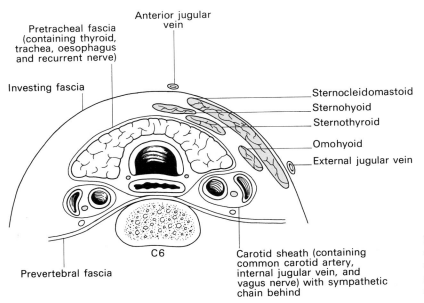

Fig. 16 The cervical part of the trachea and its environs in transverse section (through the 6th cervical vertebra).

Fig. 17 The thoracic part of the trachea and its environs in transverse section (through the 4th thoracic vertebra).

shaped cartilages. Posteriorly, where the cartilage is deficient, the trachea is flattened and its wall completed by fibrous tissue and a sheet of smooth muscle (the trachealis). Within, it is lined by a ciliated columnar epithelium with many goblet cells.

Clinical features

Radiology

Since it contains air, the trachea is more radio-translucent than the neighbouring structures and is seen in posteroanterior and lateral radiographs as a dark area passing downwards, backwards and slightly to the right. In the elderly, calcification of the tracheal rings may be a source of radiological confusion.

Displacement

The trachea may be compressed or displaced by pathological enlargement of the neighbouring structures, particularly the thyroid gland and the arch of the aorta.

'Tracheal-tug'

The intimate relationship between the arch of the aorta and the trachea and left bronchus is responsible for the physical sign known as 'tracheal-tug', characteristic of aneurysms of the aortic arch.

Tracheostomy

Tracheostomy may be required for laryngeal obstruction (diphtheria, tumours, inhaled foreign bodies), for the evacuation of excessive secretions (severe postoperative chest infection in a patient who is too weak to cough adequately), and for long-continued artificial respiration (poliomyelitis, severe chest injuries). It is important to note that respiration is further assisted by considerable reduction of the dead-space air.

The neck is extended and the head held exactly in the midline by an assistant. A vertical incision is made downwards from the cricoid cartilage, passing between the anterior jugular veins. Alternatively, a more cosmetic transverse skin crease incision, placed halfway between the cricoid and suprasternal notch, is employed. A hook is thrust under the lower border of the cricoid to steady the trachea and pull it forward. The pretracheal fascia is split longitudinally, the isthmus of the thyroid either pushed upwards or divided between clamps and the cartilage of the trachea clearly exposed. A circular opening is then made into the trachea to admit the tracheostomy tube.

In children the neck is relatively short and the left brachiocephalic vein may come up above the suprasternal notch so that dissection is rather more difficult and dangerous. This difficulty is made greater because the child's trachea is softer and more mobile than the adult's and therefore not so readily identified and isolated. Its softness means that care must be taken, in incising the child's trachea, not to let the scalpel plunge through and damage the underlying oesophagus.

In contrast, the trachea may be ossified in the elderly and small bone shears required to open into it.

The golden rule of tracheostomy — based entirely on anatomical considerations — is 'stick exactly to the midline'. If this is not done, major vessels are in jeopardy and it is possible, although the student may not credit it, to miss the trachea entirely.

The bronchi (Fig. 15)

The *right main bronchus* is wider, shorter and more vertical than the left. It is about 1 in (2.5 cm) long and passes directly to the root of the lung at T5. Before joining the lung it gives off its *upper lobe branch*, and then passes below the pulmonary artery to enter the hilum of the lung. It has two important relations: the azygos vein, which arches over it from behind to reach the superior vena cava, and the pulmonary artery which lies first below and then anterior to it.

The *left main bronchus* is nearly 2 in (5 cm) long and passes downwards and outwards below the arch of the aorta, in front of the oesophagus and descending aorta. Unlike the right, it gives off no branches until it enters the hilum of the lung, which it reaches opposite T6. The pulmonary artery spirals over the bronchus, lying first anteriorly and then above it.

Clinical features

1   The greater width and more vertical course of the right bronchus accounts for the greater tendency for foreign bodies and aspirated material to pass into the right bronchus (and thence especially into the middle and lower lobes of the right lung) rather than into the left.
2   The inner aspect of the whole of the trachea, the main and lobar bronchi and the commencement of the first segmental divisions can be seen at bronchoscopy.
3   Widening and distortion of the angle between the bronchi (the carina) as seen at bronchoscopy is a serious prognostic sign, since it usually indicates carcinomatous involvement of the tracheobronchial lymph nodes around the bifurcation of the trachea.

The lungs (Figs 18 and 19)

Each lung is conical in shape, having a blunt apex which reaches

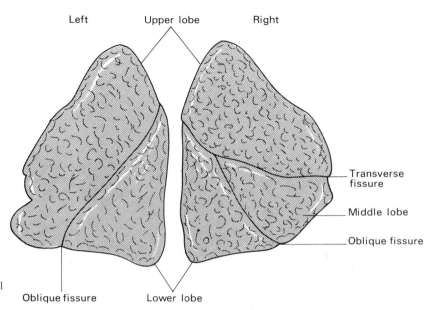

Left   Upper lobe   Right

Transverse
fissure

Middle lobe

Oblique fissure

Fig. 18 The lungs, lateral
aspects.

Oblique fissure   Lower lobe

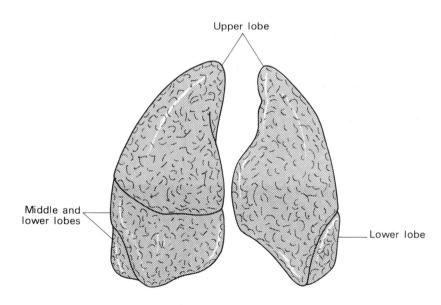

Upper lobe

Middle and
lower lobes

Lower lobe

Fig. 19 The lungs,
anterior aspects.

above the sternal end of the 1st rib, a concave base overlying the
diaphragm, an extensive costovertebral surface moulded to the form
of the chest wall and a mediastinal surface which is concave to accommodate the pericardium.

The right lung is slightly larger than the left and is divided into
three lobes—upper, middle and lower, by the oblique and horizontal

fissures. The left lung has only an oblique fissure and hence only two lobes.

Blood supply

Mixed venous blood is returned to the lungs by the pulmonary arteries; the air passages are themselves supplied by the bronchial arteries, which are small branches of the descending aorta.

Lymphatic drainage

The lymphatics of the lung drain centripetally from the pleura towards the hilum. From the *bronchopulmonary lymph nodes* in the hilum, efferent lymph channels pass to the *tracheobronchial nodes* at the bifurcation of the trachea, thence to the *paratracheal nodes* and the mediastinal lymph trunks to drain usually directly into the brachiocephalic veins or, rarely, indirectly via the thoracic or right lymphatic duct.

Nerve supply

The pulmonary plexuses derive fibres from both the vagi and the sympathetic trunk. They supply efferents to the bronchial musculature and receive afferents from the mucous membrane of the bronchioles and from the alveoli.

The bronchopulmonary segments of the lungs
(Figs 20 and 21)

A knowledge of the finer arrangement of the bronchial tree is an

Table 1 The named divisions of the main bronchi.

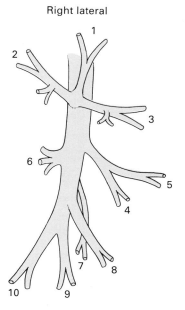

Right lateral   Left lateral

Fig. 20 The named divisions of the main bronchi.

Right
Upper lobe
1 Apical bronchus
2 Posterior bronchus
3 Anterior bronchus

Middle lobe
4 Lateral bronchus
5 Medial bronchus

Lower lobe
6 Apical bronchus
7 Medial basal (cardiac) bronchus
8 Anterior basal bronchus
9 Lateral basal bronchus
10 Posterior basal bronchus

Left
Upper lobe
1 } Apicoposterior bronchus
2
3 Anterior bronchus

Lingula
4 Superior bronchus
5 Inferior bronchus

Lower lobe
6 Apical bronchus

8 Anterior basal bronchus

9 Lateral basal bronchus

10 Posterior basal bronchus

essential prerequisite to intelligent appreciation of lung radiology, to interpretation of bronchoscopy and to the surgical resection of lung segments. Each lobe of the lung is subdivided into a number of bronchopulmonary segments, each of which is supplied by a segmental bronchus, artery and vein. These segments are wedge-shaped with their apices at the hilum and bases at the lung surface; if excised accurately along their boundaries (which are marked by intersegmental veins), there is little bleeding or alveolar air leakage from the raw lung surface.

The names and arrangements of the bronchi are given in Table 1;

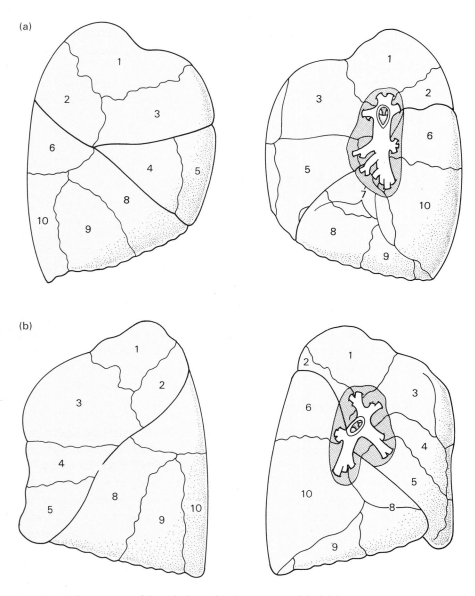

Fig. 21 (a) The segments of the right lung. (b) The segments of the left lung.

each bronchopulmonary segment takes its title from that of its supplying segmental bronchus (listed in the right-hand column of the table).

The left upper lobe has a *lingular segment*, supplied by the *lingular bronchus* from the main upper lobe bronchus. This lobe is equivalent to the *right middle lobe* whose bronchus arises as a branch from the main bronchus. Apart from this, differences between the two sides are very slight; on the left, the upper lobe bronchus gives off a combined apicoposterior segmental bronchus and an anterior branch, whereas all three branches are separate on the right side.

On the right also there is a small medial (or cardiac) lower lobe bronchus which is absent on the left, the lower lobes being otherwise mirror images of each other.

The mediastinum

The mediastinum is defined as 'the space which is sandwiched between the two pleural sacs'. For descriptive purposes the mediastinum is divided by a line drawn horizontally from the sternal angle to the lower border of T4 (angle of Louis) into superior and inferior mediastinum. The inferior mediastinum is further subdivided into the anterior in front of the pericardium, a middle mediastinum containing the pericardium itself with the heart and great vessels, and posterior mediastinum between the pericardium and the lower eight thoracic vertebrae (Fig. 22).

The pericardium

The heart and the roots of the great vessels are contained within the conical *fibrous pericardium*, the apex of which is fused with the adventitia of the great vessels and the base with the central tendon of the diaphragm. Anteriorly it is related to the body of the sternum, the 3rd–6th costal cartilages and the anterior borders of the lungs; posteriorly, to the oesophagus, descending aorta, and vertebra T5–T8, and on either side to the roots of the lungs, the mediastinal pleura and the phrenic nerves.

The inner aspect of the fibrous pericardium is lined by the *parietal layer of serous pericardium*. This, in turn, is reflected around the roots of the great vessels to become continuous with the *visceral layer* or *epicardium*. The lines of pericardial reflexion are marked on the posterior surface of the heart (Fig. 23) by the *oblique sinus*, bounded by the inferior vena cava and the four pulmonary veins, which form a recess between the left atrium and the pericardium, and the *transverse sinus* between the superior vena cava and left atrium behind and the pulmonary trunk and aorta in front.

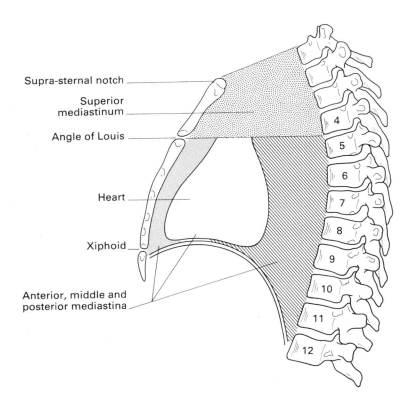

Supra-sternal notch

Superior mediastinum

Angle of Louis

Heart

Xiphoid

Anterior, middle and posterior mediastina

4

5

6

7

8

9

10

11

12

Fig. 22 The subdivisions of the mediastinum.

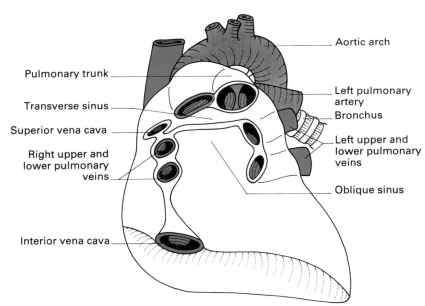

Pulmonary trunk

Transverse sinus

Superior vena cava

Right upper and lower pulmonary veins

Interior vena cava

Aortic arch

Left pulmonary artery

Bronchus

Left upper and lower pulmonary veins

Oblique sinus

Fig. 23 The transverse and oblique sinuses of the pericardium. In this illustration the heart has been removed from the pericardial sac, which is seen in anterior view.

The heart (Fig. 24)

Its great importance means no excuse need be offered for dealing with
the heart in considerable detail.

 The heart is irregularly conical in shape, and it is placed obliquely

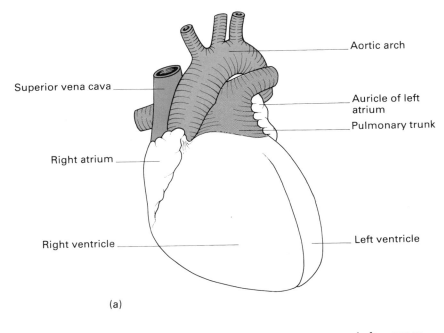

Aortic arch

Superior vena cava

Auricle of left
atrium

Pulmonary trunk

Right atrium

Right ventricle

Left ventricle

(a)

Left common
carotid artery

Left subclavian artery

Brachiocephalic artery

Aortic arch

Vena azygos

Superior vena cava

Left pulmonary veins

Right pulmonary artery
and

Veins

Left atrium

Left ventricle

Right atrium

Coronary sinus

Inferior vena cava

(b)

Fig. 24 The heart, (a) anterior and (b) posterior aspects.

in the middle mediastinum. Viewed from the front, portions of all the heart chambers can be seen. The right border is formed entirely by the right atrium, the left border partly by the auricular appendage of the left atrium but mainly by the left ventricle, and the inferior border chiefly by the right ventricle but also by the lower part of the right atrium and the apex of the left ventricle.

The bulk of the anterior surface is formed by the right ventricle which is separated from the right atrium by the vertical atrioventricular groove, and from the left ventricle by the anterior interventricular groove.

The inferior or *diaphragmatic surface* consists of the right and left ventricles separated by the posterior interventricular groove and the portion of the right atrium which receives the inferior vena cava.

The base or *posterior surface* is quadrilateral in shape and is formed mainly by the left atrium with the openings of the pulmonary veins and, to a lesser extent, by the right atrium.

Chambers of the heart

Right atrium (Fig. 25)

The right atrium receives the superior vena cava in its upper and posterior part, the inferior vena cava and coronary sinus in its lower part, and the anterior cardiac vein (draining much of the front of the heart)

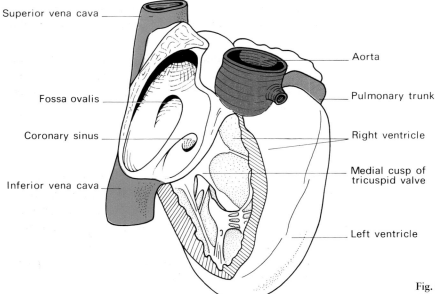

Superior vena cava

Fossa ovalis

Coronary sinus

Inferior vena cava

Aorta

Pulmonary trunk

Right ventricle

Medial cusp of
tricuspid valve

Left ventricle

Fig. 25 The interior of the right atrium and ventricle.

anteriorly. Running more or less vertically downwards between the venae cavae is a distinct muscular ridge, the *crista terminalis* (indicated on the outer surface of the atrium by a shallow groove — the *sulcus terminalis*). This ridge separates the smooth-walled posterior part of the atrium, derived from the sinus venosus, from the rough-walled anterior portion which is prolonged into the *auricular appendage* and which is derived from the true fetal atrium.

The openings of the inferior vena cava and the coronary sinus are guarded by rudimentary valves; that of the inferior vena cava being continuous with the *annulus ovalis* around the shallow depression on the atrial septum, the *fossa ovalis*, which marks the site of the fetal *foramen ovale*.

Right ventricle (Fig. 25)

The right ventricle is joined to the right atrium by the way of the vertically disposed tricuspid valve, and with the pulmonary trunk through the pulmonary valve. A muscular ridge, the *infundibuloventricular crest*, between the atrioventricular and pulmonary orifices, separates the 'inflow' and 'outflow' tracts of the ventricle. The inner aspect of the inflow tract path is marked in the presence of a number of irregular muscular elevations (*trabeculae carneae*) from some of which the *papillary muscles* project into the lumen of the ventricle and find attachment to the free borders of the cusps of the tricuspid valve by way of the *chordae tendineae*. The *moderator band* is a muscular bundle crossing the ventricular cavity from the interventricular septum to the anterior wall and is of some importance since it conveys the right branch of the atrioventricular bundle to the ventricular muscle.

The outflow tract of the ventricle or *infundibulum* is smooth-walled and is directed upwards and to the right towards the pulmonary trunk. The pulmonary orifice is guarded by the *pulmonary valves*, comprising three semilunar cusps.

Left atrium

The left atrium is rather smaller than the right but has somewhat thicker walls. On the upper part of its posterior wall it presents the openings of the four pulmonary veins and on its septal surface there is a shallow depression corresponding to the fossa ovalis of the right atrium. As on the right side, the main part of the cavity is smooth-walled but the surface of the auricle is marked by a number of ridges due to the underlying pectinate muscles.

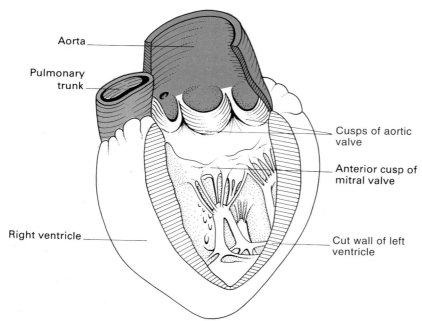

Fig. 26 The interior of the left ventricle.

Left ventricle (Fig. 26)

The left ventricle communicates with the left atrium by way of the *mitral valve*, which possesses a large anterior and a smaller posterior cusp attached to papillary muscles by chordae tendineae. With the exception of the fibrous *vestibule* immediately below the aortic orifice, the wall of the left ventricle is marked by thick trabeculae carneae.

The aortic orifice is guarded by the three semilunar cusps of the *aortic valve*, immediately above which are the dilated *aortic sinuses*. The mouths of the right and left coronary arteries are seen in the anterior and left posterior sinus respectively.

The conducting system of the heart

This consists of specialized cardiac muscle found in the sinuatrial node and in the atrioventricular node and bundle. The heart-beat is initiated in the *sinuatrial node* (the 'pacemaker of the heart'), situated in the upper part of the crista terminalis just to the right of the opening of the superior vena cava into the right atrium. From there the cardiac impulse spreads throughout the atrial musculature to reach the *atrioventricular node* lying in the atrial septum immediately above the opening of the coronary sinus. The impulse is then conducted to the ventricles by way of the specialized tissue of the *atrio-*

ventricular bundle (of His). This bundle divides at the junction of the membranous and muscular parts of the interventricular septum into its right and left branches which run immediately beneath the endocardium to activate all parts of the ventricular musculature.

The blood supply to the heart (Fig. 27)

The heart's blood supply is derived from the right and left coronary arteries whose main branches lie in the interventricular and atrioventricular grooves.

The *right coronary artery* arises from the anterior aortic sinus and passes forwards between the pulmonary trunk and the right atrium to descend in the right part of the atrioventricular groove. At the inferior border of the heart it continues along the atrioventricular groove to anastomose with the left coronary at the posterior interventricular groove. It gives off a *marginal branch* along the lower border of the heart and an interventricular branch which runs forward in the inferior interventricular groove and to anastomose near the apex of the heart with the corresponding branch of the left coronary artery.

The *left coronary artery*, which is larger than the right, rises from the left posterior aortic sinus. Passing first behind and then to the left of the pulmonary trunk, it reaches the left part of atrioventricular groove in which it runs laterally round the left border of the heart to reach the posterior interatrial groove. Its most important branch is the *anterior interventricular artery* which supplies the anterior aspect of both ventricles and passes around the apex of the heart to anastomose with the posterior interventricular branch of the right coronary. Note that the sinuatrial node is usually supplied by the right coronary

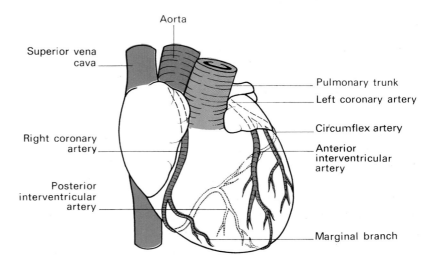

Fig. 27 The coronary arteries. (Dotted vessels lie posteriorly.)

Aorta

Superior vena cava

Pulmonary trunk

Left coronary artery

Circumflex artery

Right coronary artery

Anterior interventricular artery

Posterior interventricular artery

Marginal branch

artery, although the left coronary artery takes over this duty in about one-third of subjects.

Although anastomoses occur between the terminations of the right and left coronary arteries, these are usually inefficient. Thrombosis in one or other of these vessels leads to death of the area of heart muscle supplied (a myocardial infarction).

The venous drainage of the heart (Fig. 28)

About two-thirds of the venous drainage of the heart is achieved by veins which accompany the coronary arteries and which open into the right atrium. The rest of the blood drains by means of small veins (*venae cordis minimae*) directly into the cardiac cavity.

The coronary sinus lies in the posterior atrioventricular groove and opens into the right atrium just to the left of the mouth of the inferior vena cava.

It receives:
1   the *great cardiac vein* in the anterior interventricular groove;
2   the *middle cardiac vein* the inferior interventricular groove;
3   the *small cardiac vein* — accompanying the marginal artery along the lower border of the heart;
4   the *oblique vein* — descends obliquely on the posterior aspect of the left atrium.

The *anterior cardiac veins* (up to three or four in number) cross the anterior atrioventricular groove, drain much of the anterior surface of the heart and open directly into the right atrium.

Nerve supply

The nerve supply of the heart is derived from the vagus (cardio-

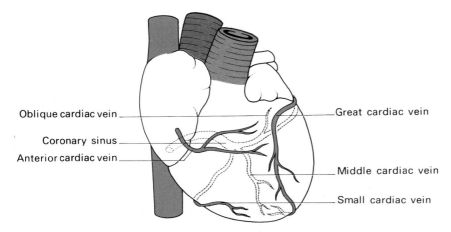

Oblique cardiac vein

Coronary sinus

Anterior cardiac vein

Great cardiac vein

Middle cardiac vein

Small cardiac vein

Fig. 28 The coronary veins. (Dotted vessels lie posteriorly.)

inhibitor) and the cervical and upper thoracic sympathetic ganglia (cardio-accelerator) by way of superficial and deep cardiac plexuses.

The development of the heart

The primitive heart is a single tube which soon shows grooves demarcating the *sinus venosus*, *atrium*, *ventricle* and *bulbus cordis* from behind forwards. As this tube enlarges it kinks so that its caudal end, receiving venous blood, comes to lie behind its cephalic end with its emerging arteries (Fig. 29).

The sinus venosus later absorbs into the atrium and the bulbus becomes incorporated into the ventricle so that, in the fully developed heart, the atria and great veins come to lie posterior to the ventricles and the roots of the great arteries.

The boundary tissue between the primitive single atrial cavity and single ventricle grows out as a *dorsal* and a *ventral endocardial cushion* which meet in the midline, thus dividing the common atrio-ventricular orifice into a right (tricuspid) and left (mitral) orifice.

The division of the primitive atrium into two is a complicated process but an important one in the understanding of congenital septal defects (Fig. 30). A partition, the *septum primum*, grows downwards from the posterior and superior walls of the primitive common atrium to fuse with the endocardial cushions. Before fusion is com-

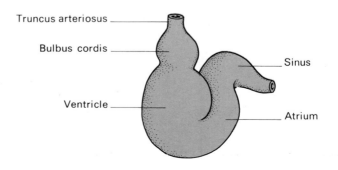

Fig. 29 The coiling of the primitive heart tube into its definitive form.

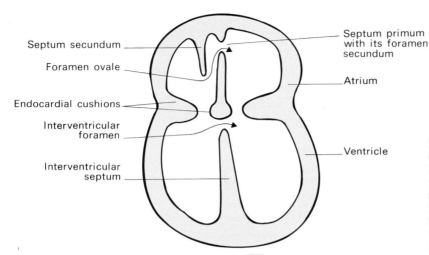

Septum secundum

Foramen ovale

Endocardial cushions

Interventricular foramen

Interventricular septum

Septum primum with its foramen secundum

Atrium

Ventricle

Fig. 30 The development of the chambers of the heart. (Note the septum primum and septum secundum which form the interatrial septum, leaving the foramen ovale as a valve-like opening passing between them.)

plete, however, a hole appears in the upper part of this septum which is termed *the foramen secundum in the septum primum.*

A second membrane, the *septum secundum,* then develops to the right of the primum but this is never complete; it has a free lower edge which does, however, extend low enough for this new septum to overlap the foramen secundum in the septum primum and hence to close it.

The two overlapping defects in the septa form the valve-like *foramen ovale* which shunts blood from the right to left heart in the fetus (see 'fetal circulation' below). After birth, this foramen usually becomes completely fused leaving only the fossa ovalis on the septal wall of the right atrium as its memorial. In about 10% of adult subjects, however, a probe can still be insinuated through an anatomically patent, although functionally sealed foramen.

Division of the ventricle is commenced by the upgrowth of a fleshy septum from the apex of the heart towards the endocardial cushions. This stops short of dividing the ventricle completely and thus it has an upper free border, forming a temporary interventricular foramen. At the same time, the single truncus arteriosus is divided into aorta and pulmonary trunk by a spiral septum (hence the spiral relations of these two vessels), which grows downwards to the ventricle and fuses accurately with the upper free border of the ventricular septum. This contributes the small *pars membranacea septi,* which completes the separation of the ventricle in such a way that blood on the left of the septum flows into the aorta and on the right into the pulmonary trunk.

The primitive *sinus venosus* absorbs into the right atrium so that the venae cavae draining into the sinus come to open separately into this atrium. The smooth-walled part of the adult atrium represents the contribution of the sinus venosus, the pectinate part represents the portion derived from the primitive atrium.

Rather similarly, the adult left atrium has a double origin. The original single pulmonary venous trunk entering the left atrium becomes absorbed into it, and donates the smooth-walled part of this chamber with the pulmonary veins entering as four separate openings; the trabeculated part of the definitive left atrium is the remains of the original atrial wall.

The development of the aortic arches and their derivatives (Fig. 31)

Emerging from the bulbus cordis is a common arterial trunk termed the *truncus arteriosus*, from which arise six pairs of aortic arches, equivalent to the arteries supplying the gill clefts of the fish. These arteries curve dorsally around the pharynx on either side and join to form two longitudinally placed dorsal aortae which fuse distally into the descending aorta.

The 1st and 2nd arches disappear; the 3rd arches become the carotids. The 4th arch on the right becomes the brachiocephalic and

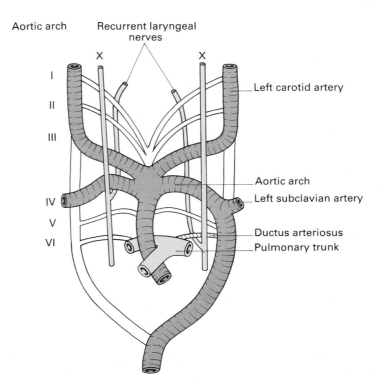

Fig. 31 The aortic arches and their derivatives. This diagram explains the relationship of the right recurrent laryngeal nerve to the right subclavian artery and the left nerve to the aortic arch and the ligamentum arteriosum (or to a patent ductus arteriosus).

right subclavian artery; on the left, it differentiates into the definitive aortic arch, gives off the left subclavian artery and links up distally with the descending aorta.

The 5th arch artery is rudimentary and disappears.

When the truncus arteriosus splits longitudinally to form the ascending aorta and pulmonary trunk, the 6th arch, unlike the others, remains linked with the latter and forms the right and left pulmonary arteries. On the left side this arch retains its connection with the dorsal aorta to form the *ductus arteriosus* (the ligamentum arteriosum of adult anatomy).

This asymmetrical development of the aortic arches accounts for the different course taken by the recurrent laryngeal nerve on each side. In the early fetus the vagus nerve lies lateral to the primitive pharynx, separated from it by the aortic arches. What are to become the recurrent laryngeal nerves pass medially, caudal to the aortic arches, to supply the developing larynx. With elongation of the neck and caudal migration of the heart, the recurrent nerves are caught up and dragged down by the descending aortic arches. On the right side the 5th and distal part of the 6th arch absorb, leaving the nerve to hook round the 4th arch (i.e. the right subclavian artery). On the left side, the nerve remains looped around the persisting distal part the 6th arch (the ligamentum arteriosum) which is overlapped and dwarfed by the arch of the aorta.

The fetal circulation (Fig. 32)

The circulation of the blood in the embryo is a remarkable example of economy in nature and results in the shunting of well-oxygenated blood from the placenta to the brain and the heart, leaving relatively desaturated blood for less essential structures.

Blood is returned from the placenta by the umbilical vein to the inferior vena cava and thence the right atrium, most of it by-passing the liver in the *ductus venosus*. Relatively little mixing of oxygenated and deoxygenated blood occurs in the right atrium since the valve overlying the orifice of the inferior vena cava serves to direct the flow of oxygenated blood from that vessel through the *foramen ovale* into the left atrium, while the deoxygenated stream from the superior vena cava is directed through the tricuspid valve into the right ventricle. From the left atrium the oxygenated blood (together with a small amount of deoxygenated blood from the lungs) passes into the left ventricle and hence into the ascending aorta for the supply of the brain and heart via the vertebral, carotid and coronary arteries.

As the lungs of the fetus are inactive, most of the deoxygenated blood from the right ventricle is short-circuited by way of the *ductus*

Brachiocephalic artery

Right pulmonary artery

Aorta

Superior vena cava

Septum II

Foramen ovale

Septum I

Inferior vena cava

Left common carotid artery

Left subclavian artery

Ductus arteriosus

Left pulmonary artery

Pulmonary trunk

Aorta

Umbilical arteries

Fig. 32 The fetal circulation.

arteriosus from the pulmonary trunk into the descending aorta. This blood supplies the abdominal viscera and the lower limbs and is shunted to the placenta, for oxygenation, along the *umbilical arteries* arising from the internal iliac arteries.

At birth, expansion of the lungs leads to an increased blood flow in the pulmonary arteries; the resulting pressure changes in the two atria bring the *septum primum* and *septum secundum* into apposition and effectively close off the foramen ovale. At the same time active contraction of the muscular wall of the ductus arteriosus results in a functional closure of this arterial shunt and, in the course of the next 2–3 months, its complete obliteration. Similarly, ligature of the umbilical cord is followed by thrombosis and obliteration of the umbilical vessels.

Congenital abnormalities of the heart and great vessels

The complex development of the heart and major arteries accounts for the multitude of congenital abnormalities which may affect these structures, either alone or in combination.

Dextro-rotation of the heart means that this organ and its emerging vessels lie as a mirror-image to the normal anatomy. It may be associated with reversal of all the intra-abdominal organs; I have seen a student correctly diagnose acute appendicitis as the cause of a patient's severe *left* iliac fossa pain because he found that the apex beat of the heart was on the right side!

Septal defects

At birth, the septum primum and septum secundum are forced together, closing the flap valve of the foramen ovale. Fusion usually takes place about 3 months after birth. In about 10% of subjects, this fusion may be incomplete. However, the two septa overlap and this patency of the foramen ovale is of no functional significance. If the septum secundum is too short to cover the foramen secundum in the septum primum, an *atrial septal defect* persists after the septum primum and septum secundum are pressed together at birth. This results in an *ostium secundum defect*, which allows shunting of blood from the left to the right atrium. This defect lies high up in the atrial wall and is relatively easy to close surgically. A more serious atrial septal defect results if the septum primum fails to fuse with the endocardial cushions. This *ostium primum defect* lies immediately above the atrioventricular boundary and may be associated with a defect of the pars membranacea septi of the ventricular septum. In such a case, the child is born with both an atrial and ventricular septal defect.

Occasionally the ventricular septal defect is so huge that the ventricles form a single cavity, giving a *trilocular heart*.

Congenital pulmonary stenosis may affect the trunk of the pulmonary artery, its valve or the infundibulum of the right ventricle. If stenosis occurs in conjunction with a septal defect, the compensatory hypertrophy of the right ventricle (developed to force blood through the pulmonary obstruction) develops a sufficiently high pressure to shunt blood through the defect into the left heart; this mixing of the deoxygenated right heart blood with the oxygenated left-sided blood results in the child being *cyanosed at birth*.

The commonest combination of congenital abnormalities causing cyanosis is *Fallot's tetralogy* (Fig. 33). This results from unequal division of the truncus arteriosus by the spinal septum, resulting in a stenosed pulmonary trunk and a wide aorta which overrides the orifices of both the ventricles. The displaced septum is unable to close the interventricular septum, which results in a ventricular septal defect. Right ventricular hypertrophy develops as a consequence of the pulmonary stenosis. Cyanosis results from the shunting of large amounts of unsaturated blood from the right ventricle through the

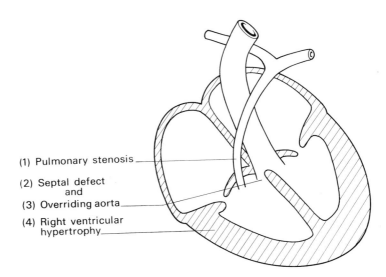

(1) Pulmonary stenosis

(2) Septal defect
and

(3) Overriding aorta

(4) Right ventricular
hypertrophy

Fig. 33 The tetralogy of
Fallot.

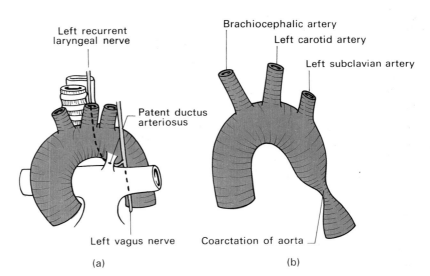

Left recurrent
laryngeal nerve

Patent ductus
arteriosus

Left vagus nerve

Brachiocephalic artery

Left carotid artery

Left subclavian artery

Coarctation of aorta

Fig. 34 (a) Patent ductus
arteriosus—showing its
close relationship to the
left recurrent laryngeal
nerve. (b) Coarctation of
the aorta.

(a)

(b)

ventricular septal defect into the left ventricle and also directly into
the aorta.

A *patent ductus arteriosus* (Fig. 34a) is a relatively common congenital defect. If left uncorrected, it causes progressive work hypertrophy of the left heart and pulmonary hypertension.

Aortic coarctation (Fig. 34b) is thought to be due to an abnormality of the obliterative process which normally occludes the ductus
arteriosus. There may be an extensive obstruction of the aorta from
the left subclavian artery to the ductus, which is widely patent and
maintains the circulation to the lower parts of the body; often there

are multiple other defects and frequently infants so afflicted die at an early age. More commonly there is a short segment involved in the region of the ligamentum arteriosum or still patent ductus. In these cases, circulation to the lower limb is maintained via collateral arteries around the scapula anastomosing with the intercostal arteries, and via the link-up between the internal thoracic and inferior epigastric arteries.

Clinically, this circulation may be manifest by enlarged vessels being palpable around the scapular margins; radiologically, dilatation of the engorged intercostal arteries results in notching of the inferior borders of the ribs.

Abnormal development of the primitive aortic arches may result in the aortic arch being on the right or actually being double. An abnormal right subclavian artery may arise from the dorsal aorta and pass behind the oesophagus—a rare cause of difficulty in swallowing (dysphagia lusoria).

Rarely, the division of the truncus into aorta and pulmonary artery is incomplete, leaving an *aorta–pulmonary window*, the most unusual congenital fistula between the two sides of the heart.

The superior mediastinum

This is bounded in front by the manubrium sterni and behind the first four thoracic vertebrae (Fig. 22). Above, it is in direct continuity with the root of the neck and below it is continuous with the three compartments of the inferior mediastinum. Its principal contents are: the great vessels, trachea, oesophagus, thoracic duct, vagi, left recurrent laryngeal nerve and the phrenic nerves (Fig. 17).

The *arch of the aorta* is directed anteroposteriorly, its three great branches, the *brachiocephalic*, *left carotid* and *left subclavian arteries*, ascend to the thoracic inlet, the first two forming a V around the trachea. The *brachiocephalic veins* lie in front of the arteries, the left running almost horizontally across the superior mediastinum and the right vertically downwards; the two unite to form the *superior vena cava*. Posteriorly lies the trachea with the oesophagus immediately behind it lying against the vertebral column.

The oesophagus

The oesophagus, which is 10 in (25 cm) long, extends from the level of the lower border of the cricoid cartilage to the cardiac orifice of the stomach (Fig. 35).

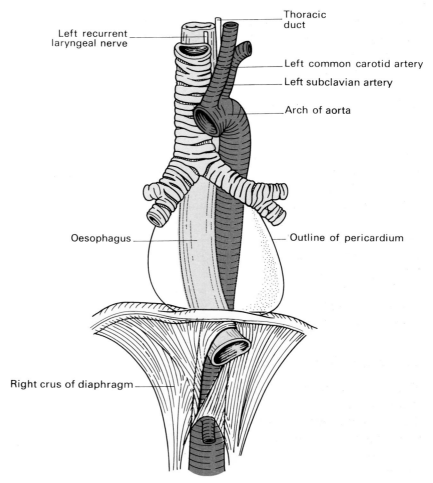

Fig. 35 The oesophagus and its relations.

Course and relations

Cervical

In the neck it commences in the median plane and deviates slightly to the left as it approaches the thoracic inlet. The trachea and the thyroid gland are its immediate anterior relations, the lower cervical vertebrae and prevertebral fascia are behind it and on either side it is related to the common carotid arteries and the recurrent laryngeal nerves. On the left side it is also related to the subclavian artery and the terminal part of the thoracic duct (Fig. 16).

Thoracic

The thoracic part traverses first the superior and then the posterior mediastinum. From being somewhat over to the left, it returns to the midline at T5 then passes downwards, forwards and to the left to reach the oesophageal opening in the diaphragm (T10). For convenience, the relations of this part are given in sequence from above downwards.

Anteriorly, it is crossed by the trachea, the left bronchus (which constricts it), the pericardium (separating it from the left atrium) and the diaphragm.

Posteriorly lie the thoracic vertebrae, the thoracic duct, the azygos vein and its tributaries and, near the diaphragm, the descending aorta.

On the left side it is related to the left subclavian artery, the terminal part of the aortic arch, the left recurrent laryngeal nerve, the thoracic duct and the left pleura. In the posterior mediastinum it relates to the descending thoracic aorta before this passes posteriorly to the oesophagus above the diaphragm.

On the right side there is the pleura and the azygos vein.

Below the root of the lung the vagi form a plexus on the oesophagus, the left vagus lying anteriorly, the right posteriorly.

In the abdomen, passing forwards through the opening in the right crus of the diaphragm, the oesophagus comes to lie in the oesophageal groove on the posterior surface of the left lobe of the liver, covered by peritoneum on its anterior and left aspects. Behind it is the left crus of the diaphragm.

Structure

The oesophagus is made of:

1   an outer connective tissue sheath of areolar tissue;
2   a muscular layer of external longitudinal and internal circular fibres which are striated in the upper two-thirds and smooth in the lower one-third;
3   a submucous layer containing mucous glands;
4   a mucosa of stratified epithelium passing abruptly into the columnar epithelium of the stomach.

Blood supply is from the inferior thyroid artery, branches of the descending thoracic aorta and the left gastric artery. The veins from the cervical part drain into the inferior thyroid veins, from the thoracic portion into the azygos vein and from the abdominal portion partly into the azygos and partly into the left gastric veins.

The *lymphatic drainage* is from a peri-oesophageal lymph plexus into the posterior mediastinal nodes, which drain both into the supra-clavicular nodes and into nodes around the left gastric vessels. It is not uncommon to be able to palpate hard, fixed supraclavicular nodes in patients with advanced oesophageal cancer.

Radiographically, the oesophagus is studied by X-rays taken after a barium swallow, in which it is seen lying in the retrocardiac space just in front of the vertebral column. Anteriorly, the normal oesophagus is indented from above downwards by the three most important structures that cross it, the arch of the aorta, the left bronchus and the left atrium.

Clinical features

1   For oesophagoscopy, measurements are made from the upper incisor teeth; the three important levels 7 in (17 cm), 11 in (28 cm) and 17 in (43 cm) corresponding to the commencement of the oesophagus, the point at which it is crossed by the left bronchus and its termination respectively.

2   These three points also indicate the narrowest parts of the oesophagus: the sites at which, as might be expected, swallowed foreign bodies are most likely to become impacted and strictures to occur after swallowing corrosive fluids.

3   The anastomosis between the azygos (systemic) and left gastric (portal) venous tributaries in the oesophageal veins is of great importance. In portal hypertension these veins distend into large collateral channels, *oesophageal varices*, which may then rupture with severe haemorrhage (probably as a result of peptic ulceration of the overlying mucosa).

4   Use is made of the close relationship between the oesophagus and the left atrium in determining the degree of left atrial enlargement in mitral stenosis; a barium swallow may show marked backward displacement of the oesophagus caused by the dilated atrium.

5   The oesophagus is crossed solely by the vena azygos on the right side. This is therefore the side of election to approach the oesophagus surgically.

Development of the oesophagus

The oesophagus develops from the distal part of the primitive fore-gut. From the floor of the fore-gut also differentiate the larynx and trachea, first as a groove (the laryngotracheal groove) which then converts into a tube, a bud on each side of which develops and ramifies into the lung.

This close relationship between the origins of the oesophagus and trachea accounts for the relatively common malformation in which

← T4

Fig. 36 The usual form of oesophageal
stenosis. The upper oesophagus ends
blindly; the lower oesophagus
communicates with the trachea at the level
of the 4th thoracic vertebra.

the upper part of the oesophagus ends blindly while the lower part
opens into the lower trachea at the level of T4 (*oesophageal atresia
with tracheo-oesophageal fistula*). Less commonly, the upper part of
the oesophagus opens into the trachea, or oesophageal atresia occurs
without concomitant fistula into the trachea. Rarely, there is a
tracheo-oesophageal fistula without atresia (Fig. 36).

The thoracic duct (Figs 37 and 213)

The *cisterna chyli* lies between the abdominal aorta and right crus of
the diaphragm. It drains lymphatics from the abdomen and the lower
limbs, then passes upwards through the aortic opening to become the
thoracic duct. This ascends behind the oesophagus, inclines to the left
of the oesophagus at the level of T5, then runs upwards behind the
carotid sheath, descends over the subclavian artery and drains into
the commencement of the left brachiocephalic vein (see Fig. 213).

The left *jugular, subclavian* and *mediastinal lymph trunks*, draining the left side of the head and neck, upper limb and thorax respectively, usually join the thoracic duct, although they may open directly
into the adjacent large veins at the root of the neck.

The thoracic duct thus usually drains the whole lymphatic field
below the diaphragm and the left half of the lymphatics above it.

On the right side, the right subclavian, jugular and mediastinal
trunks may open independently into the great veins. Usually the subclavian and jugular trunks first join into a *right lymphatic duct* and
this may be joined by the mediastinal trunk so that all three then have
a common opening into the origin of the right brachiocephalic vein.

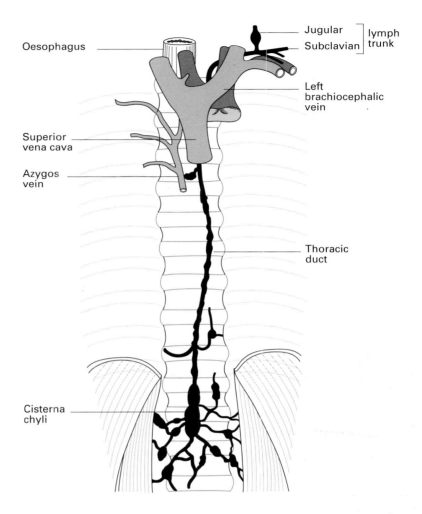

Oesophagus

Jugular ⎤ lymph
Subclavian ⎦ trunk

Left
brachiocephalic
vein

Superior
vena cava

Azygos
vein

Thoracic
duct

Cisterna
chyli

Fig. 37 The course of the
thoracic duct.

Clinical features

1 The lymphatics may become blocked by infection and fibrosis due
to the *Microfilaria bancrofti*. This usually results in lymphoedema of
the legs and scrotum but occasional involvement of the main channels
of the trunk and thorax is followed by chylous ascites, chyluria and
chylous pleural effusion.

2 The thoracic duct may be damaged during block dissection of the
neck. If noticed at operation, the injured duct should be ligated;
lymph then finds its way into the venous system by anastomosing
channels. If the accident is missed, there follows an unpleasant
chylous fistula in the neck.

3 Tears of the thoracic duct have also been reported as a complication of fractures of the thoracic vertebrae to which, in its lower part,

the duct is closely related. Such injuries are followed by a chylothorax.

The thoracic sympathetic trunk (Fig. 38)

The sympathetic chain lies immediately lateral to the mediastinum behind the parietal pleura.

Descending from the cervical chain, it crosses:

- the neck of the first rib;
- the heads of the 2nd to 10th ribs;
- the bodies of the 11th and 12th thoracic vertebrae.

It then passes behind the medial arcuate ligament of the diaphragm to continue as the lumbar sympathetic trunk.

The thoracic chain bears a ganglion for each spinal nerve; the first frequently joins the inferior cervical ganglion to form the *stellate ganglion*. Each ganglion receives a white ramus communicans containing preganglionic fibres from its corresponding spinal nerve and donates back a grey ramus, bearing postganglionic fibres.

Branches

1   Sympathetic fibres are distributed to the skin with each of the thoracic spinal nerves.
2   Postganglionic fibres from T1–5 are distributed to the thoracic viscera—the heart and great vessels, the lungs and the oesophagus.
3   Mainly preganglionic fibres from T5–12 form the *splanchnic nerves*, which pierce the crura of the diaphragm and pass to the coeliac and renal ganglia from which they are relayed as postganglionic fibres to the abdominal viscera. These splanchnic nerves are the:

- *greater splanchnic* (T5–10);
- *lesser splanchnic* (T10–11);
- *least splanchnic* (T12).

They lie medial to the sympathetic trunk on the bodies of the thoracic vertebra and are quite easily visible through the parietal pleura.

Clinical features

1   Bilateral thoraco-abdominal sympathectomy is now rarely performed in cases of severe essential hypertension, since the majority of patients can be controlled medically. The sympathetic chain is removed from T5 to L2 and the splanchnic nerves excised. This results in considerable dilatation of the splanchnic vascular bed, released from sympathetic control, with consequent fall in blood pressure.

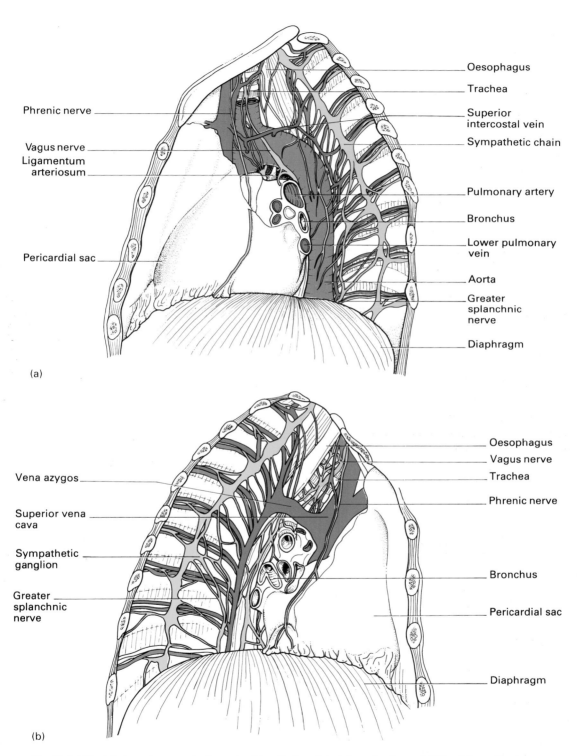

Oesophagus

Trachea

Superior
intercostal vein

Sympathetic chain

Phrenic nerve

Vagus nerve
Ligamentum
arteriosum

Pulmonary artery

Bronchus

Lower pulmonary
vein

Pericardial sac

Aorta

Greater
splanchnic
nerve

Diaphragm

(a)

Oesophagus

Vagus nerve

Trachea

Phrenic nerve

Vena azygos

Superior vena
cava

Sympathetic
ganglion

Bronchus

Greater
splanchnic
nerve

Pericardial sac

Diaphragm

(b)

Fig. 38 (a) The left thoracic sympathetic trunk with a display of the left mediastinum. (b) The right mediastinum.

2   A high spinal anaesthetic will produce temporary hypotension by paralysing the sympathetic (vasoconstrictor) preganglionic outflow from spinal segment T5 downwards, passing to the abdominal viscera.

On the examination of a chest radiograph

The following features should be examined in every radiograph of the chest.

Centering and density of film

The sternal ends of the two clavicles should be equidistant from the shadow of the vertebral spines. The assessment of the density of the film can only be learned by experience, but in a 'normal' film the bony cage should be clearly outlined and the larger vessels in the lung fields clearly visible.

General shape

Any abnormalities in the general form of the thorax (scoliosis, kyphosis and the barrel chest of emphysema, for example) should always be noted before other abnormalities are described.

Bony cage

The thoracic vertebrae should be examined first, then each of the ribs in turn (counting conveniently from their posterior ends and comparing each one with its fellow of the opposite side), and finally clavicles and scapulae. Unless this procedure is carried out systematically, important diagnostic clues (e.g. the presence of a cervical rib, or notching of the ribs by enlarged anastomotic vessels) are liable to be missed.

The domes of the diaphragm

These should be examined for height and symmetry and the nature of the cardiophrenic and costophrenic angles observed.

The mediastinum

The outline of the mediastinum should be traced systematically. Special note should be made of the size of the heart, of mediastinal shift and of the vessels and nodes at the hilum of the lung.

Lung fields

Again, systematic examination of the lung fields visible in each inter-costal space is necessary if slight differences between the two sides are not to be overlooked.

Abnormalities

When this scheme has been carefully followed, any abnormalities in the bony cage, the mediastinum or lung fields should now be appar-ent. They should then be defined anatomically as accurately as possi-ble and checked, where necessary, by reference to a film taken from a different angle.

Radiographic appearance of the heart

For the appearance of the heart as seen at fluoroscopy, reference should be made to a standard work in radiology or cardiology. In the present account, only the more important features of the heart and great vessels which can be seen in standard posteroanterior and oblique lateral radiographs of the chest will be described.

The heart and great vessels in anteroposterior radiographs (Fig. 39)

The greater part of the 'mediastinal shadow' in an anteroposterior film of the chest is formed by the heart and great vessels. These should be examined as follows.

Size and shape of the heart

Normally the transverse diameter should not exceed half the total width of the chest, but since it varies widely with bodily build and the position of the heart, these factors must also be assessed. The *shape* of the cardiac shadow also varies a good deal with the position of the heart, being long and narrow in a vertically disposed heart and broad and rounded in the so-called horizontal heart.

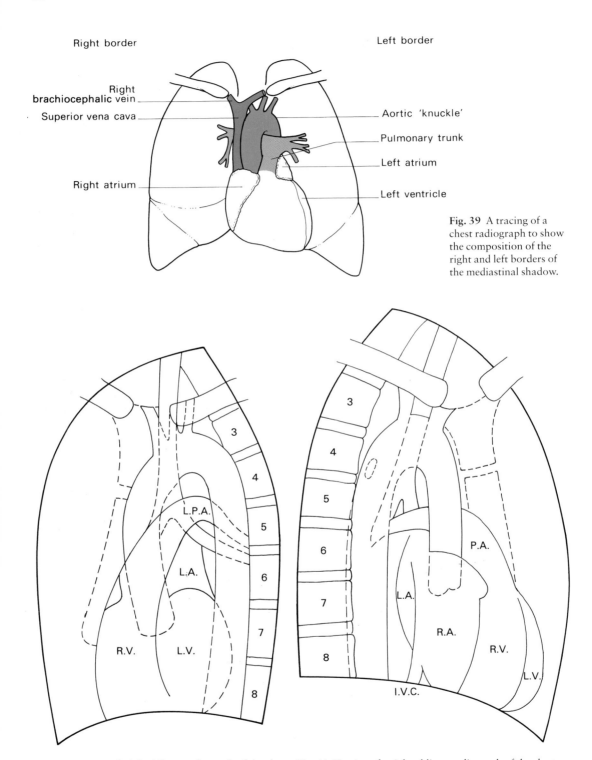

Right border

Left border

Right
brachiocephalic vein

Superior vena cava

Right atrium

Aortic 'knuckle'

Pulmonary trunk

Left atrium

Left ventricle

Fig. 39 A tracing of a chest radiograph to show the composition of the right and left borders of the mediastinal shadow.

Fig. 40 Tracing of a left oblique radiograph of the chest.   **Fig. 41** Tracing of a right oblique radiograph of the chest.

The cardiac outline

Each 'border' of the cardiac shadow should be examined in turn. The *right border* of the mediastinal shadow is formed from above downwards by the right brachiocephalic vein, the superior vena cava and the right atrium. Immediately above the heart, the *left border* of the mediastinal shadow presents a well-marked projection, the *aortic knuckle*, which represents the arch of the aorta seen 'end-on'. Beneath this there are, successively, the shadows due to the pulmonary trunk (or the infundibulum of the right ventricle), the auricle of the left atrium, and the left ventricle. The shadow of the inferior border of the heart blends centrally with that of the diaphragm, but on either side the two shadows are separated by the well-defined cardiophrenic angles.

The heart and great vessels in anterior oblique radiographs

The left oblique view (Fig. 40)

The greater part of the mediastinal shadow in this view is formed by the right and left ventricles, above which the relation of the arch of the aorta and the pulmonary trunk to the translucent trachea can be seen.

The right oblique view (Fig. 41)

Almost all of the cardiac shadow in this view is due to the right ventricle. It is particularly useful for the assessment of the size of the left atrium since its posterior wall forms the upper half of the posterior border of the cardiac shadow. This border can be defined more accurately by giving the patient barium paste to swallow; the outlined oesophagus is indented by an enlarged left atrium.

Part 2
The Abdomen and Pelvis

Surface anatomy and surface markings

Be able to identify these landmarks on yourself or the patient (Fig. 42).

The *xiphoid*. The *costal margin* extends from the 7th costal cartilage at the xiphoid to the tip of the 12th rib (although the latter is often difficult to feel); this margin bears a distinct step, which is the tip of the 9th costal cartilage.

The *iliac crest* ends in front at the *anterior superior spine* from which the *inguinal ligament* (*Poupart's ligament*) passes downwards and medially to the *pubic tubercle*. Identify this tubercle by direct palpation and also by running the fingers along the adductor longus tendon (tensed by flexing, abducting and externally rotating the thigh) to its origin at the tubercle.

Feel the firm *vas deferens* between the finger and thumb as it lies within the spermatic cord at the scrotal neck. Trace the vas upwards and note that it passes medially to the pubic tubercle and thence through the *external inguinal ring*, which can be felt by invaginating the scrotal skin with the fingertip.

Vertebral levels (Fig. 42)

- T9—*the xiphoid*.
- L1—*the transpyloric plane of Addison* lies half-way between the suprasternal notch and the pubis, or approximately a hand's breadth

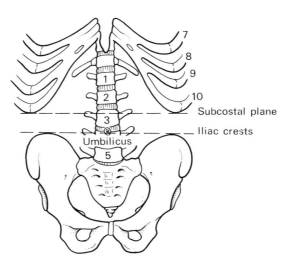

Fig. 42 Boundaries, bony landmarks and vertebral levels of the abdomen.

below the xiphoid. This plane passes through the pylorus, the pancreatic neck, the duodenojejunal flexure, the fundus of the gall-bladder, the tip of the 9th costal cartilage (felt as a distinct 'step'), and the hila of the kidneys. It also corresponds to the level of termination of the spinal cord.

• L3 — the *subcostal plane*, a line joining the lowest point of the thoracic cage on each side, which is the inferior margin of the 10th rib.

• L4 — the *plane of the iliac crests*. This corresponds to the level of the bifurcation of the aorta. It is also a useful landmark in performing a lumbar puncture, since it is well below the level of the termination of the spinal cord, which is approximately at L1 (see page 364).

• The *umbilicus* is an inconstant landmark. In the healthy adult it lies at the junction of L3 and L4 vertebrae. It is lower in the infant and, naturally, when the abdomen is pendulous.

Surface markings

The abdominal viscera are inconstant in their position but the surface markings of the following structures are of clinical value.

Liver

The lower border of the liver extends along a line from the tip of the right 10th rib to just below the left nipple; it may just be palpable in the normal subject, especially on deep inspiration. The upper border follows a line passing just below the nipple on each side.

Spleen

This underlies the 9th, 10th and 11th ribs posteriorly on the left side.

Gall-bladder

The fundus of the gall-bladder corresponds to the point where the lateral border of the rectus abdominis cuts the costal margin; this is at the tip of the 9th costal cartilage, easily detected as a distinct 'step' when the fingers are run along the costal margin.

Pancreas

The transpyloric plane defines the level of the neck of the pancreas which overlies the vertebral column. From this landmark, the head can be imagined passing downward and to the right, the body and tail passing upwards and to the left.

Aorta

This terminates just to the left of the midline at the level of the iliac crest at L4; a pulsatile swelling below this level may thus be an iliac, but cannot be an aortic, aneurysm.

Kidneys

The lower pole of the normal right kidney may be felt in the thin subject. Anteriorly, the hilum of the kidney lies on the transpyloric plane four finger breadths from the midline. Posteriorly, the upper pole of the kidney lies deep to the 12th rib. The right kidney normally extends about 1 in (2.5 cm) lower than the left. Using these landmarks, the kidney outlines can be projected on to either the anterior or posterior aspects of the abdomen.

In some perfectly normal thin people, especially women, it is possible to palpate the lower border of the liver, the lower pole of the right kidney, the aorta, the caecum and the sigmoid colon.

The fasciae and muscles of the abdominal wall

Fasciae of the abdominal wall

There is no deep fascia over the trunk, only the superficial fascia. (If there were, we would presumably be unable to take a deep breath or enjoy a large meal!) This, in the lower abdomen, forms a *superficial fatty layer* (*of Camper*) and a deeper *fibrous layer* (*of Scarpa*). The fatty layer is continuous with the superficial fat of the rest of the body, but the fibrous layer blends with the deep fascia of the upper thigh, extends into the penis and scrotum (or labia majora), and into the perineum as *Colles' fascia*. In the perineum it is attached behind to the perineal body and posterior margin of the perineal membrane and, laterally, to the rami of the pubis and ischium. It is because of these attachments that a rupture of the urethral bulb may be followed by extravasation of blood and urine into the scrotum, perineum and penis and then into the lower abdomen deep to the fibrous fascial plane, but not by extravasation downwards into the lower limb, from which the fluid is excluded by the attachment of the fascia to the deep fascia of the upper thigh.

Nerve supply

The segmental nerve supply of the abdominal muscles and the overlying skin is derived from T7 to L1. This distribution can be mapped out approximately if it is remembered that the umbilicus is supplied by T10 and the groin and scrotum by L1 (via the ilio-inguinal and iliohypogastric nerves—see Fig. 140).

The muscles of the anterior abdominal wall

These are of considerable practical importance because their anatomy forms the basis of abdominal incisions.

The *rectus abdominis* (Fig. 43) arises on a 3 in (7.5 cm) horizontal line from the 5th, 6th and 7th costal cartilages and is inserted for a length of 1 in (2.5 cm) into the crest of the pubis. At the tip of the xiphoid, at the umbilicus and half-way between, are three constant *transverse tendinous intersections*; below the umbilicus there is sometimes a fourth. These intersections are seen only on the anterior aspect of the muscle and here they adhere to the anterior rectus sheath. Posteriorly they are not in evidence and, in consequence, the

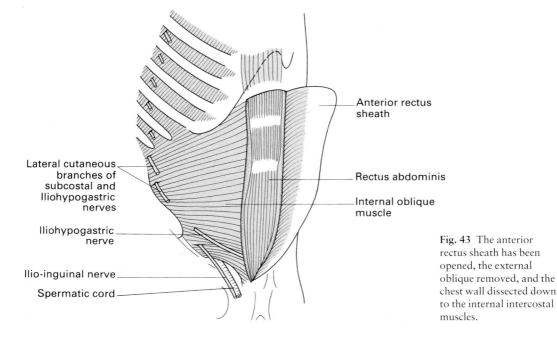

Lateral cutaneous branches of subcostal and Iliohypogastric nerves

Iliohypogastric nerve

Ilio-inguinal nerve

Spermatic cord

Anterior rectus sheath

Rectus abdominis

Internal oblique muscle

Fig. 43 The anterior rectus sheath has been opened, the external oblique removed, and the chest wall dissected down to the internal intercostal muscles.

Fig. 44 The composition
of the rectus sheath
shown in transverse
section (a) above the
costal margin, (b) above
the arcuate line and (c)
below the arcuate line.

rectus muscle is completely free behind. At each intersection, vessels from the superior epigastric artery and vein pierce the rectus.

The sheath in which the rectus lies is formed, to a large extent, by the aponeurotic expansions of the lateral abdominal muscles (Fig. 44).

(a) Above the costal margin, the anterior sheath comprises the external oblique aponeurosis only; posteriorly lie the costal cartilages.

(b) From the costal margin to a point half-way between umbilicus and pubis, the external oblique and the anterior part of the internal oblique aponeurosis form the anterior sheath. Posteriorly lie the posterior part of this split internal oblique aponeurosis and the aponeurosis of transversus abdominis.

(c) Below a point half-way between umbilicus and pubis, all the aponeuroses pass in front of the rectus so that the anterior sheath here comprises the tendinous expansions of all three oblique muscles blended together. The posterior wall at this level is made up of the only other structures available — the transversalis fascia and peritoneum.

The posterior junction between (b) and (c) is marked by the *arcuate line of Douglas*, which is the lower border of the posterior aponeurotic part of the rectus sheath. At this point the inferior epigastric artery and vein (from the external iliac vessels) enter the sheath, pass upwards and anastomose with the superior epigastric vessels which are terminal branches of the internal thoracic artery and vein.

The rectus sheaths fuse in the midline to form the *linea alba* stretching from the xiphoid to the pubic symphysis.

The lateral muscles of the abdominal wall comprise the external and internal oblique and the transverse muscles. They are clinically important in making up the rectus sheath and the inguinal canal, and also because they must be divided in making lateral abdominal incisions.

Their attachments can be remembered when one bears in mind that they fill the space between the costal margin above, the iliac crest below, and the lumbar muscles covered by lumbar fascia behind. Medially, as already noted, they constitute the rectus sheath and thence blend into the linea alba from xiphoid to pubic crest.

The *obliquus externus abdominis* (external oblique) arises from the outer surfaces of the lower eight ribs and fans out into the xiphoid, linea alba, the pubic crest, pubic tubercle and the anterior half of the iliac crest.

From the pubic tubercle to the anterior superior iliac spine its lower border forms the aponeurotic *inguinal ligament of Poupart*.

The *obliquus internus abdominis* (internal oblique) arises from the lumbar fascia, the anterior two-thirds of the iliac crest and the lateral two-thirds of the inguinal ligament. It is inserted into the lowest six costal cartilages, linea alba and the pubic crest.

The *transversus abdominis* arises from the lowest six costal cartilages (interdigitating with the diaphragm), the lumbar fascia, the anterior two-thirds of the iliac crest and the lateral one-third of the inguinal ligament; it is inserted into the linea alba and the pubic crest.

Note that the external oblique passes downwards and forwards, the internal oblique upwards and forwards and the transversus transversely. Note also that the external oblique has its posterior border free but the deeper two muscles both arise posteriorly from the lumbar fascia.

The anatomy of abdominal incisions

Incisions to expose the intraperitoneal structures represent a compromise on the part of the operator. On the one hand he requires maximum access; on the other hand he wishes to leave a scar which lies, if possible, in an unobtrusive crease, and which will have done minimal damage to the muscles of the abdominal wall and to their nerve supply.

The nerve supply to the lateral abdominal muscles forms a richly communicating network so that cuts across the lines of fibres of these muscles, with division of one or two nerves, produce no clinical ill-effects. The segmental nerve supply to the rectus, however, has little cross-communication and damage to these nerves must, if possible, be avoided.

The copious anastomoses between the blood vessels supplying the abdominal muscles make damage to these by operative incisions of no practical importance.

Midline incision

The midline incision is made through the linea alba. Superiorly, this is a relatively wide fibrous structure, but below the umbilicus it becomes almost hair-line and the surgeon may experience difficulty in finding the exact point of cleavage between the recti at this level. Being made of fibrous tissue only, it provides an almost bloodless line along which the abdomen can be opened rapidly and, if necessary, from Dan in the North to Beersheba in the South!

Paramedian incision

The paramedian incision is placed 1 in (2.5 cm) to 1.5 in (4 cm) lateral, and parallel, to the midline; the anterior rectus sheath is opened, the rectus displaced laterally and the posterior sheath, together with peritoneum, then incised. This incision has the advantage that, on suturing the peritoneum, the rectus slips back into place to cover and protect the peritoneal scar.

The adherence of the anterior sheath to the rectus muscle at its tendinous intersections means that the sheath must be dissected off the muscle at each of these sites, and at each of these a segmental vessel requires division. Having done this, the rectus is easily slid laterally from the posterior sheath from which it is quite free. The posterior sheath and the peritoneum form a tough membrane down to half-way between pubis and umbilicus, but it is much thinner and more fatty below this where, as we have seen, it loses its aponeurotic component and is made up of only transversalis fascia and peritoneum. The inferior epigastric vessels may be seen passing under the arcuate line of Douglas in the posterior sheath and usually require division in a low paramedian incision.

The transrectus incision

Occasionally, the rectus muscle is split in the line of the paramedian incision. The rectus receives its nerve supply laterally and the muscle medial to the incision must, in consequence, be deprived of its innervation and undergo atrophy; it is an incision therefore best avoided.

Pararectus incision

The pararectus (Battle) incision opens the rectus sheath and displaces the rectus medially. It was once popular but has the disadvantage that extending this incision longitudinally may damage the nerves running across segmentally to supply the rectus.

Subcostal incision

The subcostal (Kocher) incision is used on the right side in biliary surgery and, on the left, in exposure of the spleen. The skin incision commences at the midline and extends parallel to, and 1 in (2.5 cm) below, the costal margin.

The anterior rectus sheath is opened, the rectus cut and the posterior sheath with underlying adherent peritoneum incised. The small 8th intercostal nerve branch to the rectus is sacrificed but the larger and more important 9th nerve, in the lateral part of the wound, is preserved. The divided rectus muscle is held by the intersections above and below and retracts very little. It subsequently heals by fibrous tissue. This incision is valuable in the patient with the wide subcostal angle. Where this angle is narrow, the paramedian incision is usually preferred.

The muscle split or gridiron approach to the appendix

The oblique skin incision centred at *McBurney's point* (two-thirds of the way laterally along the line from the umbilicus to the anterior superior iliac spine) is now less popular than an almost transverse incision in the line of the skin crease forwards from, and 1 in (2.5 cm) above, the anterior spine.

The aponeurosis of the external oblique is incised in the line of its fibres (obliquely downwards and medially); the internal oblique and transversus muscles are then split in the line of their fibres, and retracted without their having to be divided. On closing the incision, these muscles snap together again, leaving a virtually undamaged abdominal wall.

Transverse and oblique incisions

Incisions cutting through the lateral abdominal muscles do not damage their richly anastomosing nerve supply and heal without weakness. They are useful, for example, in exposing the sigmoid colon or the caecum or, by displacing the peritoneum medially, extraperitoneal structures such as the ureter, sympathetic chain and the external iliac vessels.

Thoraco-abdominal incisions

An upper paramedian or upper oblique abdominal incision can be extended through the 8th or 9th intercostal space, the diaphragm incised and an extensive exposure achieved of both upper abdomen and thorax. This is used, for example, in removing growths of the upper stomach or lower oesophagus and in resection of the right lobe of the liver.

Paracentesis abdominis

Intraperitoneal fluid collections can be evacuated via a cannula inserted through the abdominal wall. The bladder having been first emptied with a catheter, the cannula is introduced on a trocar either through the midline (where the linea alba is relatively bloodless) or lateral to McBurney's point (where there is no danger of wounding the inferior epigastric vessels). The coils of gut are not in danger in this procedure because they are mobile and are pushed away by the tip of the trocar.

The inguinal canal (Fig. 45)

This canal represents the oblique passage taken through the lower abdominal wall by the testis and cord (the round ligament in the female).

Questions on the anatomy of this region are probably asked more often than any other in examinations because of its importance in diagnosis and treatment of hernias.

The canal is 1.5 in (4 cm) long. It passes downwards and medially from the internal to the external inguinal rings and lies parallel to and immediately above, the inguinal ligament.

Relations

- Anteriorly — the skin, superficial fascia and the external oblique aponeurosis cover the full length of the canal; the internal oblique covers its lateral one-third.
- Posteriorly — the *conjoint tendon* forms the posterior wall of the canal medially, the transversalis fascia laterally. (The conjoint tendon represents the fused common insertion of the internal oblique and transversus into the pubic crest.)

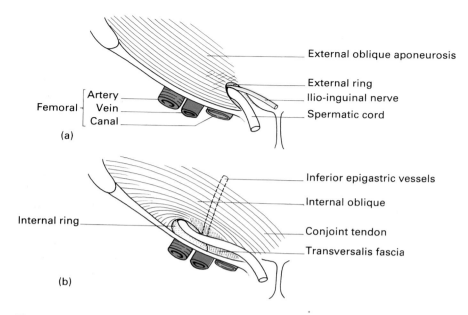

External oblique aponeurosis

External ring
Ilio-inguinal nerve
Spermatic cord

Femoral { Artery
Vein
Canal

(a)

Inferior epigastric vessels

Internal oblique

Internal ring

Conjoint tendon

Transversalis fascia

(b)

Fig. 45 The right inguinal canal (a) with the external oblique aponeurosis intact, (b) with the aponeurosis laid open.

- Above — arch the lowest fibres of the internal oblique and transversus abdominis.
- Below — lies the inguinal ligament.

The *internal (or deep) ring* represents the point at which the spermatic cord pushes through the transversalis fascia, dragging from it a covering which forms the *internal spermatic fascia*. This ring is demarcated medially by the inferior epigastric vessels passing upwards from the external iliac artery and vein.

The *external (or superficial) ring* is a V-shaped defect in the external oblique aponeurosis and lies immediately above and medial to the pubic tubercle. As the cord traverses this opening, it carries the *external spermatic fascia* from the ring's margins.

The inguinal canal transmits the spermatic cord and the ilio-inguinal nerve in the male and the round ligament and ilio-inguinal nerve in the female.

The *spermatic cord* comprises (Fig. 46):
- three layers of fascia — the *external spermatic*, from the external oblique aponeurosis; the *cremasteric*, from the internal oblique aponeurosis (containing muscle fibres termed the *cremaster muscle*); the *internal spermatic*, from the transversalis fascia;
- three arteries — the testicular (from the aorta); the cremasteric (from the inferior epigastric artery); the artery of the vas (from the inferior vesical artery);

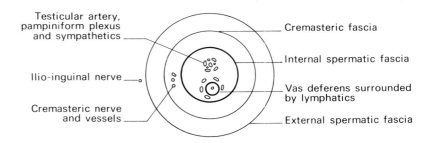

Fig. 46 Scheme of the spermatic cord and its contents, in transverse section.

- three nerves—the nerve to the cremaster (from the genito-femoral nerve); sympathetic fibres; the ilio-inguinal nerve (strictly, *on* and not *in* the cord);
- three other structures—the vas deferens; the pampiniform plexus of veins (draining the right testis into the inferior vena cava and the left testis into the renal vein); lymphatics which drain the testis to the aortic lymph nodes.

Clinical features

An *indirect inguinal hernia* passes through the internal ring, along the canal and then, if large enough, emerges through the external ring and descends into the scrotum. If reducible, such a hernia can be completely controlled by pressure with the fingertip over the internal ring, which lies 0.5 in (12 mm) above the point where the femoral artery passes under the inguinal ligament, i.e. 0.5 in (12 mm) above the femoral pulse. This pulse can be felt at the mid-inguinal point, half-way between the anterior superior iliac spine and the symphysis pubis (see Fig. 153).

If the hernia protrudes through the external ring, it can be felt to lie above and medial to the pubic tubercle, and is thus differentiated from a femoral hernia emerging from the femoral canal, which lies below and lateral to this landmark (see Fig. 176).

A *direct inguinal hernia* pushes its way directly forwards through the posterior wall of the inguinal canal. Since it lies medial to the internal ring, it is not controlled by digital pressure applied immediately above the femoral pulse. Occasionally, a direct hernia becomes large enough to push its way through the external ring and then into the neck of the scrotum. This is so unusual that one can usually assume that a scrotal hernia is an indirect hernia.

The only certain way of determining the issue is at operation; the inferior epigastric vessels demarcate the medial edge of the internal ring, therefore an indirect hernia sac will pass lateral and a direct hernia medial to these vessels. Quite often both a direct and an indirect hernia coexist; they bulge through on each side of the inferior epigastric vessels like the legs of a pair of pantaloons.

Peritoneal cavity

The endothelial lining of the primitive coelomic cavity of the embryo becomes the thoracic pleura and the abdominal peritoneum. Each is invaginated by ingrowing viscera which thus come to be covered by a serous membrane and to be packed snugly into a serous-lined cavity, the visceral and parietal layer respectively.

In the male, the peritoneal cavity is completely closed, but in the female it is perforated by the openings of the uterine tubes which constitute a possible pathway of infection from the exterior.

To revise the complicated attachments of the peritoneum, it is best to start at one point and trace this membrane in an imaginary round-trip of the abdominal cavity, aided by Figs 47 and 48. A convenient point of departure is the parietal peritoneum of the anterior abdominal wall below the umbilicus. At this level the membrane is smooth apart from the shallow ridges formed by the median umbilical cord (the obliterated fetal urachus passing from the bladder to the umbilicus), the medial umbilical cords (the obliterated umbilical arteries

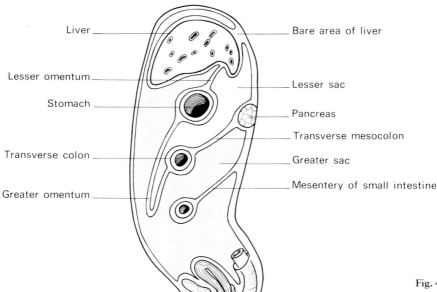

Fig. 47 The peritoneal cavity in longitudinal section (female).

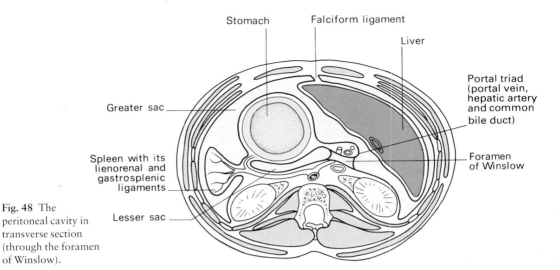

Stomach   Falciform ligament   Liver

Portal triad
(portal vein,
hepatic artery
and common
bile duct)

Greater sac

Spleen with its
lienorenal and
gastrosplenic
ligaments

Foramen
of Winslow

Lesser sac

Fig. 48 The peritoneal cavity in transverse section (through the foramen of Winslow).

passing to the umbilicus from the internal iliac arteries) and the lateral umbilical cords (the peritoneum covering the inferior epigastric vessels).

A cicatrix can usually be felt and seen at the posterior aspect of the umbilicus, and from this the *falciform ligament* sweeps upwards and slightly to the right of the midline to the liver. In the free border of this ligament lies the *ligamentum teres* (the obliterated fetal left umbilical vein) which passes into the groove between the quadrate lobe and left lobe of the liver.

Elsewhere, the peritoneum sweeps over the inferior aspect of the diaphragm, to be reflected on to the liver (leaving a bare area demarcated by the *upper* and *lower coronary ligaments* of the liver) and on to the right margin of the abdominal oesophagus. After enclosing the liver (for further details, see pages 99 and 100), the peritoneum descends from the porta hepatis as a double sheet, the *lesser omentum*, to the lesser curve of the stomach. Here it again splits to enclose this organ, reforms at its greater curve, then loops downwards, then up again to attach to the length of the transverse colon, forming the apron-like *greater omentum*.

The transverse colon, in turn, is enclosed within this peritoneum which then passes upwards and backwards as the transverse mesocolon to the posterior abdominal wall, where it is attached along the anterior aspect of the pancreas.

At the base of the transverse mesocolon, this double peritoneal sheet divides once again; the upper leaf passes upwards over the posterior abdominal wall to reflect on to the liver (at the bare area), the lower leaf passes over the lower part of the posterior abdominal wall to cover the pelvic viscera and to link up once again with the peritoneum of the anterior wall. This posterior layer is, however, interrupted by its being reflected along an oblique line running from the duodenojejunal flexure to the ileocaecal junction to form the *mesentery of the small intestine*.

The mesentery of the small intestine, the lesser and greater omenta and mesocolon all carry the vascular supply and lymph drainage of their contained viscera.

The *lesser sac* (Fig. 48) is the extensive pouch lying behind the lesser omentum and the stomach and projecting downwards (although usually this space is obliterated) between the layers of the greater omentum. Its left wall is formed by the spleen attached by the *gastrosplenic* and *lienorenal ligaments*. The right extremity of the sac opens into the main peritoneal cavity via the *epiploic foramen or foramen of Winslow* (Fig. 49), whose boundaries are as follows:

• anteriorly — the free edge of lesser omentum, containing the common bile duct to the right, hepatic artery to the left and portal vein posteriorly;
• posteriorly—the inferior vena cava;
• inferiorly — the 1st part of the duodenum, over which runs the hepatic artery before this ascends into the anterior wall of the foramen;
• superiorly—the caudate process of the liver.

Clinical features

1   Occasionally a loop of intestine passes through the foramen of Winslow into the lesser sac and becomes strangulated by the edges of the foramen. Notice that none of these important boundaries can be incised to release the strangulation; the bowel must be decompressed by a needle to allow its reduction.

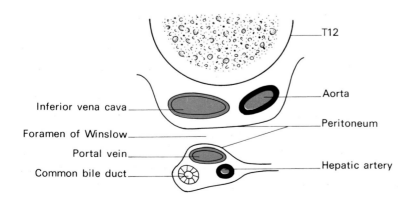

Fig. 49 The foramen of Winslow in transverse section.

2   It is important to the surgeon that the hepatic artery can be compressed between his index finger within the foramen of Winslow and his thumb on its anterior wall. If the cystic artery is torn during cholecystectomy, haemorrhage can be controlled by this manoeuvre which then enables the damaged vessel to be identified and secured.

Intraperitoneal fossae

A number of fossae occur within the peritoneal cavity into which loops of bowel may become caught and strangulated. Those of importance are:

1   the *lesser sac* via the foramen of Winslow, which we have just described;

2   *paraduodenal fossa*—between the duodenojejunal flexure and the inferior mesenteric vessels;

3   *retrocaecal fossa*—in which the appendix frequently lies;

4   *intersigmoid fossa*—formed by the inverted V attachment of the mesosigmoid.

The subphrenic spaces (Fig. 50)

Below the diaphragm are a number of potential spaces formed in relation to the attachments of the liver. One or more of these spaces may become filled with pus (a subphrenic abscess) walled off inferiorly by adhesions. There are five subdivisions of clinical importance.

The *right* and *left subphrenic spaces* lie between the diaphragm and the liver, separated from each other by the falciform ligament.

The *right* and *left subhepatic spaces* lie below the liver. The right is the *pouch of Morison* and is bounded by the posterior abdominal wall behind and by the liver above. It communicates anteriorly with the right subphrenic space around the anterior margin of the right lobe of the liver and below both open into the general peritoneal cavity from which infection may track, for example, from a perforated appendix or a perforated peptic ulcer. The left subhepatic space is the lesser sac which communicates with the right through the foramen of Winslow. It may fill with fluid as a result of a perforation in the posterior wall of the stomach or from an inflamed or injured pancreas to form a pseudocyst of the pancreas.

The *right extraperitoneal space* lies between the bare area of the liver and the diaphragm. It may become involved in retroperitoneal infections or directly from a liver abscess.

Posterior and anterior subphrenic abscesses are drained by an incision below, or through the bed of, the 12th rib. A finger is then passed upwards and forwards between liver and diaphragm to open

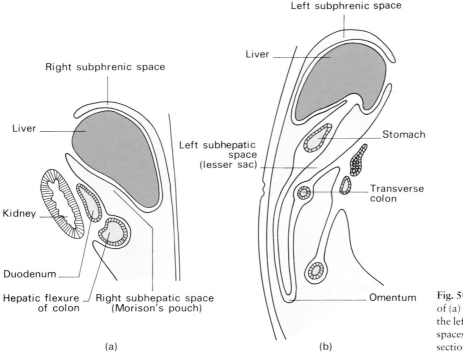

Fig. 50 The anatomy of (a) the right and (b) the left subphrenic spaces in sagittal section.

into the abscess cavity. An anteriorly placed collection of pus below the diaphragm can alternatively be drained via an incision placed below and parallel to the costal margin.

The gastrointestinal tract

The stomach

The stomach is roughly J-shaped, although its size and shape vary considerably. It tends to be high and transverse in the obese short subject and to be elongated in the asthenic individual; even in the same person, its shape depends on whether it is full or empty, on the position of the body and on the phase of respiration. The stomach has two surfaces—the anterior and posterior; two curvatures—the greater and lesser; and two orifices—the cardia and pylorus (Fig. 51).

The stomach projects to the left, above the level of the *cardia*, to form the dome-like *gastric fundus*. Between the cardia and the pylorus lies the *body* of the stomach leading to a narrow portion, immediately preceding the pylorus, which is termed the *pyloric*

Cardia

Incisura angularis

Pyloric sphincter

Fundus

Body

Pyloric antrum

Fig. 51 The stomach and its subdivisions.

antrum. The junction of the body with the pyloric antrum is marked by a distinct notch on the lesser curvature termed the *incisura angularis*. The junction of pylorus with duodenum is marked by a constriction externally and also by a *constant vein (of Mayo)* which crosses it at this level.

The thickened *pyloric sphincter* is easily felt and surrounds the lumen of the *pyloric canal*. The pyloric sphincter is an anatomical structure as well as a physiological mechanism. The cardia, on the other hand, although competent (gastric contents do not flow out of your mouth if you stand on your head), is not demarcated by a distinct anatomical sphincter. The exact nature of the cardiac sphincter action is still not fully understood, but the following mechanisms have been suggested, each supported by some experimental and clinical evidence.

1 Mucosal folds at the oesophagogastric junction act as a valve.
2 The acute angle of entry of oesophagus into stomach produces a valve-like effect.
3 The circular muscle of the lower oesophagus is a physiological, as distinct from an anatomical, sphincter.
4 The arrangement of the muscle fibres of the stomach around the cardia acts either as a sphincter or else maintains the acute angle of entry of oesophagus into stomach.
5 The right crus of the diaphragm acts as a 'pinch-cock' to the lower oesophagus as it pierces this muscle.
6 The positive intra-abdominal pressure compresses the walls of the short segment of intra-abdominal oesophagus.

Relations of the stomach (Fig. 52)

• Anteriorly — the abdominal wall, the left costal margin, the diaphragm and the left lobe of the liver.
• Posteriorly—the lesser sac, which separates the stomach from the pancreas, transverse mesocolon, left kidney, left suprarenal, the spleen and the splenic artery.
• Superiorly—the left dome of the diaphragm.

Spleen

Left kidney and suprarenal

Pancreas and splenic artery

Diaphragm

Transverse colon and mesocolon

Fig. 52 The posterior relations of the stomach; the stomach (grey tint) is superimposed upon its bed.

The lesser omentum is attached along the lesser curvature of the stomach, the greater omentum along the greater curvature. These omenta contain the vascular and lymphatic supply of the stomach.

The *arterial supply* (Fig. 53) to the stomach is extremely rich and comprises:

- the left gastric artery—from the coeliac axis;
- the right gastric artery—from the hepatic artery;
- the right gastro-epiploic artery—from the gastroduodenal branch of the hepatic artery;
- the left gastro-epiploic artery—from the splenic artery;
- the short gastric arteries—from the splenic artery.

The corresponding veins drain into the portal system.

Left gastric artery

Coeliac axis

Cystic artery

Hepatic artery

Right gastric artery

Gastroduodenal artery

Superior pancreaticoduodenal artery

Short gastric arteries

Splenic artery

Left gastro-epiploic artery

Right gastro-epiploic artery

Fig. 53 The arterial supply of the stomach.

The *lymphatic drainage* of the stomach accompanies its blood vessels. The stomach can be divided into three drainage zones (Fig. 54).

- Area I—the superior two-thirds of the stomach drain along the left and right gastric vessels to the aortic nodes.
- Area II—the right two-thirds of the inferior one-third of the stomach drain along the right gastro-epiploic vessels to the subpyloric nodes and thence to the aortic nodes.
- Area III—the left one-third of the greater curvature of the stomach drains along the short gastric and splenic vessels lying in the gastro-splenic and lienorenal ligaments, then, via the suprapancreatic nodes, to the aortic group.

This extensive lymphatic drainage and the technical impossibility of its complete removal is one of the serious problems in dealing with stomach cancer. Involvement of the nodes along the splenic vessels can be dealt with by removing spleen, gastrosplenic and lienorenal ligaments and the body and tail of the pancreas. Lymph nodes among the gastro-epiploic vessels are removed by excising the greater omentum. However, involvement of the nodes around the aorta and the head of the pancreas may render the growth incurable.

The vagal supply to the stomach (Fig. 55)

The anterior and posterior vagi enter the abdomen through the oesophageal hiatus. The anterior nerve lies close to the stomach wall

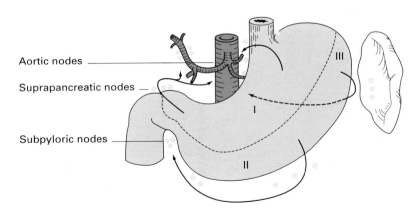

Aortic nodes

Suprapancreatic nodes

Subpyloric nodes

Fig. 54 The lymph drainage of the stomach.
- Area I drains along the right and left gastric vessels to the aortic nodes.
- Area II drains to the subpyloric and thence aortic nodes via lymphatics along the right gastro-epiploic vessels.
- Area III drains via lymphatics along the splenic vessels to the suprapancreatic nodes and thence to aortic nodes.

(a)

(b)

Fig. 55 The vagal supply to the stomach: (a) anterior vagus; (b) posterior vagus.

but the posterior, and larger, nerve is at a little distance from it. The anterior vagus supplies branches to the cardia and lesser curve of the stomach and also a large hepatic branch. The posterior vagus gives branches to both the anterior and posterior aspects of the body of the stomach but the bulk of the nerve forms the coeliac branch. This runs along the left gastric artery to the coeliac ganglion for distribution to the intestine, as far as the midtransverse colon, and the pancreas.

The exact means by which the vagal fibres reach the stomach is of considerable practical importance to the surgeon. The gastric divisions of both the anterior and posterior vagi reach the stomach at the cardia and descend along the lesser curvature between the anterior and posterior peritoneal attachments of the lesser omentum (the anterior and posterior nerves of Latarjet). The stomach is innervated by terminal branches from the anterior and posterior gastric nerves and it is, therefore, possible to divide those branches which supply the acid-secreting body of the stomach yet preserving the pyloric innervation (highly selective vagotomy, see below).

The vagus constitutes the motor and secretory nerve supply for the stomach. When divided, in the operation of *vagotomy*, the neurogenic (reflex) gastric acid secretion is abolished but the stomach is, at the same time, rendered atonic so that it empties only with difficulty; because of this, total vagotomy must always be accompanied by some sort of drainage procedure, either a pyloroplasty (to enlarge the pyloric exit and render the pyloric sphincter incompetent) or by a gastrojejunostomy (to drain the stomach into the proximal small intestine). Drainage can be avoided if the nerve of Latarjet is preserved, thus maintaining the innervation and function of the pyloric antrum (highly selective vagotomy).

Clinical features

1   A posterior gastric ulcer or cancer may erode the pancreas, giving pain referred to the back. Ulceration into the splenic artery—a direct posterior relation—may cause torrential haemorrhage.

2   There may be adhesions across the lesser sac which bring the transverse mesocolon into intimate relationship with the stomach or greater omentum. In these circumstances the middle colic vessels are in danger of damage during mobilization of the stomach for gastrectomy.

3   *Radiology of the stomach* (Fig. 56). A plain erect film of the abdomen reveals a bubble of air below the left diaphragm; this is gas in the stomach fundus. After the subject has swallowed barium sulphate paste, the stomach can be seen and its position, movements and

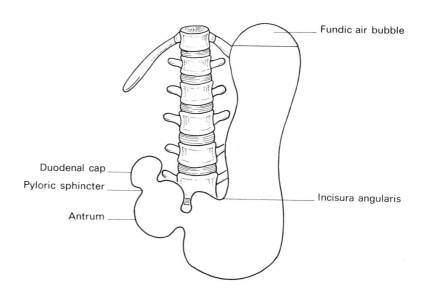

Fundic air bubble

Duodenal cap

Pyloric sphincter

Antrum

Incisura angularis

Fig. 56 Tracing of barium meal X-ray of the stomach.

outline studied. The wide variations in the position and shape of the stomach that we have already mentioned have come to light principally as a result of such investigations.

By tipping the subject head-down, the opaque meal can be made to impinge against the cardia; incompetence of this sphincter mechanism will be demonstrated by seeing barium regurgitate into the oesophagus.

4   Gastroscopy. The mucosa of the air-inflated stomach can be inspected in the living subject through the gastroscope. With the modern fibre-optic instrument the whole of the gastric mucosa can be viewed, the duodenum examined, and the common bile duct and the pancreatic duct intubated for retrograde radiological study.

The duodenum

The duodenum curves in a C around the head of the pancreas and is 10 in (25 cm) long. At its origin from the pylorus it is completely covered with peritoneum for about 1 in (2.5 cm), but then becomes a retroperitoneal organ, only partially covered by serous membrane.

Relations (Figs 57, 58)

For descriptive purposes, the duodenum is divided into four sections.

The first part (2 in (5 cm)) ascends from the gastroduodenal junction, overlapped by the liver and gall-bladder. Immediately posterior to it lie the portal vein, common bile duct and gastroduodenal artery which separate it from the inferior vena cava.

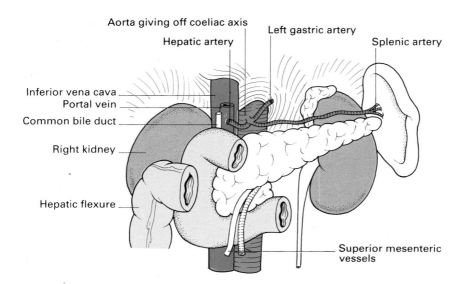

Fig. 57 The relations of the duodenum.

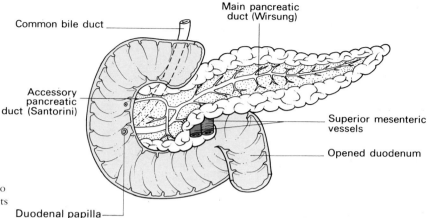

Common bile duct

Main pancreatic
duct (Wirsung)

Accessory
pancreatic
duct (Santorini)

Superior mesenteric
vessels

Opened duodenum

Fig. 58 The duodenum
and pancreas dissected to
show the pancreatic ducts
and their orifices.

Duodenal papilla

The second part (3 in (7.5 cm)) descends in a curve around the head of the pancreas. It is crossed by the transverse colon and lies on the right kidney and ureter. Half-way along, its posteromedial aspect enters the common opening of the bile duct and *main pancreatic duct (of Wirsung)* on to an eminence called the *duodenal papilla*. This common opening is guarded by the *sphincter of Oddi*. The *subsidiary pancreatic duct (of Santorini)* opens into the duodenum a little above the papilla.

The third part (4 in (10 cm)) runs transversely to the left, crossing the inferior vena cava, the aorta and the third lumbar vertebra. It is itself crossed anteriorly by the root of the mesentery and the superior mesenteric vessels. Its upper border hugs the pancreatic head.

The fourth part (1 in (2.5 cm)) ascends upwards and to the left to end at the duodenojejunal junction. It is surprisingly easy for the surgeon to confuse this with the ileocaecal junction, a mistake which may be disastrous. He confirms the identity of the duodenal termination by the presence of the *suspensory ligament of Treitz*, which is a well-marked peritoneal fold descending from the right crus of the diaphragm to the duodenal termination, and by visualizing the inferior mesenteric vein which descends from behind the pancreas immediately to the left of the duodenojejunal junction.

Blood supply of the duodenum

The superior pancreaticoduodenal artery arises from the gastroduodenal artery; the inferior pacreaticoduodenal artery originates as the first branch of the superior mesenteric artery. These vessels both lie in the curve between the duodenum and the head of the pancreas, supplying both structures. Interestingly, their anastomosis represents the site of junction of the fore-gut (supplied by the coeliac artery), and the mid-gut (supplied by the superior mesenteric artery), at the level of the duodenal papilla (see page 92).

Clinical features

1   The first part of the duodenum is overlapped by the liver and gall-bladder, either of which may become adherent to, or even ulcerated by, a duodenal ulcer. Moreover, a gallstone may ulcerate from the fundus of the gall-bladder into the duodenum. The gallstone may then impact in the lower ileum as it traverses the gut to produce intestinal obstruction (*gallstone ileus*).
2   The pancreas, as the duodenum's most intimate relation, is readily invaded by a posterior duodenal ulcer. This should be suspected if the patient's pain radiates into the dorsolumbar region. Erosion of the gastroduodenal artery by such an ulcer results in severe haemorrhage.
3   Extensive dissection of a duodenum, scarred by severe ulceration, may damage the common bile duct which passes behind the first part of the duodenum about 1 in (2.5 cm) from the pylorus.
4   The hepatic flexure of the colon crosses the second part of the duodenum and the latter may be damaged during the right hemi-colectomy. Similarly, the right kidney lies directly behind this part of the duodenum, which may be injured in performing a right nephrectomy.
5   *Radiology of the duodenum.* Within a few minutes of swallowing a barium meal, the first part of the duodenum becomes visible as a tri-angular shadow termed the *duodenal cap*. Every few seconds the duo-denum contracts, emptying this cap, which promptly proceeds to fill again. It is in this region that the great majority of duodenal ulcers occur; an actual ulcer crater may be visualized, filled with barium, or deformity of the cap, produced by scar tissue, may be evident. The rest of the duodenum can also be seen, the shadow being floccular due to the rugose arrangement of the mucosa.

Small intestine

The *length* of the small intestine varies from 10 to 33 feet (3–10 m) in different subjects; the average is some 24 feet (6.5 m). Resection of up to one-third or even half of the small intestine is compatible with a perfectly normal life, and survival has been reported with only 18 in (45 cm) of small intestine preserved.

The *mesentery* of the small intestine has a 6 in (15 cm) origin from the posterior abdominal wall, which commences at the duodenojejunal junction to the left of the 2nd lumbar vertebra, and passes obliquely downwards to the right sacro-iliac joint; it contains the superior mesenteric vessels, the lymph nodes draining the small gut and autonomic nerve fibres.

The upper half of the small intestine is termed the *jejunum*, the remainder is the *ileum*. There is no sharp distinction between the two and this division is a conventional one only. The bowel does, however, change its character from above downwards, the following points enabling the surgeon to determine the level of a loop of small intestine at operation.

1   The jejunum has a thicker wall as the circular folds of mucosa (*valvulae conniventes*) are larger and thicker more proximally.

2   The proximal small intestine is of greater diameter than the distal.

3   The jejunum tends to lie at the umbilical region, the ileum in the hypogastrium and pelvis.

4   The mesentery becomes thicker and more fat-laden from above downwards.

5   The mesenteric vessels from only one or two arcades to the jejunum, with long and relatively infrequent terminal branches passing to the gut wall. The ileum is supplied by shorter and more numerous terminal vessels arising from complete series of three, four or even five arcades (Fig. 59).

Large intestine

The large intestine is subdivided, for descriptive purposes, into:
* caecum with the appendix vermiformis;
* ascending colon (5–8 in (12–20 cm));
* hepatic flexure;
* transverse colon (18 in (45 cm));

Fig. 59 The simple arterial arcades of the jejunum (a) compared with the complex arcades of the ileum (b).

(a) Jejunum

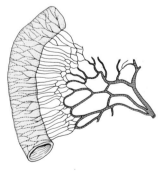

(b) Ileum

- splenic flexure;
- descending colon (9–12 in (22–30 cm));
- sigmoid colon (5–30 in (12–75 cm), average 15 in (37 cm));
- rectum (5 in (12 cm));
- anal canal (1.5 in (4 cm)).

The large bowel may vary considerably in length in different subjects; the average is approximately 5 feet (1.5 m).

The colon (but not the appendix, caecum or rectum), bears characteristic fat-filled peritoneal tags called *appendices epiploicae* scattered over its surface. These are especially numerous in the sigmoid colon.

The colon and caecum (but not the appendix or rectum) are marked by the *taeniae coli*. These are three flattened bands commencing at the base of the appendix and running the length of the large intestine to end at the rectosigmoid junction. They represent the great bulk of the longitudinal muscle of the large bowel; because the taeniae are about a foot shorter than the gut to which they are attached, the colon becomes condensed into its typical sacculated shape. These sacculations may be seen in a plain radiograph of the abdomen when the large bowel is distended and appear as incomplete septa projecting into the gas shadow. The radiograph of distended small intestine, in contrast, characteristically has complete transverse lines across the bowel shadow due to the transverse mucosal folds of the valvulae conniventes.

Peritoneal attachments

The transverse colon and sigmoid are completely peritonealized (the former being readily identified by its attachment to the greater omentum). The ascending and descending colon have no mesocolon but adhere directly to the posterior abdominal wall (although exceptionally the ascending colon has a mesocolon). The caecum may or may not be completely peritonealized, and the appendix, although usually free within its own mesentery, occasionally lies extraperitoneally behind caecum and ascending colon or adheres to the posterior wall of these structures.

The rectum is extraperitoneal on its posterior aspect in its upper third, posteriorly and laterally in its middle third and completely in its lower third as it sinks below the pelvic peritoneum.

The appendix

The appendix arises from the posteromedial aspect of the caecum about 1 in (2.5 cm) below the ileocaecal valve; its length ranges from

0.5 in (12 mm) to 9 in (22 cm). In the fetus it is a direct outpouching of the caecum, but differential overgrowth of the lateral caecal wall results in its medial displacement.

The position of the appendix is extremely variable—more so than that of any other organ (Fig. 60). Most frequently (75% of cases) the appendix lies behind the caecum. The appendix is usually quite free in this position although occasionally it lies beneath the peritoneal covering of the caecum. If the appendix is very long, it may actually extend behind the ascending colon and abut against the right kidney or the duodenum; in these cases its distal portion lies extraperitoneally.

In about 20% of cases, the appendix lies just below the caecum or else hangs down into the pelvis. Less commonly, it passes in front of or behind the terminal ileum, or lies in front of the caecum or in the right paracolic gutter.

A long appendix has been known to ulcerate into the duodenum or perforate into the *left* paracolic gutter. It may well be said that 'the appendix is the only organ in the body that has no anatomy'.

The mesentery of the appendix, containing the appendicular branch of the ileocolic artery, descends behind the ileum as a triangular fold (Fig. 61). Another peritoneal sheet, the *ileocaecal fold*, passes to the appendix or to the base of the caecum from the front of the ileum. The ileocaecal fold is termed the *bloodless fold of Treves* although, in fact, it often contains a vessel and, if cut, proves far from bloodless.

Clinical features

1   The lumen of the appendix is relatively wide in the infant and is frequently completely obliterated in the elderly. Since obstruction of

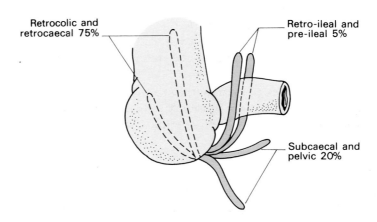

Retrocolic and retrocaecal 75%

Retro-ileal and pre-ileal 5%

Subcaecal and pelvic 20%

Fig. 60 The positions in which the appendix may lie, together with their approximate incidence.

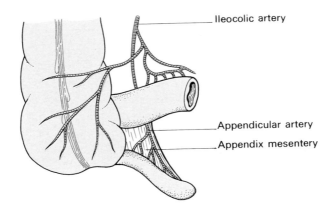

Fig. 61 The blood supply
of the appendix.

the lumen is the usual precipitating cause of acute appendicitis it is
not unnatural, therefore, that appendicitis should be uncommon at
the two extremes of life.

2   The appendicular artery represents the entire vascular supply of
the appendix. It runs first in the edge of the appendicular mesentery
and then, distally, along the wall of the appendix. Acute infection of
the appendix may result in thrombosis of this artery with rapid development of gangrene and subsequent perforation. This is in contrast to
acute cholecystitis, where the rich collateral vascular supply from the
liver bed ensures the rarity of gangrene of the gall-bladder even if the
cystic artery becomes thrombosed.

3   *Appendicectomy* is usually performed through a muscle-splitting
incision in the right iliac fossa (see 'abdominal incisions', page 66).
The caecum is delivered into the wound and, if the appendix is not
immediately visible, it is located by tracing the taeniae coli along the
caecum—they fuse at the base of the appendix. When the caecum is
extraperitoneal it may be difficult to bring the appendix up into the
incision; this is facilitated by first mobilizing the caecum by incising
the almost avascular peritoneum along its lateral and inferior
borders.

The appendix mesentery, containing the appendicular vessels, is
firmly tied and divided, the appendix base tied, the appendix removed
and its stump invaginated into the caecum.

The rectum

The rectum is 5 in (12 cm) in length. It commences anterior to the
third segment of the sacrum and ends at the level of the apex of the
prostate or at the lower quarter of the vagina, where it leads into
the anal canal.

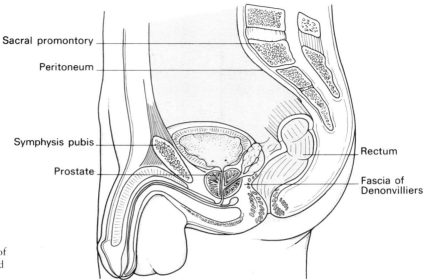

Sacral promontory

Peritoneum

Symphysis pubis

Prostate

Rectum

Fascia of
Denonvilliers

Fig. 62 Sagittal section of
the rectum and its related
viscera in the male.

The rectum is straight in lower mammals (hence its name) but is curved in man to fit into the sacral hollow. Moreover, it presents a series of three lateral inflexions, capped by the *valves of Houston*, projecting left, right and left from above downwards.

Relations (Figs 62, 63)

The main relations of the rectum are important. They must be visualized in carrying out a rectal examination, they provide the key to the

Sacral
promontory

Ovary
and tube

Peritoneum

Bladder

Symphysis
pubis

Urethra

Uterus

Rectum

Vagina

Fig. 63 Sagittal section of
the rectum and its related
viscera in the female.

local spread of rectal growths and they are important in operative removal of the rectum.

Posteriorly lie sacrum and coccyx and the middle sacral artery, which are separated from it by extraperitoneal connective tissue containing the rectal vessels and lymphatics. The lower sacral nerves, emerging from the anterior sacral foramina, may be involved by growth spreading posteriorly from the rectum, resulting in severe sciatic pain.

Anteriorly, the upper two-thirds of the rectum are covered by peritoneum and relate to coils of small intestine which lie in the cul-de-sac of the pouch of Douglas between the rectum and the bladder or the uterus. In front of the lower one-third lie the prostate, bladder base and seminal vesicles in the male, or the vagina in the female. A layer of fascia (*Denonvilliers*) separates the rectum from the anterior structures and forms the plane of dissection which must be sought after in excision of the rectum.

Laterally, the rectum is supported by the levator ani.

The anal canal (Fig. 64)

The anal canal is 1.5 in (4 cm) long and is directed downwards and backwards from the rectum to end at the anal orifice. The mid-anal canal represents the junction between endoderm and ectoderm.

1   The lower half is lined by squamous epithelium and the upper half by columnar epithelium; the latter presents vertical columns of mucosa (*the columns of Morgagni*) connected at their distal extremities by valve-like folds (*the valves of Ball*). A carcinoma of the upper anal canal is thus an adenocarcinoma, whereas that arising from the lower part is a squamous tumour.

2   The blood supply of the upper half of the anal canal is from the superior rectal vessels, whereas that of the lower half is the blood supply of the surrounding anal skin, the inferior rectal vessels, which

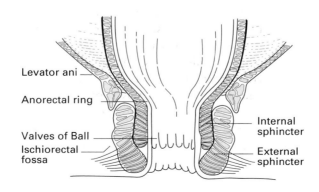

Fig. 64 The sphincters of the anus.

derive from the internal pudendal, and ultimately the internal iliac vessels. The two venous systems communicate and therefore form one of the anastomoses between the portal and systemic circulations.

3   The lymphatics above this mucocutaneous junction drain along the superior rectal vessels to the lumbar nodes whereas, below this line, drainage is to the inguinal nodes. A carcinoma of the rectum which invades the lower anal canal may thus metastasize to the groin nodes.

4   The nerve supply to the upper anal canal is via the autonomic plexuses, the lower part is supplied by the somatic inferior rectal nerve, a terminal branch of the pudendal nerve (see Fig. 99b). (The lower canal is therefore sensitive to the prick of a hypodermic needle, whereas injection of an internal haemorrhoid with sclerosant fluid, by passing a needle through the mucosa of the upper part of the canal, is painless.)

The anal sphincter

Forming the walls of the anal canal is a rather complicated muscle arrangement which constitutes a powerful sphincter mechanism (Fig. 64). This comprises:

• *the internal anal sphincter*, of involuntary muscle, which continues above with the circular muscle coat of the rectum;

• *the external anal sphincter*, of voluntary muscle, which surrounds the internal sphincter and which extends further downwards and curves medially to occupy a position below and slightly lateral to the lower rounded edge of the internal sphincter, close to the skin of the anal orifice. The lowermost, or subcutaneous, portion of the external sphincter is traversed by a fan-shaped expansion of the longitudinal muscle fibres of the anal canal which continue above with the longitudinal muscle of the rectal wall. At its upper end the external sphincter fuses with the fibres of levator ani.

In carrying out a digital rectal examination, the ring of muscle on which the flexed finger rests just over an inch from the anal margin is the *anorectal ring*. This represents the deep part of the external sphincter where this blends with the internal sphincter and levator ani, and demarcates the junction between anal canal and rectum.

The anal canal is related posteriorly to the fibrous tissue between it and the coccyx (anococcygeal body), laterally to the ischiorectal fossae containing fat, and anteriorly to the perineal body separating it from the bulb of the urethra in the male or the lower vagina in the female.

Rectal examination

The following structures can be palpated by the finger passed per
rectum in the normal patient:

1   both sexes—the anorectal ring (see above), coccyx and sacrum,
ischiorectal fossae, ischial spines;

2   male—prostate, rarely the healthy seminal vesicles;

3   female—perineal body, cervix, occasionally the ovaries.

Abnormalities which can be detected include:

1   within the lumen—faecal impaction, foreign bodies;

2   in the wall—rectal growths, strictures, granulomata, etc., but *not*
haemorrhoids unless these are thrombosed;

3   outside the rectal wall—pelvic bony tumours, abnormalities of the
prostate or seminal vesicle, distended bladder, uterine or ovarian
enlargement, collections of fluid or neoplastic masses in the pouch of
Douglas.

Do not be deceived by foreign objects placed in the vagina. The
commonest are a tampon or a pessary.

During parturition, dilatation of the cervical os can be assessed by
rectal examination since it can be felt quite easily through the rectal
wall.

Clinical features

Haemorrhoids

Haemorrhoids (piles) are dilatations of the superior rectal veins. Initially contained within the anus (1st degree), they gradually enlarge
until they prolapse on defaecation (2nd degree) and finally remain
prolapsed through the anal orifice (3rd degree).

The superior rectal vessels are asymmetrical—there is one branch
to the left side of the rectum and two to the right, thus giving the well-
known arrangement of piles at 3, 7, and 11 o'clock, viewing the
patient in the lithotomy position.

Anatomically, each pile comprises: a varicose venous plexus
draining into one of the superior rectal veins; terminal branches of the
corresponding superior rectal artery; and a covering of anal canal
mucosa and submucosa.

The so-called 'thrombosed external pile' is a small tense
haematoma at the anal margin caused by rupture of a subcutaneous
vein and is much better termed a *perianal haematoma*.

Perianal abscesses (Fig. 65)

These may be localized beneath the anal mucosa (submucous), be

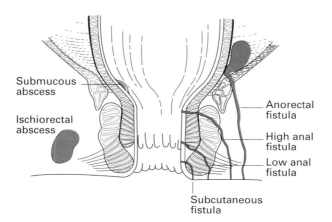

Fig. 65 The anatomy of perianal fistulae and abscess.

beneath perianal skin (subcutaneous) or occupy the ischiorectal fossa (ischiorectal). Occasionally, abscesses lie in the pelvirectal space above levator ani, alongside the rectum and deep to the pelvic peritoneum.

Fistulae (Fig. 65)

Anal fistulae usually result from rupture of perianal abscesses. They are classified anatomically and may be:
• submucous—confined to the tissues immediately below the anal mucosa;
• subcutaneous—confined to the perianal skin;
• low-level — passing through the lower part of the superficial sphincter (most common);
• high-level — passing through the deeper part of the superficial sphincter;
• anorectal—which has its track passing above the anorectal ring and which may or may not open into the rectum.

In laying open *fistulae in ano*, it is essential to preserve the anorectal ring if faecal incontinence is to be avoided. The lower part of the sphincter, on the other hand, can be divided quite safely without this risk.

Fissure *in ano*

This is a tear in the anal mucosa; over 90% occur posteriorly in the midline. The anatomical basis for this probably lies in the insertion of the superficial component of the external anal sphincter posteriorly into the coccyx; between the two limbs of the V thus formed, the mucosa is relatively unsupported and may therefore be torn by a hard faecal mass at this site.

Arterial supply of the intestine

The alimentary tract develops from the fore-, mid- and hind-gut; the arterial supply to each is discrete, although anastomosing with its neighbour. The fore-gut comprises stomach and duodenum as far as the entry of the bile duct and is supplied by branches of the *coeliac axis* which arises from the aorta at T12 vertebral level (see Fig. 53). The mid-gut extends from mid-duodenum to the distal transverse colon and is supplied by the *superior mesenteric artery* (Fig. 66) arising from the aorta at L1. Its branches are:

1 the inferior pancreaticoduodenal artery;
2 jejunal and ileal branches — supplying the bulk of the small intestine;
3 the ileocolic artery, supplying terminal ileum, caecum and commencement of ascending colon and giving off an *appendicular branch* to the appendix—the most commonly ligated intra-abdominal artery;
4 the right colic artery—supplying the ascending colon;
5 the middle colic artery—supplying the transverse colon.

The hind-gut receives its supply from the *inferior mesenteric artery* (Fig. 66), arising from the aorta at L3 and giving the following branches:

1 the left colic artery—supplying the descending colon;
2 the sigmoid branches—supplying the sigmoid;
3 the superior rectal artery—supplying the rectum.

Each branch of the superior and inferior mesenteric artery anastomoses with its neighbour above and below so that there is, in fact, a

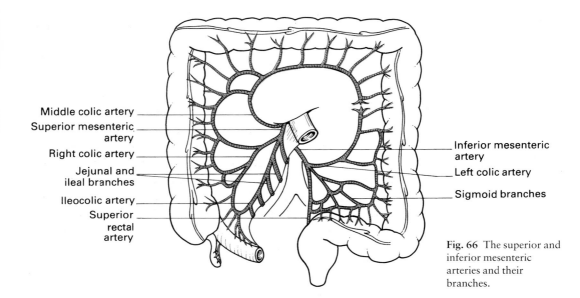

Middle colic artery
Superior mesenteric artery
Right colic artery
Jejunal and ileal branches
Ileocolic artery
Superior rectal artery

Inferior mesenteric artery
Left colic artery
Sigmoid branches

Fig. 66 The superior and inferior mesenteric arteries and their branches.

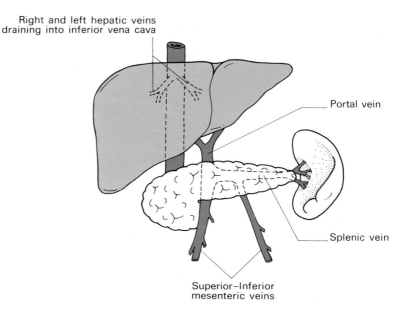

Right and left hepatic veins
draining into inferior vena cava

Portal vein

Splenic vein

Superior–Inferior
mesenteric veins

Fig. 67 The composition
of the portal system.

continuous vascular arcade along the whole length of the gastrointestinal canal.

The portal system of veins

The portal venous system drains blood to the liver from the abdominal part of the alimentary canal (excluding the anus), the spleen, the pancreas and the gall-bladder and its ducts.

The distal tributaries of this system correspond to, and accompany, the branches of the coeliac and the superior and inferior mesenteric arteries enumerated above; only proximally (Fig. 67) does the arrangement differ.

The *inferior mesenteric vein* ascends above the point of origin of its artery to enter the splenic vein behind the pancreas.

The *superior mesenteric vein* joins the splenic vein behind the neck of the pancreas in the transpyloric plane to form the *portal vein*, which ascends behind the first part of the duodenum into the anterior wall of the foramen of Winslow and thence to the porta hepatis. Here the portal vein divides into right and left branches and breaks up into capillaries running between the lobules of the liver. These capillaries drain into the radicles of the hepatic vein through which they empty into the inferior vena cava.

Connections between the portal and systemic venous systems

Normally, portal venous blood traverses the liver as described above and empties into the systemic venous circulation via the hepatic vein and inferior vena cava. This pathway may be blocked by a variety of causes which are classified into:

- prehepatic — e.g. thrombosis or congenital obliteration of the portal vein;
- hepatic—e.g. cirrhosis of the liver;
- posthepatic—e.g. congenital stenosis of the hepatic vein.

If obstruction from any of these causes occurs, the portal venous pressure rises (*portal hypertension*) and collateral pathways open up between the portal and systemic venous systems.

These communications are:

1   between the oesophageal branch of the left gastric vein and the oesophageal veins of the azygos system (these *oesophageal varices* are the cause of the severe haematemeses that may occur in portal hypertension);

2   between the superior rectal branch of the inferior mesenteric vein and the inferior rectal veins draining into the internal iliac vein via its internal pudendal tributary;

3   between the portal tributaries in the mesentery and mesocolon and retroperitoneal veins communicating with the renal, lumbar and phrenic veins;

4   between the portal branches in the liver and the veins of the abdominal wall via veins passing along the falciform ligament to the umbilicus (which may result in the formation of a cluster of dilated veins which radiate from the navel and which are called the *caput Medusae*);

5   between the portal branches in the liver and the veins of the diaphragm across the bare area of the liver.

A striking feature of operations upon patients with portal hypertension is the extraordinary dilatation of every available channel between the two systems which renders such procedures tedious and bloody.

Lymph drainage of the intestine (Fig. 68)

The arrangement of lymph nodes is relatively uniform throughout the small and large intestine. Numerous small nodes lying near, or even on, the bowel wall drain to intermediately placed and rather larger nodes along the vessels in the mesentery or mesocolon and thence to clumps of nodes situated near the origins of the superior and inferior

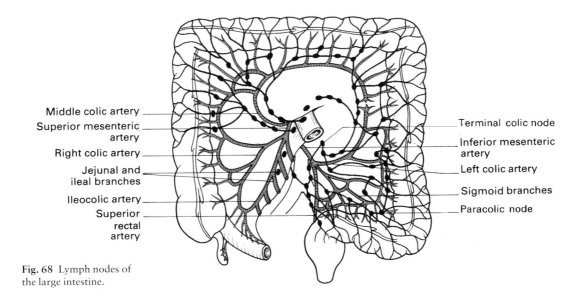

Middle colic artery
Superior mesenteric artery
Right colic artery
Jejunal and ileal branches
Ileocolic artery
Superior rectal artery

Terminal colic node
Inferior mesenteric artery
Left colic artery
Sigmoid branches
Paracolic node

Fig. 68 Lymph nodes of the large intestine.

mesenteric arteries. From these, efferent vessels link up to drain into the cisterna chyli.

The lymphatic drainage field of each segment of bowel corresponds fairly accurately to its blood supply. High ligation of the vessels to the involved segment of bowel with removal of a wide surrounding segment of mesocolon will, therefore, remove the lymph nodes draining the area. Division of the middle colic vessels and a resection of a generous wedge of transverse mesocolon, for example, would be performed for a growth of transverse colon.

The structure of the alimentary canal

The alimentary canal is made up of mucosa demarcated by the *muscularis mucosae* from the *submucosa*, the *muscle coat* and the *serosa* —the last being absent where the gut is extraperitoneal.

The oesophageal mucosa and that of the lower anal canal is squamous; elsewhere it is columnar. At the cardio-oesophageal junction this transition is quite sharp, although occasionally columnar epithelium may line the lower oesophagus.

The gastric mucosa bears simple crypt-like glands projecting down to the muscularis mucosae. The pyloric antrum secretes an alkaline juice containing mucus and the hormone gastrin. The body of the stomach secretes pepsin and also HCl, the latter from the *oxyntic cells* lying sandwiched deeply between the surface cells. The stomach mucosa also produces intrinsic factor.

The mucosa of the duodenum and small intestine, as well as

bearing crypt-like glands, projects into the bowel lumen in villous processes which greatly increase its surface area. The duodenum is distinguished by its crypts extending deep through the muscularis mucosae and opening into an extensive system of acini in the submucosa termed *Brunner's glands*.

The mucosa of the large intestine is lined almost entirely by mucus-secreting goblet cells; there are no villi.

The muscle coat of the alimentary tract is made up of an inner circular layer and an outer longitudinal layer. In the upper two-thirds of the oesophagus and at the anal margin this muscle is voluntary; elsewhere it is involuntary. The stomach wall is reinforced by an innermost oblique coat of muscle and the colon is characterized by the condensation of its longitudinal layer into three *taeniae coli*.

The nerve plexuses of Meissner and Auerbach lie respectively in the submucosal layer and between the circular and longitudinal muscle coats.

The development of the intestine and its congenital abnormalities (Fig. 69)

The primitive endodermal tube of the gut is divided into:
1   the fore-gut (supplied by the coeliac axis) extending as far as the entry of the bile duct into the duodenum;
2   the mid-gut (supplied by the superior mesenteric artery) continuing as far as the distal transverse colon;
3   the hind-gut (supplied by the inferior mesenteric artery) extending thence to the ectodermal part of the anal canal.

At an early stage rapid proliferation of the gut wall obliterates its lumen and this is followed by subsequent recanalization.

The fore-gut becomes rotated with the development of the lesser sac so that the original right wall of the stomach comes to form its posterior surface and the left wall its anterior surface. The vagi rotate with the stomach and therefore lie anteriorly and posteriorly to it at the oesophageal hiatus.

This rotation swings the duodenum to the right and the mesentery of this organ then blends with the peritoneum of the posterior abdominal wall—this blending process is termed *zygosis*.

The mid-gut enlarges rapidly in the 5-week fetus, becomes too large to be contained within the abdomen and herniates into the umbilical cord. The apex of this herniated bowel is continuous with the vitello-intestinal duct and the yolk sac, but this connection, even at this early stage of fetal life, is already reduced to a fibrous strand.

The axis of this herniated loop of gut is formed by the superior mesenteric artery, which demarcates a cephalic and a caudal limb.

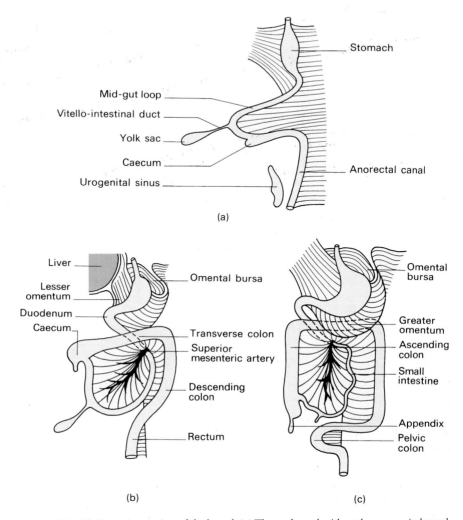

Fig. 69 Stages in rotation of the bowel. (a) The prolapsed mid-gut loop, seen in lateral view. (b) The mid-gut returns to the abdomen. (c) The caecum descends to its definitive position. Note the completion of stomach-rotation with the formation of the lesser sac (omental bursa).

The cephalic element develops into the proximal small intestine; the caudal segment differentiates into the terminal 2 feet (62 cm) of ileum, the caecum and the colon as far as the junction of the middle and left thirds of the transverse colon.

A bud which develops on the caudal segment indicates the site of subsequent formation of the caecum; it may well be that this bud delays the return of the caudal limb in favour of the cephalic gut during the subsequent reduction of the herniated bowel.

At 10 weeks this return of the bowel into the abdominal cavity commences. The mid-gut loop first rotates anti-clockwise through 90° so that the cephalic limb now lies to the right and the caudal limb to the left.

The cephalic limb returns first, passing upwards and to the left into the space left available by the bulky liver. In doing so, this mid-gut passes *behind* the superior mesenteric artery (which thus comes to cross the third part of the duodenum) and also pushes the hind-gut— the definitive distal colon—over to the left.

When the caudal limb returns, it lies in the only space remaining to it, superficial to, and above, the small intestine with the caecum lying immediately below the liver.

The caecum then descends into its definitive position in the right iliac fossa, dragging the colon with it. The transverse colon thus comes to lie in front of the superior mesenteric vessels and the small intestine.

Finally, the mesenteries of the ascending and descending parts of the colon blend with the posterior abdominal wall peritoneum by zygosis.

Numerous anomalies may occur in the highly complex developmental process.

1 *Atresia* or *stenosis of the bowel* may result from failure of recanalization of the lumen. Another cause of this may be damage to the blood supply to the bowel within the fetal umbilical hernia with consequent ischaemic changes.

2 *Meckel's diverticulum* represents the remains of the embryonic vitello-intestinal duct (communication between the primitive mid-gut and yolk sac) and is, therefore, always on the anti-mesenteric border of the bowel. As an approximation to the truth it can be said to occur in 2% of subjects, twice as often in males as females, to be situated at 2 feet (62 cm) from the ileocaecal junction and to be 2 in (5 cm) long. In fact, it may occur anywhere from 6 in (15 cm) to 12 feet (3.5 m) from the terminal ileum and vary from a tiny stump to a 6 in (15 cm) long sac. Occasionally the diverticulum ends in a whip-like solid strand.

As well as a diverticulum—the commonest form—this duct may persist as a fistula or band connecting the intestine to the umbilicus, as a cyst hanging from the anti-mesenteric border of the ileum or as a 'raspberry tumour' at the umbilicus, formed by the red mucosa of a persistent umbilical extremity of the diverticulum pouting at the navel (Fig. 70).

The mucosa lining the diverticulum may contain islands of peptic epithelium with oxyntic (acid-secreting) cells. Peptic ulceration of adjacent intestinal epithelium may then occur with haemorrhage or perforation.

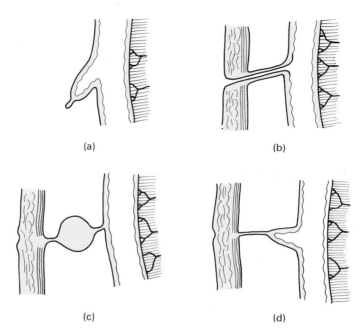

(a)

(b)

(c)

(d)

Fig. 70 Abnormalities associated with persistence of the vitello-intestinal tract. (a) Meckel's diverticulum. (b) Patent vitellointestinal duct. (c) Cyst within a fibrous cord passing from the anti-mesenteric border of the intestine to the umbilicus. (d) Meckel's diverticulum with terminal filament passing to umbilicus.

3 The caecum may fail to descend; the peritoneal fold which normally seals it in the right iliac fossa passes, instead, across the duodenum and causes a neonatal intestinal obstruction. The mesentery of the small intestine in such a case is left as a narrow pedicle, which allows volvulus of the whole small intestine to occur (*volvulus neonatorum*).

4 Occasionally, reversed rotation occurs, in which the transverse colon comes to lie *behind* the superior mesenteric vessels with the duodenum in front of them; this may again be accompanied by extrinsic duodenal obstruction due to a peritoneal fold.

5 *Exomphalos* is persistence of the mid-gut herniation at the umbilicus after birth.

The gastrointestinal adnexae: liver, gall-bladder and its ducts, the pancreas and spleen

The liver (Fig. 71)

This is the largest organ in the body. It is related by its domed upper surface to the diaphragm, which separates it from pleura, lungs, pericardium and heart. Its postero-inferior (or visceral) surface abuts

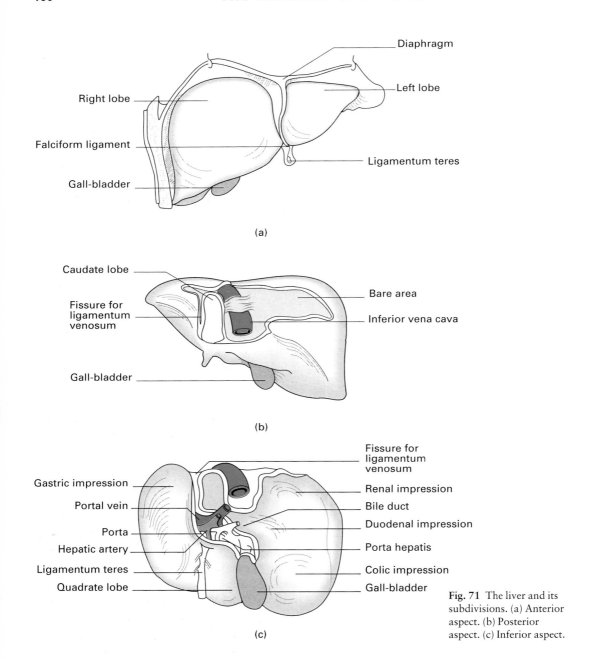

Fig. 71 The liver and its subdivisions. (a) Anterior aspect. (b) Posterior aspect. (c) Inferior aspect.

against the abdominal oesophagus, the stomach, duodenum; hepatic flexure of colon and the right kidney and suprarenal, as well as carrying the gall-bladder.

The liver is divided into a larger right and small left lobe, separated superiorly by the falciform ligament and postero-inferiorly by an H-shaped arrangement of fossae (Fig. 71c):

- anteriorly and to the right—the fossa for the gall-bladder;
- posteriorly and to the right—the groove in which the inferior vena cava lies embedded;
- anteriorly and to the left—the fissure containing the ligamentum teres;
- posteriorly and to the left — the fissure for the ligamentum venosum.

The cross-bar of the H is the *porta hepatis*. Two subsidiary lobes are marked out on the visceral aspect of the liver between the limbs of this H—the *quadrate lobe* in front and the *caudate lobe* behind.

The *ligamentum teres* is the obliterated remains of the left umbilical vein which, in utero, brings blood from the placenta back into the fetus. The *ligamentus venosum* is the fibrous remnant of the fetal *ductus venosus* which shunts blood from this left umbilical vein to the inferior vena cava, short-circuiting the liver. It is easy enough to realize, then, that the grooves for the ligamentum teres, ligamentum venosum and inferior vena cava, representing as they do the pathway of a fetal venous trunk, are continuous in the adult.

Lying in the porta hepatis (which is 2 in (5 cm) long) are:

1 the common hepatic duct—anteriorly;
2 the hepatic artery—in the middle;
3 the portal vein—posteriorly.

As well as these, autonomic nerve fibres (sympathetic from the coeliac axis and parasympathetic from the vagus), lymphatic vessels and lymph nodes are found there.

Peritoneal attachments

The liver is enclosed in peritoneum except for a small posterior bare area, demarcated by the peritoneum from the diaphragm reflected on to it as the upper and lower layers of the *coronary ligament*. To the right, these fuse to form the *right triangular ligament*.

The *falciform ligament* ascends to the liver from the umbilicus, somewhat to the right of the midline, and bears the ligamentum teres in its free border. The ligamentum teres passes into its fissure in the inferior surface of the liver while the falciform ligament passes over the dome of the liver and then divaricates. Its right limb joins the upper layer of the coronary ligament and its left limb stretches out as the long narrow *left triangular ligament* which, when traced posteriorly and to the right, joins the lesser omentum in the upper end of the fissure for the ligamentum venosum.

The *lesser omentum* arises from the fissures of the porta hepatis and the ligamentum venosum and passes as a sheet to be attached along the lesser curvature of the stomach.

Structure

The liver is made up of lobules, each with a solitary central vein which is a tributary of the hepatic vein which, in turn, drains into the inferior vena cava. In spaces between the lobules, termed *portal canals*, lie branches of the hepatic artery (bringing systemic blood) and the portal vein, both of which drain into the central vein by means of sinusoids traversing the lobule.

Branches of the *hepatic duct* also lie in the portal canals and receive fine bile capillaries from the liver lobules.

Segmental anatomy

The gross anatomical division of the liver into a right and left lobe, demarcated by a line passing from the attachment of the falciform ligament on the anterior surface to the fissures for the ligamentum teres and ligamentum venosum on its posterior surface, is simply a gross anatomical descriptive term with no morphological significance. Studies of the distribution of the hepatic blood vessels and ducts have indicated that the true morphological and physiological division of the liver is into right and left lobes demarcated by a plane which passes through the fossa of the gall-bladder and the fossa of the inferior vena cava. Although these two lobes are not differentiated by any visible line on the dome of the liver, each has its own arterial and portal venous blood supply and separate biliary drainage. This morphological division lies to the right of the gross anatomical plane and in this the quadrate lobe comes to be part of the left morphological lobe of the liver while the caudate lobe divides partly to the left and partly to the right lobe (Fig. 72).

The right and left morphological lobes of the liver can be further subdivided into a number of segments, four for each lobe (Fig. 72c). The student need not learn the details of these, but of course to the hepatic surgeon, carrying out a partial resection of the liver, knowledge of these segments, with their individual blood supply and biliary drainage, is of great importance.

At the hilum of the liver, the hepatic artery, portal vein and bile duct each divide into right and left branches and there is little or no anastomosis between the divisions on the two sides (Fig. 73). From the region of the porta hepatis, the branches pass laterally and spread upwards and downwards throughout the liver substance, defining the morphological left and right lobes.

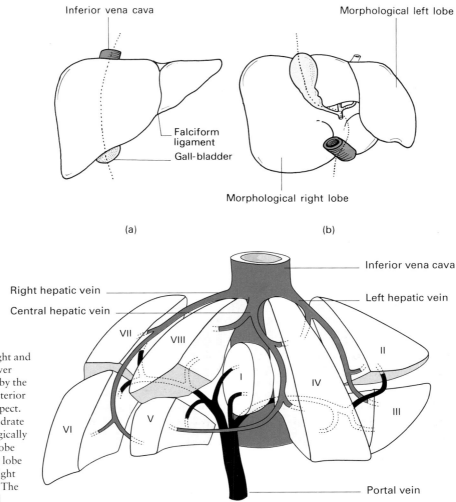

Fig. 72 The morphological right and left lobes of the liver shown separated by the dotted line: (a) anterior and (b) ventral aspect. Note that the quadrate lobe is morphologically a part of the left lobe while the caudate lobe belongs to both right and left lobes. (c) The further segmental divisions of the liver.

The hepatic veins

These veins are massive and their distribution is somewhat different from that of the portal, hepatic arterial and bile duct systems already described. There are three major hepatic veins, comprising a right, a central and a left. These pass upwards and backwards to drain into the inferior vena cava at the superior margin of the liver. Their terminations are somewhat variable but usually the central hepatic vein enters the left hepatic vein near its termination. In other specimens it may drain directly into the cava. In addition, small hepatic venous tributaries run directly backwards from the substance of the liver to enter the vena cava more distally to the main hepatic veins. Although

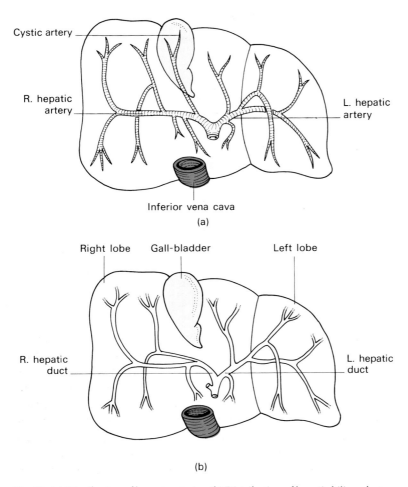

Cystic artery

R. hepatic
artery

L. hepatic
artery

Inferior vena cava

(a)

Right lobe   Gall-bladder   Left lobe

R. hepatic
duct

L. hepatic
duct

(b)

Fig. 73 (a) Distribution of hepatic arteries. (b) Distribution of hepatic biliary ducts.
Note that the quadrate lobe is supplied exclusively by the left hepatic artery and
drained by the left hepatic duct. The caudate lobe is supplied by each.

these are not of great functional importance they obtrude upon the
surgeon during the course of a right hepatic lobectomy.

The three principal hepatic veins have three zones of drainage corresponding roughly to the right, the middle and left thirds of the liver.
The plane defined by the falciform ligament corresponds to the
boundary of the zones drained by the left and middle hepatic veins.
Unfortunately for the surgeon, the middle hepatic vein lies just at the
line of the principal plane of the liver between its right and left morphological lobes and it is this fact which complicates the operation of
right hepatic resection (Fig. 74).

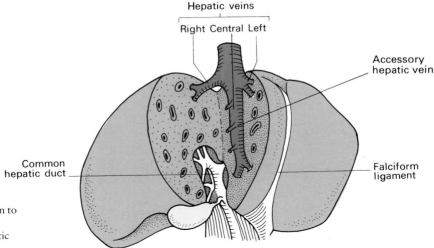

Fig. 74 Liver split open to demonstrate the tributaries of the hepatic vein.

The biliary system (Fig. 75)

The *right* and *left hepatic ducts* fuse in the porta hepatis to form the *common hepatic duct* (1.5 in (4 cm)). This joins with the *cystic duct* (1.5 in (4 cm)), draining the gall-bladder, to form the *common bile duct* (4 in (10 cm)). The common bile duct commences about 1 in (4 cm) above the duodenum, then passes behind it to open at a papilla on the medial aspect of the second part of the duodenum. In this course the common duct lies either in a groove in the posterior aspect of the head of the pancreas or is actually buried in its substance.

As a rule, the common duct termination joins that of the main pancreatic duct (of Wirsung) in a dilated common vestibule, the *ampulla of Vater*, whose opening in the duodenum is guarded by the *sphincter of Oddi*. Occasionally, the bile and pancreatic ducts open separately into the duodenum.

The common hepatic duct and the supraduodenal part of the common bile duct lie in the free edge of the lesser omentum where they are related as follows (Fig. 49):
• bile duct—anterior to the right;
• hepatic artery—anterior to the left;
• portal vein—posterior;
• inferior vena cava—still more posterior, separated from the portal vein by the foramen of Winslow.

The gall-bladder (Fig. 75)

The gall-bladder normally holds about 50 ml of bile and acts as a bile

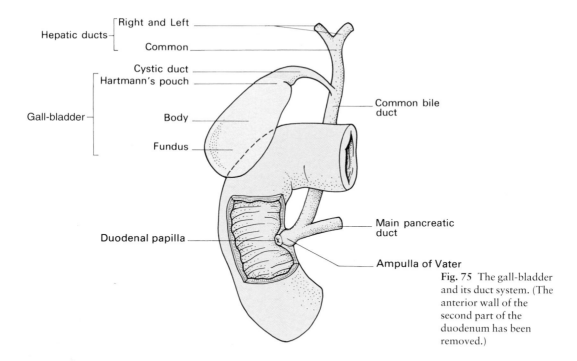

Fig. 75 The gall-bladder and its duct system. (The anterior wall of the second part of the duodenum has been removed.)

concentrator and reservoir. It lies in a fossa separating the right and quadrate lobes of the liver and is related inferiorly to the duodenum and transverse colon. (An inflamed gall-bladder may occasionally ulcerate into either of these structures.)

For descriptive purposes, the organ is divided into fundus, body and neck, the latter opening into the cystic duct. In dilated and patho-logical gall-bladders there is frequently a pouch present on the ventral aspect just proximal to the neck termed *Hartmann's pouch* in which gallstones may become lodged.

Blood supply (Fig. 76)

The gall-bladder is supplied by the cystic artery (a branch usually of the right hepatic artery) which lies in the triangle made by the liver, the cystic duct and the common hepatic duct. Other vessels derived from the hepatic artery pass to the gall-bladder from its bed in the liver. Interestingly, there is no accompanying vein to the cystic artery. Small veins pass from the gall-bladder through its bed directly into the liver.

Structure

The gall-bladder wall and the sphincter of Oddi contain muscle, but

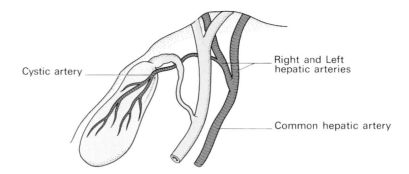

Cystic artery

Right and Left hepatic arteries

Common hepatic artery

Fig. 76 The arterial supply of the gall-bladder.

there are only scattered muscle fibres throughout the remaining biliary duct system. The mucosa is lined throughout by columnar cells and bears mucus-secreting glands.

Development

The gall-bladder and ducts are subject to numerous anatomical variations which are best understood by considering their embryological development. A diverticulum grows out from the ventral wall of the duodenum which differentiates into the hepatic ducts and the liver (see Fig. 78). Another diverticulum from the side of the hepatic duct bud forms the gall-bladder and cystic duct.

Some variations are shown in a series of diagrams (Fig. 77).

Clinical features

1   Errors in gall-bladder surgery are frequently the result of failure to appreciate the variations in the anatomy of the biliary system; it is important, therefore, before dividing any structures and removing the gall-bladder, to have all the three biliary ducts clearly identified, together with the cystic and hepatic arteries.

2   Haemorrhage during cholecystectomy may be controlled by compressing the hepatic artery (which gives off the cystic branch) between the finger and thumb where it lies in the anterior wall of the foramen of Winslow (Pringle's manoeuvre).

3   Gangrene of the gall-bladder is rare because even if the cystic artery becomes thrombosed in acute cholecystitis there is a rich secondary blood supply coming in from the liver bed. Gangrene may occur in the unusual event of a gall-bladder on an abnormally long mesentery undergoing torsion, which will destroy both its sources of blood supply.

4   Stones in the common duct can usually be removed through an

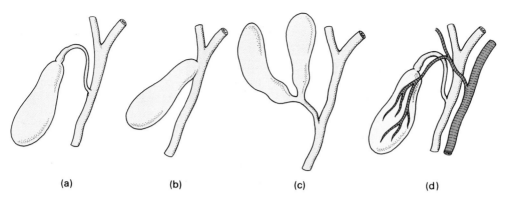

Fig. 77 Some variations in biliary anatomy (a) A long cystic duct joining the hepatic duct low down behind the duodenum. (b) Absence of the cystic duct—the gall-bladder opens directly into the common hepatic duct. (c) A double gall-bladder, the result of a rare bifid embryonic diverticulum from the hepatic duct. (d) The right hepatic artery crosses *in front* of the common hepatic duct; this occurs in 25 per cent of cases.

incision in the supraduodenal part of the common duct. Occasionally, the duodenum and pancreatic head must be mobilized by incising the peritoneum along the lateral duodenal margin to expose the duct in its retropancreatic course (Kocher's manoeuvre). Sometimes a stone impacted at the ampulla of Vater must be approached via an incision in the second part of the duodenum. This last approach is also used when it is necessary to divide the sphincter of Oddi or to remove a tumour arising at the termination of the common bile duct.

The pancreas (Fig. 57)

The pancreas lies retroperitoneally in roughly the transpyloric plane. For descriptive purposes it is divided into head, neck, body and tail.

Relations

The head lies in the C-curve of the duodenum and sends out the *uncinate process* which hooks posteriorly to the superior mesenteric vessels as these travel from behind the pancreas into the root of the mesentery.

Posteriorly lie the inferior vena cava, the commencement of the portal vein, aorta, superior mesenteric vessels, the crura of diaphragm, coeliac plexus, the left kidney and suprarenal gland. The splenic artery runs along the upper border of the pancreas. The splenic vein runs behind the gland, receives the inferior mesenteric vein and joins the superior mesenteric to form the portal vein behind the pancreatic neck (Fig. 67).

To complete this list of important posterior relationships, the common bile duct lies either in a groove in the right extremity of the gland or embedded in its substance, as it passes to open into the second part of the duodenum.

Anteriorly lies the stomach separated by the lesser sac. To the left, the pancreatic tail lies against the hilum of the spleen.

The blood supply

Blood is supplied from the splenic and the pancreaticoduodenal arteries; the corresponding veins drain into the portal system.

The lymphatics

The lymphatics drain into nodes which lie along its upper border, in the groove between its head and the duodenum, and along the root of the superior mesenteric vessels.

Structure

The pancreas macroscopically is lobulated and is contained within a fine capsule; these lobules are made up of alveoli of serous secretory cells draining via their ductules into the principal ducts. Between these alveoli lie the insulin-secreting *islets of Langerhans*.

The main duct of the pancreas (Wirsung) (Fig. 58) runs the length of the gland and usually opens at the ampulla of Vater in common with the common bile duct; occasionally it drains separately into the duodenum.

The *accessory duct (of Santorini)* passes from the lower part of the head in front of the main duct, communicates with it, and then opens into the duodenum above it. Occasionally it is absent.

Development (Fig. 78)

The pancreas develops from a larger dorsal diverticulum from the duodenum and a smaller ventral outpouching from the side of the common bile duct. The ventral pouch swings round posteriorly to fuse with the lower aspect of the dorsal diverticulum, trapping the superior mesenteric vessels between the two parts.

The ducts of the two formative segments of the pancreas then communicate; that of the smaller takes over the main pancreatic flow to form the main duct, leaving the original duct of the larger portion of the gland as the accessory duct.

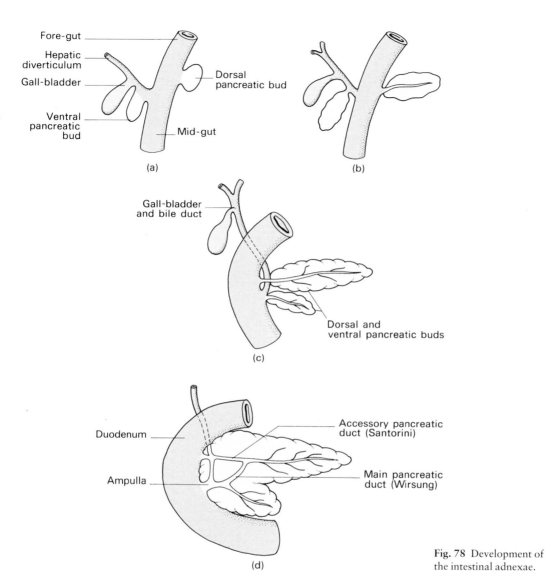

Fig. 78 Development of the intestinal adnexae.

Clinical features

1   Rarely, the two developing segments of the pancreas completely surround the second part of the duodenum ('annular pancreas') and may produce duodenal obstruction.

2   Note from the posterior relations of the pancreas that a neoplasm of the head of the pancreas will produce obstructive jaundice by compressing the common bile duct. An extensive growth in the body of the gland may cause portal or inferior vena caval obstruction.

3   Anterior to the pancreas lies the stomach, separated from it by the lesser sac. This sac may become closed off and distended with fluid either from perforation of a posterior gastric ulcer or from the outpouring of fluid in acute pancreatitis, forming a *pseudo-cyst of the pancreas*. Such a collection may almost fill the abdominal cavity.

The spleen

The spleen is about the size of the cupped hand. If forms the left lateral extremity of the lesser sac. Passing from it are the *gastrosplenic ligament* to the greater curvature of stomach (carrying the short gastric and left gastro-epiploic vessels) and the *lienorenal ligament* to the posterior abdominal wall (carrying the splenic vessels and tail of the pancreas).

Relations (Fig. 79)

• Posteriorly—the left diaphragm, separating it from the pleura, left lung and the 9th, 10th and 11th ribs.
• Anteriorly—the stomach.
• Inferiorly—the splenic flexure of the colon.
• Medially—the left kidney.
   The tail of the pancreas abuts against the hilum of the spleen through which vessels and nerves enter and leave this organ.

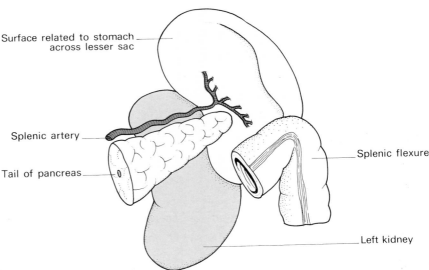

Surface related to stomach across lesser sac

Splenic artery

Tail of pancreas

Splenic flexure

Left kidney

Fig. 79 The spleen and its immediate relations.

Blood supply

The splenic artery is one of the three main branches of the coeliac axis. The splenic vein is joined by the superior mesenteric to form the portal vein.

Structure

The spleen represents the largest reticulo-endothelial accumulation in the body. It has a thin fibrous capsule, to which the peritoneum adheres intimately. The fibrous tissue of the capsule extends into the spleen to form a series of trabeculae between which lies the splenic pulp.

Clinical features

1   In performing a splenectomy the close relation of the pancreatic tail to the hilum and splenic pedicle must be remembered; it is easily wounded.

2   Note the close proximity of the lower ribs, lowest part of the left lung and pleural cavity, left diaphragm, left kidney and the spleen; injuries to the left upper abdomen may damage any combination of these structures. Similarly, a stab wound of the posterior left chest may penetrate the diaphragm and tear the spleen. The spleen, with its thin tense capsule, is the commonest intra-abdominal viscus to be ruptured by blunt trauma.

3   Accessory spleens (one or more) may occur most commonly near the hilum, but also in the tail of pancreas, the mesentery of the spleen, the omentum, small bowel mesentery, ovary and even testis. They occur in about one in ten subjects and, if left behind, may result in persistence of symptoms following splenectomy for congenital acholuric jaundice or thrombocytopenic purpura.

The urinary tract

The kidneys

The kidneys lie retroperitoneally on the posterior abdominal wall; the right kidney is 0.5 in (12 mm) lower than the left, presumably because of its downward displacement by the bulk of the liver. Each measures approximately 4.5 in (11 cm) long, 2.5 in (6 cm) wide and 1.5 in (4 cm) thick.

Relations (Figs 80, 81)

• Posteriorly — the diaphragm (separating pleura), quadratus lumborum, psoas, transversus abdominis, the 12th rib and three nerves — the subcostal (T12), iliohypogastric and ilio-inguinal (L1).

• Anteriorly — the right kidney is related to the liver, the 2nd part of the duodenum (which may be opened accidentally in performing a right nephrectomy), and the ascending colon. In front of the left kidney lie the stomach, the pancreas and its vessels, the spleen, and the descending colon. The suprarenals sit on each side as a cap on the kidney's upper pole.

The medial aspect of the kidney presents a deep vertical slit, the *hilum*, which transmits, from before backwards, the renal vein, renal artery, pelvis of the ureter and, usually, a subsidiary branch of the renal artery. Lymphatics and nerves also enter the hilum, the latter being sympathetic, mainly vasomotor, fibres.

The *pelvis of the ureter* is subject to considerable anatomical variations (Fig. 82); it may lie completely outside the substance of the kidney (even to the extent of having part of the major calyces extrarenal) or may be almost buried in the renal hilum. All gradations exist between these extremes. If a calculus is lodged in the pelvis of the ureter, its removal is comparatively simple when this is extrarenal, and it is correspondingly difficult when the pelvis is hidden within the substance of the kidney.

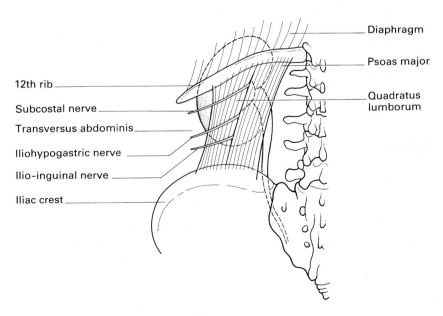

Fig. 80 The posterior relations of the kidney.

Liver

Duodenum

Right colon
(hepatic flexure)

Inferior
vena cava

Aorta

Spleen

Stomach

Pancreas

Left colon
(splenic flexure)

Small gut

Fig. 81 The anterior
relations of the kidneys.

Within the kidney, the pelvis of the ureter divides into two or three
major calyces, each of which divides into a number of *minor calyces*.
Each of these, in turn, is indented by a papilla of renal tissue and it is
here that the collecting tubules of the kidney discharge urine into the
ureter.

The kidneys lie in an abundant fatty cushion (*perinephric fat*) contained in the *renal fascia* (Fig. 83). Above, the renal fascia blends with
the fascia over the diaphragm, leaving a separate compartment for the
suprarenal (which is thus easily separated and left behind in performing a nephrectomy). Medially, the fascia blends with the sheaths of
the aorta and inferior vena cava. Laterally it is continuous with the
transversalis fascia. Only inferiorly does it remain relatively open —
tracking around the ureter into the pelvis.

The kidney has, in fact, three capsules:

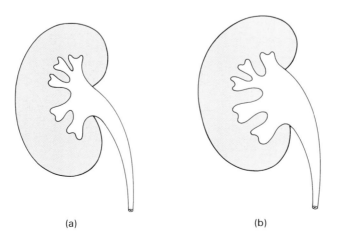

(a)   (b)

Fig. 82 Variations in the
renal pelvis. (a) The pelvis
is buried within the renal
parenchyma—
pyelolithotomy difficult.
(b) The pelvis protrudes
generously—
pyelolithotomy easy.

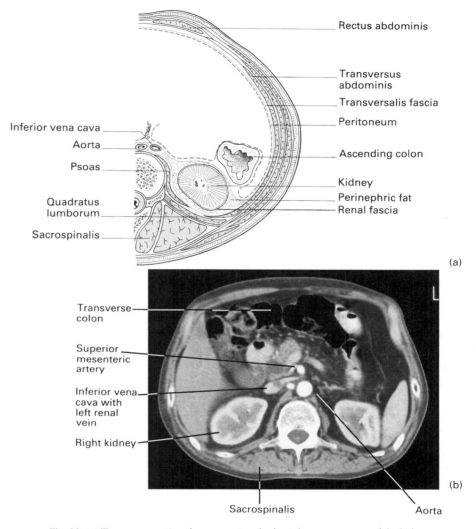

Fig. 83 (a) Transverse section demonstrating the fascial compartments of the kidney. (b) CT scan of the same region. Note that CT scans, by convention, are viewed from below, so that the aorta, for example, is seen on the right side. The blood vessels have been enhanced by an intravenous injection of contrast.

1 fascial (renal fascia);

2 fatty (perinephric fat);

3 true — the fibrous capsule which strips readily from the normal kidney surface but adheres firmly to an organ that has been inflamed.

Blood supply

The *renal artery* is derived directly from the aorta. The *renal vein* drains directly into the inferior vena cava. The *left renal vein* passes in front of the aorta immediately below the origin of the superior mesenteric artery. The *right renal artery* passes behind the inferior vena cava.

Lymph drainage

Lymphatics drain directly to the para-aortic lymph nodes.

Clinical features

1   Blood from a ruptured kidney or pus in a perinephric abscess first distend the renal fascia, then force their way within the fascial compartment downwards into the pelvis. The midline attachment of the renal fascia prevents extravasation to the opposite side.
2   In hypermobility of the kidney ('floating kidney'), this organ can be moved up and down in its fascial compartment but not from side to side. To a lesser degree, it is in this plane that the normal kidney moves during respiration.
3   *Exposure of the kidney via the loin.* An oblique incision is usually favoured midway between the 12th rib and the iliac crest, extending laterally from the lateral border of erector spinae. Latissimus dorsi and serratus posterior inferior are divided and the free posterior border of external oblique is identified, enabling this muscle to be split along its fibres. Internal oblique and transversus abdominis are then divided, revealing peritoneum anteriorly, which is pushed forward. The renal fascial capsule is then brought clearly into view and is opened. The subcostal nerve and vessels are usually encountered in the upper part of the incision and are preserved. If more room is required, the lateral edge of quadratus lumborum may be divided and also the 12th rib excised, care being taken to push up, but not to open, the pleura, which crosses the medial half of the rib.

The ureter

The ureter is 10 in (25 cm) long and comprises the pelvis of the ureter (see above) and its abdominal, pelvic and intravesical portions.

The *abdominal ureter* lies on the medial edge of psoas major (which separates it from the tips of the transverse processes of L2–L5) and then crosses into the pelvis at the bifurcation of the common iliac artery in front of the sacroiliac joint. Anteriorly, the right ureter is covered at its origin by the second part of the duodenum and then lies lateral to the inferior vena cava and behind the posterior peritoneum. It is crossed by the testicular (or ovarian), right colic, and ileocolic vessels. The left ureter is crossed by the testicular (or ovarian) and left colic vessels and then passes above the pelvic brim, behind the mesosigmoid and sigmoid colon.

The *pelvic ureter* runs on the lateral wall of the pelvis in front of the internal iliac artery to just in front of the ischial spine; it then turns forwards and medially to enter the bladder. In the male it lies above the seminal vesicle near its termination and is crossed superficially by the vas deferens (see Fig. 87). In the female, the ureter passes above the lateral fornix of the vagina 0.5 in (12 mm) lateral to the supravaginal portion of the cervix and lies below the broad ligament and uterine vessels (see Fig. 104).

The *intravesical ureter* passes obliquely through the wall of the bladder for 0.75 in (2 cm); the vesical muscle and obliquity of this course produce respectively a sphincteric and valve-like arrangement at the termination of this duct.

Blood supply

The ureter receives a rich segmental blood supply from all available arteries along its course: the aorta, and the renal, testicular (or ovarian), internal iliac and inferior vesical arteries.

Clinical features

1   The ureter is readily identified in life by its thick muscular wall which is seen to undergo worm-like (vermicular) writhing movements, particularly if gently stroked or squeezed.

2   Throughout its abdominal and the upper part of its pelvic course, it adheres to the overlying peritoneum (through which it can be seen in the thin subject), and this fact is used in exposing the ureter—as the parietal peritoneum is dissected upwards, the ureter comes into view sticking to its posterior aspect.

3   The ureter is relatively narrowed at three sites:
- at the junction of the pelvis of ureter with its abdominal part,
- at the pelvic brim, and
- at the ureteric orifice (narrowest of all).

   A ureteric calculus is likely to lodge at one of these three levels.

4   In searching for a ureteric stone on a plain radiograph of the abdomen, one must imagine the course of the ureter in relation to the bony skeleton (Fig. 84). It lies along the tips of the transverse processes, crosses in front of the sacroiliac joint, swings out to the ischial spine and then passes medially to the bladder. An opaque shadow along this line is suspicious of calculus. This course of the ureter is readily studied by examining a radiograph showing a radio-opaque ureteric catheter *in situ*.

Fig. 84 Drawing from an intravenous pyelogram to show the relationship of the ureters to the bony landmarks.

The embryology and congenital abnormalities of the kidney and ureter (Fig. 85)

The kidney and ureter are mesodermal in origin and develop in an unusual manner of considerable interest to the comparative anatomist.

The *pronephros*, of importance in primitive animals, is transient in humans, but the distal part of its duct receives the tubules of the next renal organ to develop, the *mesonephros*, and now becomes the *mesonephric* or *Wolffian duct*. The mesonephros itself then disap-

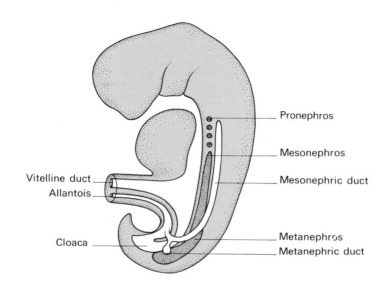

Fig. 85 Development of the pro-, meso-, and metanephric systems (after Langman).

pears except for some of its ducts which form the efferent tubules of the testis.

A diverticulum then appears at the lower end of the mesonephric duct which develops into the *metanephric duct*; on top of the latter a cap of tissue differentiates to form the definitive kidney or *metanephros*. The metanephric duct develops into the ureter, pelvis, calyces and collecting tubules, the metanephros into the glomeruli and the proximal part of the renal duct system.

The mesonephric duct now loses its renal connection, atrophies in the female (remaining only as the epoöphoron) but persists in the male, to become the epididymis and vas deferens.

The kidney first develops in the pelvis and then migrates upwards. Its blood supply is first obtained from the common iliac artery but, during migration, a series of vessels form to supply it, only to involute again when the renal artery takes over this duty.

Developmental abnormalities (Fig. 86)

1 It is common for one or more distally placed arteries to persist (*aberrant renal arteries*) and one may even run to the kidney from the common iliac artery.

(a)

(b)

(c)

(d)

Fig. 86 Renal abnormalities.
(a) Polycystic kidney.
(b) Horseshoe kidney.
(c) Pelvic kidney and double ureter.
(d) Aberrant renal artery and associated hydronephrosis.

2   Occasionally the kidney will fail to migrate cranially, resulting in a persistent *pelvic kidney*.

3   The two metanephric masses may fuse in development, forming a *horseshoe kidney* linked across the midline.

4   In 1 in 2400 births there is complete failure of development of one kidney (*congenital absence of the kidney*).

5   *Congenital polycystic kidneys* (which are nearly always bilateral) are believed to result from failure of metanephric tissue to link up with some of the metanephric duct collecting tubules; blind ducts therefore form which subsequently become distended with fluid. This theory of origin does not explain their occasional association with multiple cysts of the liver, pancreas, lung and ovary.

6   The mesonephric duct may give off a double metanephric bud so that two ureters may develop on one or both sides. These ureters may fuse into a single duct anywhere along their course or open separately into the bladder (where the *upper* ureter enters *below* the *lower* ureter).

Rarely, the extra ureter may open ectopically into the vagina or urethra resulting in urinary incontinence.

The bladder (Figs 62, 63, 87)

The urinary bladder of a normal subject is uncomfortably distended by half a pint of fluid. When fully distended, the adult bladder projects from the pelvic cavity into the abdomen, stripping the peritoneum upwards from the anterior abdominal wall. The surgeon utilizes this fact in carrying out an extraperitoneal incision into the bladder. In children up to the age of about 3 years, the pelvis is relatively small and the bladder is, in fact, intra-abdominal although still extraperitoneal.

Posterior aspect of bladder

Vas deferens

Ureter

Seminal vesicle

Prostate (with its median groove)

Fig. 87 The prostate, seminal vesicles and vasa shown in a posterior view of the bladder.

Relations

- Anteriorly—the pubic symphysis.
- Superiorly—the bladder is covered by peritoneum with coils of small intestine and sigmoid colon lying against it. In the female, the body of the uterus flops against its posterosuperior aspect.
- Posteriorly—in the male the rectum, the termination of the vasa deferentia and the seminal vesicles; in the female, the vagina and the supravaginal part of the cervix.
- Laterally—the levator ani and obturator internus.

The neck of the bladder fuses with the prostate in the male; in the female it lies directly on the pelvic fascia surrounding the short urethra.

The muscle coat of bladder is formed by a criss-cross arrangement of bundles; when these undergo hypertrophy in chronic obstruction (due to an enlarged prostate, for example) they account for the typical trabeculated 'open weave' appearance of the bladder wall, readily seen through a cystoscope.

The circular component of the muscle coat condenses as an (involuntary) *internal urethral sphincter* around the internal orifice. This can be destroyed without incontinence providing the external sphincter remains intact (as occurs in prostatectomy).

Cystoscopy

The interior of the bladder and its three orifices (the internal meatus and the two ureters) are easily inspected by means of a cystoscope. The ureteric orifices lie 1 in (2.5 cm) apart in the empty bladder, but when this is distended for cystoscopic examination, the distance increases to 2 in (5 cm). The submucosa and mucosa of most of the bladder are only loosely adherent to the underlying muscle and are thrown into folds when the bladder is empty, smoothing out during distension of the organ. Over the *trigone*, the triangular area bounded by the ureteric orifices and the internal meatus, the mucosa is adherent and remains smooth even in the empty bladder.

Between the ureters, a raised fold of mucosa can be seen called the *interureteric ridge* which is produced by an underlying bar of muscle.

Blood supply

Blood is supplied from the superior and inferior vesical branches of the internal iliac artery. The vesical veins form a plexus which drains into the internal iliac vein.

Lymph drainage

Lymphatics drain alongside the vesical blood vessels to the iliac and then para-aortic nodes.

Nerve supply

Efferent parasympathetic fibres from S2 to S4 accompany the vesical arteries to the bladder. They convey motor fibres to the muscles of the bladder wall and inhibitory fibres to its internal sphincter. Sympathetic efferent fibres are said to be inhibitory to the bladder muscles and motor to its sphincter, although they may be mainly vasomotor in function, so that normal filling and emptying of the bladder are probably controlled exclusively by its parasympathetic innervation. The external sphincter is made up of striated muscle. It is also concerned in the control of micturition and is supplied by the pudendal nerve. Sensory fibres from the bladder, which are stimulated by distension, are conveyed in both the sympathetic and parasympathetic nerves, the latter pathway being the more important.

The urethra

The male urethra

The male urethra is 8 in (20 cm) long and is divided into the prostatic, membranous and spongy parts.

The *prostatic urethra* (1.25 in (3 cm)), as its name implies, traverses the prostate. Its posterior wall bears a longitudinal elevation termed the *urethral crest*, on each side of which is a shallow depression, the *prostatic sinus*, into which the 15–20 prostatic ducts empty. At about the middle of the crest is a prominence termed the *colliculus seminalis (verumontanum)* into which opens the *prostatic utricle*. This is a blind tract, about 5 mm long, running downwards from the substance of the median lobe of the prostate. It is believed to represent the male equivalent of the vagina (see page 160). On either side of the orifice of the prostatic utricle open the *ejaculatory ducts*, formed by the union of the duct of the seminal vesicle and the terminal part of the vas deferens.

The *membranous urethra* (0.75 in (2 cm)) pierces the external sphincter urethrae (the voluntary sphincter of the bladder) and the fascial perineal membrane which covers the superficial aspect of the sphincter.

The *spongy urethra* (6 in (15 cm)) traverses the corpus spongiosum of the penis. It first passes upwards and forwards to lie below the pubic symphysis and then in its flaccid state bends downwards and forwards.

Clinical features

1   Where the urethra passes beneath the pubis is a common site for it to be ruptured by a fall astride a sharp object, which crushes it against the edge of the symphysis.

2   The external orifice is the narrowest part of the urethra and a calculus may lodge there. Immediately within the meatus, the urethra dilates into a terminal fossa whose roof bears a mucosal fold (the *lacuna magna*) which may catch the tip of a catheter. Instruments should always be introduced into the urethra beak downwards for this reason.

The female urethra

The female urethra is 1.5 in (4 cm) long; it traverses the sphincter urethrae and lies immediately in front of, indeed embedded in the wall of, the vagina. Its external meatus opens 1 in (2.5 cm) behind the clitoris. The sphincter urethrae in the female is a tenuous structure and vesical control appears to depend mainly on the intrinsic sphincter of condensed circular muscle fibres of the bladder.

The mucosa of the urinary tract

The pelvis, ureter, bladder and urethra are lined by a transitional epithelium as far as the entry of the ejaculatory ducts in the prostatic urethra. This is conveniently termed the *uroepithelium* since it has a uniform appearance and is subject to the same pathological processes —for example, the development of papillomata. The remainder of the urethra has a columnar lining except at its termination, where the epithelium becomes squamous.

Radiology of the urinary tract

The renal contours can often be identified on a soft tissue radiograph of the abdomen. Intravenous injection of iodine-containing compounds excreted by the kidney will produce an outline of the calyces and the ureter (intravenous pyelogram). When the injection medium enters the bladder, a cystogram is obtained (Fig. 84).

Further information can be obtained by passing a catheter up the ureter through a cystoscope and injecting radio-opaque fluid to fill the pelvis and calyx system (retrograde pyelogram). Similarly, injection of such fluid into the urethra or bladder may be used for the radio-graphic study of these viscera.

The male genital organs

The prostate (Fig. 87)

This is a pyramidal-shaped, fibromuscular and glandular organ, 1.25 in (3 cm) long, which surrounds the prostatic urethra. It resembles the size and shape of a chestnut.

Relations (Fig. 62)

- Superiorly — the prostate is continuous with the neck of the bladder. The urethra enters the upper aspect of the prostate near it anterior border.
- Inferiorly—the apex of the prostate rests on the external sphincter of the bladder which lies within the deep perineal pouch.
- Anteriorly — lies the pubic symphysis separated by the extraperi-toneal fat of the *cave of Retzius* or *retropubic space*. Close against the prostate in this space lies the prostatic plexus of veins. Near the apex of the prostate, the *puboprostatic ligament* (a condensation of fibrous tissue) passes forward to the pubis.
- Posteriorly—lies the rectum separated by *the fascia of Denon-villiers*.
- Laterally—lies levator ani.

The ejaculatory ducts enter the upper posterior part of the gland to open into the urethra at the *colliculus seminalis* or verumontanum, one on either side of the prostatic utricle, dividing off a *median pro-static lobe* lying between these three ducts. A shallow posterior *median groove* (which can be felt on rectal examination) further divides the prostate into left and right *lobes*. Anterior to the urethra, the prostate consists of a narrow isthmus only.

The prostatic capsules

These are normally two, pathologically three, in number.
1 The true capsule — a thin fibrous sheath which surrounds the gland.
2 The false capsule — condensed extraperitoneal fascia which continues into the fascia surrounding the bladder and with the fascia of

Denonvilliers posteriorly. Between layers 1 and 2 lies the prostatic venous plexus.

3   The pathological capsule—when benign 'adenomatous' hypertrophy of the prostate takes place, the normal peripheral part of the gland becomes compressed into a capsule around this enlarging mass (Fig. 88). In performing an enucleation of the prostate, the plane between the adenomatous mass and this compressed peripheral tissue is entered, the 'tumour' enucleated and a condensed rim of prostate tissue, lying deep to the true capsule, left behind. The prostatic venous plexus, lying external to this, is thus undisturbed.

Blood supply

The arterial supply is derived from the inferior vesical artery (a branch of the internal iliac artery), a branch entering the prostate on each side at its lateral extremity.

The veins form a prostatic plexus which receives the dorsal vein of the penis and drains into the internal iliac vein on each side. Some of the venous drainage passes to the plexus of veins lying in front of the vertebral bodies and within the neural canal. These veins are valveless and constitute the *valveless vertebral veins of Batson*. This communication may explain the readiness with which carcinoma of the prostate spreads to the pelvis and vertebrae.

Clinical features

1   *Approaches to the prostate.* 'Prostatectomy' usually means the enucleation of a hyperplastic mass of glandular tissue from the surrounding prostatic gland which is compressed into a thin rim around it (Fig. 88). The prostate can be approached, for this purpose, either across the bladder (*transvesically*), or through the prostatic capsule itself (*retropubically*), or, rarely, across the perineum, dissecting in front of the rectum and reaching the prostate through the fascia of Denonvilliers (*the perineal approach*). In obstructive conditions other than benign hypertrophy—that is to say, carcinoma or fibrosis of the prostate — this plane of cleavage does not exist and the obstructing mass of prostatic tissue may be removed *transurethrally* by means of an operating cystoscope. In fact, the modern instrument is so efficient that this route is now used routinely for benign disease unless the gland is greatly enlarged. In transurethral resection the verumontanum (colliculus seminalis) is an important landmark; the surgeon keeps proximal to this point in order not to damage the urethral sphincter mechanism.

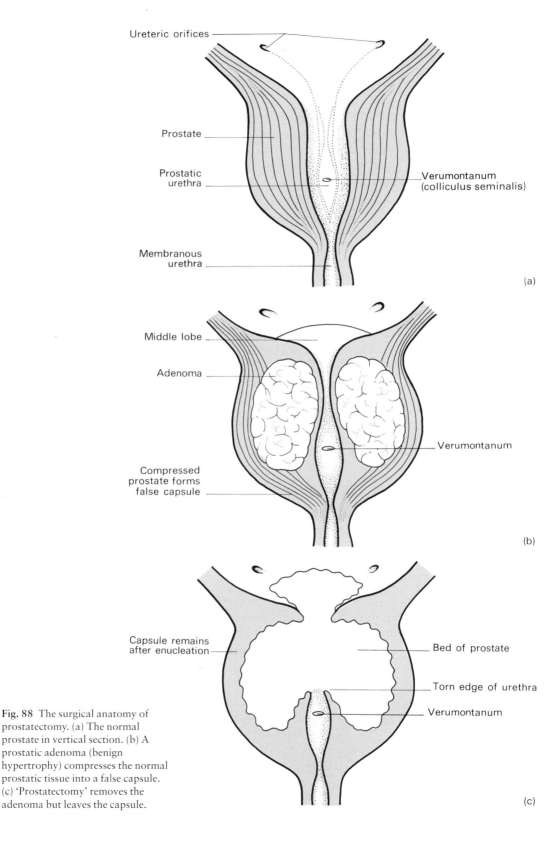

Ureteric orifices

Prostate

Prostatic
urethra

Verumontanum
(colliculus seminalis)

Membranous
urethra

(a)

Middle lobe

Adenoma

Verumontanum

Compressed
prostate forms
false capsule

(b)

Capsule remains
after enucleation

Bed of prostate

Torn edge of urethra

Verumontanum

Fig. 88 The surgical anatomy of
prostatectomy. (a) The normal
prostate in vertical section. (b) A
prostatic adenoma (benign
hypertrophy) compresses the normal
prostatic tissue into a false capsule.
(c) 'Prostatectomy' removes the
adenoma but leaves the capsule.

(c)

2   After the age of 45 years some degree of prostatic hypertrophy is all but invariable; it is as much a sign of ageing as greying of the hair. Usually the lateral lobes are affected and such enlargement is readily detected on rectal examination. The median lobe may also be involved in this process or may be enlarged without the lateral lobes being affected. In such an instance, symptoms of prostatic obstruction may occur (because of the valve-like effect of this hypertrophied lobe lying over the internal urethral orifice) without prostatic enlargement being detectable on rectal examination.

Anterior to the urethra the prostate consists of a narrow fibromuscular isthmus containing little, if any, glandular tissue. Benign glandular hypertrophy of the prostate, therefore, never affects this part of the organ.

3   The fascia of Denonvilliers is important surgically; in excising the rectum it is the plane to be sought after in order to separate off the prostate and urethra without damaging these structures. A carcinoma of the prostate only rarely penetrates this fascial barrier so that ulceration into the rectum is very unusual.

The scrotum

The scrotum is the pouch in which lie the testes and their coverings. In cryptorchidism, not unnaturally, this pouch is not well developed.

The skin of the scrotum is thin, pigmented, rugose and marked by a longitudinal median raphe. It is richly endowed with sebaceous glands, and consequently a common site for sebaceous cysts, which are often multiple. The subcutaneous tissue contains no fat but does contain the involuntary *dartos muscle*.

Clinical features

The scrotal subcutaneous tissue is continuous with the fasciae of the abdominal wall and perineum and therefore extravasations of urine or blood deep to this plane will gravitate into the scrotum. The scrotum is divided by a septum into right and left compartments but this septum is incomplete superiorly so extravasations of fluid into this sac are always bilateral.

The lax tissues of the scrotum and its dependent position cause it to fill readily with oedema fluid in cardiac or renal failure. Such a condition must be carefully differentiated from extravasation or from a scrotal swelling due to a hernia or hydrocele.

Testis and epididymis (Figs 89, 90)

The left testis lies at a lower level than the right within the scrotum; rarely, this arrangement is reversed. Each testis is contained by a white fibrous capsule, the *tunica albuginea*, and each is invaginated anteriorly into a double serous covering, the *tunica vaginalis*, just as the intestine is invaginated anteriorly into the peritoneum.

Along the posterior border of the testis, rather to its lateral side, lies the *epididymis*, which is divided into an expanded head, a body and a pointed tail inferiorly. Medially, there is a distinct groove, the *sinus epididymis*, between it and the testis. The epididymis is covered by the tunica vaginalis except at its posterior margin which is free or, so to say, 'extraperitoneal'.

The testis and epididymis each bear at their upper extremities a small stalked body, termed respectively the *appendix testis* and *appendix epididymis* (*hydatid of Morgagni*). The appendix testis is a remnant of the upper end of the paramesonephric (Müllerian) duct; the appendix epididymis is a remnant of the mesonephros.

These structures, being stalked, are liable to undergo torsion.

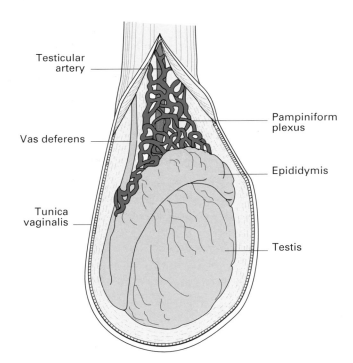

Testicular
artery

Pampiniform
plexus

Vas deferens

Epididymis

Tunica
vaginalis

Testis

Fig. 89 Testis and epididymis.

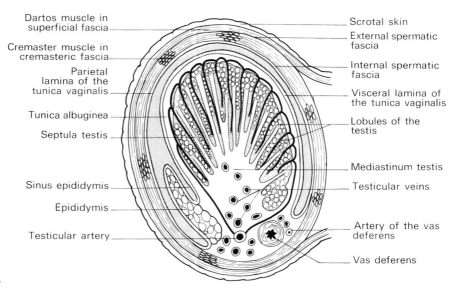

Dartos muscle in superficial fascia

Cremaster muscle in cremasteric fascia

Parietal lamina of the tunica vaginalis

Tunica albuginea

Septula testis

Sinus epididymis

Epididymis

Testicular artery

Scrotal skin

External spermatic fascia

Internal spermatic fascia

Visceral lamina of the tunica vaginalis

Lobules of the testis

Mediastinum testis

Testicular veins

Artery of the vas deferens

Vas deferens

Fig. 90 Transverse section of the testis.

Blood supply

The testicular artery arises from the aorta at the level of the renal vessels. It anastomoses with the artery to the vas, supplying the vas deferens and epididymis, which arises from the inferior vesical branch of the internal iliac artery. This cross-connection means that ligation of the testicular artery is not necessarily followed by testicular atrophy.

The pampiniform plexus of veins becomes a single vessel, the testicular vein, in the region of the internal ring. On the right this drains into the inferior vena cava, on the left into the renal vein.

Lymph drainage

The lymphatic drainage of the testis obeys the usual rule; it accompanies the venous drainage and thus passes to the para-aortic lymph nodes at the level of the renal vessels. Free communication occurs between the lymphatics on either side; there is also a plentiful anastomosis with the para-aortic intrathoracic nodes and, in turn, with the cervical nodes, so that spread of malignant disease from the testis to the root of the neck is not rare.

Nerve supply

T10 sympathetic fibres via the renal and aortic plexus.

Structure

The testis is divided into 200–300 lobules each containing one to three *seminiferous tubules*. Each tubule is some 2 feet (62 cm) in length when teased out, and is thus obviously coiled and convoluted to pack away within the testis. The tubules anastomose posteriorly into a plexus termed the *rete testis* from which about a dozen fine *efferent ducts* arise, pierce the tunica albuginea at the upper part of the testis and pass into the head of the epididymis, which is actually formed by these efferent ducts coiled within it. The efferent ducts fuse to form a considerably convoluted single tube which constitutes the body and tail of the epididymis.

Development of the testis

This is important and is the key to several features which are of clinical interest.

The testis arises from a germinal ridge of mesoderm in the posterior wall of the abdomen just medial to the mesonephros (Fig. 85), and links up with the epididymis and vas, which differentiate from the mesonephric duct. As the testis enlarges, it also undergoes a caudal migration according to the following timetable:

3rd month (of fetal life)   reaches the iliac fossa;
7th month   traverses the inguinal canal;
8th month   reaches the external ring;
9th month   descends into the scrotum.

A mesenchymal strand, the *gubernaculum testis*, extends from the caudal end of the developing testis along the course of its descent to blend into the scrotal fascia. The exact role of this structure in the descent of the testis is not known; theories are that it acts as a guide (*gubernaculum* = rudder) or that its swelling dilates the inguinal canal and scrotum.

In the third fetal month, a prolongation of the peritoneal cavity invades the gubernacular mesenchyme and projects into the scrotum as the *processus vaginalis*. The testis slides into the scrotum posterior to this, projects into it and is therefore clothed front and sides with peritoneum. About the time of birth this processus obliterates, leaving the testis covered by the tunica vaginalis. Very rarely, fragments of adjacent developing organs — spleen or suprarenal — are caught up and carried into the scrotum along with the testis.

Clinical features

1   The testis arises at the level of the mesonephros at the level of L2/3

vertebrae and drags its vascular, lymphatic and nerve supply from this region. Pain from the kidney is often referred to the scrotum and, conversely, testicular pain may radiate to the loin.

2   When searching for secondary lymphatic spread from a neoplasm of the testis, the upper abdomen must be palpated carefully for enlarged para-aortic nodes; because of cross-communications, these may be present on either side. Mediastinal and cervical nodes may also become involved. It is the beginner's mistake to feel for nodes in the groin; these are only involved if the tumour has ulcerated the scrotal skin and hence invaded scrotal lymphatics which drain to the inguinal nodes.

3   Rarely, a rapidly developing *varicocele* (dilatation of the pampiniform plexus of veins) is said to be a presenting sign of a tumour of the left kidney which, by invading the renal vein, blocks the drainage of the left testicular vein. Most examples of varicocele are idiopathic; why the vast majority are on the left side is unknown, but theories are that the left testicular vein is compressed by a loaded sigmoid colon, obstructed by angulation at its entry into the renal vein or even that it is put into spasm by adrenalin-rich blood entering the renal vein from the suprarenal vein!

4   The testis may fail to descend and may rest anywhere along its course—intra-abdominally, within the inguinal canal, at the external ring or high in the scrotum. Failure to descend must be carefully distinguished from retraction of the testis; it is common in children for contraction of the cremaster muscle to draw the testis up into the superficial inguinal pouch—a potential space deep to the superficial fascia over the external ring. Gentle pressure from above, or the relaxing effect of a hot bath, coaxes the testis back into the scrotum in such cases.

Occasionally the testis descends, but into an unusual (*ectopic*) position; most commonly it passes laterally after leaving the external ring to lie superficial to the inguinal ligament, but it may be found in front of the pubis, in the perineum or in the upper thigh. In these cases (unlike the undescended testis), the cord is long and replacement into the scrotum without tension presents no surgical difficulty.

5   Abnormalities of the obliteration of the processus vaginalis lead to a number of extremely common surgical conditions of which the *indirect inguinal hernia* is the most important.

This variety of hernia may be present at birth or develop in later life; in the latter circumstances it is probable that the processus vaginalis has persisted as a narrow empty sac and that development of the hernia results from some sudden strain due to a cough, straining at micturition or at stool, which forces abdominal contents into this peritoneal recess.

| | | | |
| (a) | (b) | (c) | (d) |

In infants, the sac frequently has the testis lying in its wall (congenital inguinal hernia) but this is unusual in older patients.

The closed-off tunica vaginalis may become distended with fluid to form a *hydrocele* which may be idiopathic (primary) or secondary to disease in the underlying testis. The anatomical classification of hydroceles is into the following groups (Fig. 91):

- *Vaginal*—confined to the scrotum and so called because it distends the tunica vaginalis.
- *Congenital*—communicating with the peritoneal cavity.
- *Infantile*—extending upwards to the internal ring.
- *Hydrocele of the cord*—confined to the cord.

Notice that, from the anatomical point of view, a hydrocele (apart from one of the cord) must surround the front and sides of the testis since the tunica vaginalis bears this relationship to it. A cyst of the epididymis, in contrast, arises from the efferent ducts of the epididymis and must therefore lie above and behind the testis. This point enables the differential diagnosis between these two common scrotal cysts to be made confidently.

Fig. 91 Types of hydrocele. (a) Vaginal hydrocele, (b) congenital hydrocele, (c) infantile hydrocele, (d) hydrocele of the cord. (The tube at the upper end of each diagram represents the internal inguinal ring.)

Vas deferens (Ductus deferens)

This tube is 18 in (45 cm) long (a distance which one may remember is also the length of the thoracic duct, the spinal cord and the femur, and the distance from the incisor teeth to the cardiac end of the stomach).

The vas passes from the tail of the epididymis to traverse the scrotum, inguinal canal and so comes to lie upon the side wall of the pelvis. Here, it lies immediately below the peritoneum of the lateral wall, extends almost to the ischial tuberosity then turns medially to the base of the bladder. Here it joins the more laterally placed seminal vesicle to form the *ejaculatory duct* which traverses the prostate to open into the prostatic urethra at the verumontanum on either side of the utricle.

Clinical features

Infection may track from bladder and urethra along the vas to the epididymis, particularly from the prostatic bed after prostatectomy; division of the vasa at prostatectomy is performed by some surgeons in an attempt to decrease this complication.

The operation of bilateral vasectomy is a common procedure for male sterilization. The vas is identified by its very firm consistency which, in coaching days, was likened to whipcord but which today might, more aptly, be compared with fine plastic tubing.

The seminal vesicles

These are coiled sacculated tubes 2 in (5 cm) long which can be unravelled to three times that length. They lie, one on each side, extraperitoneally at the bladder base, lateral to the termination of the vasa. Each has common drainage with its neighbouring vas via the ejaculatory duct (Fig. 87).

Clinical features

1   The vesicles can be felt on rectal examination if enlarged; this occurs typically in tuberculous infection.
2   The seminal vesicles, as their name implies, act as stores for semen. They receive their nerve supply from the 1st lumbar sympathetic ganglion through the presacral plexus. A bilateral high lumbar sympathectomy results in sterility as ejaculation is prevented.

The bony and ligamentous pelvis

The pelvis is made up of the innominate bones, the sacrum and the coccyx, bound to each other by dense ligaments.

The os innominatum (Fig. 92)

Examine the bone and revise the following structures.

The *ilium* with its *iliac crest* running between the *anterior* and

Iliac crest

Anterior gluteal line

Anterior superior spine

Posterior gluteal line
Posterior superior spine

Inferior gluteal line

Posterior inferior spine

Anterior inferior spine

Greater sciatic notch

Acetabulum

Iliopectineal eminence
Ischial spine

Superior ramus
Lesser sciatic notch

Pubic tubercle
Obturator foramen

Body of pubis
Inferior ramus
Ischial tuberosity

Ramus of ischium

Fig. 92 Lateral view of the os innominatum.

posterior superior iliac spines; below each of these are the corresponding *inferior spines*. Well-defined ridges on its lateral surface are the strong muscle markings of the glutei. Its inner aspect bears the large *auricular surface* which articulates with the sacrum. The *iliopectineal line* runs forward from the apex of the auricular surface and demarcates the *true* from the *false pelvis*.

The *pubis* comprises the *body* and the *superior* and *inferior pubic rami*.

The *ischium* has a vertically disposed *body*, bearing the *ischial spine* on its posterior border which demarcates an upper (greater) and lower (lesser), sciatic notch. The inferior pole of the body bears the *ischial tuberosity* then projects forwards almost at right angles into the *ischial ramus* to meet the inferior pubic ramus.

The *obturator foramen* lies bounded by the body and rami of the pubis and the body and ramus of the ischium.

All three bones fuse at the *acetabulum* which forms the socket for the femoral head, for which it bears a wide crescentic articular surface.

The pelvis is tilted in the erect position so that the plane of its inlet is at an angle 60° to the horizontal. (To place a pelvis into this position, hold it against a wall so that the anterior superior spine and the top of the pubic symphysis both touch it.)

The sacrum (Fig. 93)

The sacrum is made up of five fused vertebrae and is roughly triangular. The anterior border of its upper part is termed the *sacral promontory* and is readily felt at laparotomy.

Its anterior aspect presents a *central mass*, a row of four *anterior sacral foramina* on each side (transmitting the upper four sacral anterior primary rami), and, lateral to these, the *lateral masses* of the sacrum. The superior aspect of the lateral mass on each side forms a fan-shaped surface termed the *ala*.

Note that the central mass is roughly rectangular—the triangular shape of the sacrum is due to the rapid shrinkage in size of the lateral masses of the sacrum from above down.

Posteriorly lies the *sacral canal*, continuing the vertebral canal, bounded by short pedicles, strong laminae and diminutive spinous processes. Perforating through from the sacral canal is a row of four *posterior sacral foramina* on each side. Inferiorly, the canal terminates in the *sacral hiatus*, which transmits the 5th sacral nerve. On either side of the lower extremity of the hiatus lie the *sacral cornua*. These can easily be palpated by the finger immediately above the natal cleft.

On its lateral aspect, the sacrum presents an *auricular facet* for articulation with the corresponding surface of the ilium.

The 5th lumbar vertebra may occasionally fuse with the sacrum in whole or in part; alternatively, the 1st sacral segment may be partially or completely separated from the rest of the sacrum. The posterior arch of the sacrum is occasionally bifid.

Note that the dural sheath terminates distally at the second piece

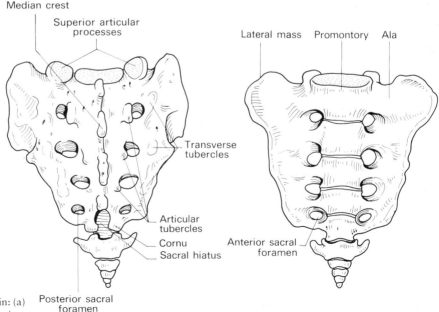

Fig. 93 The sacrum in: (a) posterior and (b) anterior views.

Median crest

Superior articular processes

Lateral mass   Promontory   Ala

Transverse tubercles

Articular tubercles

Cornu

Sacral hiatus

Anterior sacral foramen

Posterior sacral foramen

(a)

(b)

of the sacrum. Beyond this the sacral canal is filled with the fatty tissue of the extradural space, the cauda equina and the filum terminale. (For sacral block, see page 142).

The coccyx

This is made up of three to five fused vertebrae articulating with the sacrum; occasionally the first segment remains separate. It represents, of course, the tail of more primitive animals.

The functions of the pelvis

1 It protects the pelvic viscera.
2 It supports the weight of the body which is transmitted through the vertebrae, thence through the sacrum, across the sacroiliac joints to the innominate bones and then to the femora in the standing position or to the ischial tuberosities when sitting. (The sacroiliac joint is reinforced for this task as will be described below.)
3 During walking the pelvis swings from side to side by a rotatory movement at the lumbosacral articulation which occurs together with similar movements of the lumbar intervertebral joints. Even if the hip joints are fixed, this swing of the pelvis enables the patient to walk reasonably well.
4 As with all but a few small bones in the hand and foot, the pelvis provides attachments for muscles.
5 In the female it provides bony support for the birth canal.

Joints and ligamentous connections of the pelvis

The *symphysis pubis* is the name given to the cartilaginous joint between the pubic bones. Each pubic bone is covered by a layer of hyaline cartilage and is connected across the midline by a dense layer of fibrocartilage. The centre of the latter may break down to form a cleft-like joint space which, however, is not seen before the first decade and which is not lined by a synovial membrane.

The joint is surrounded and strengthened by fibrous ligaments, especially above and below.

The *sacroiliac joints* are the articulations between the auricular surfaces of the sacrum and ilium on each side and are true synovium-lined and cartilage-covered joints.

The sacrum 'hangs' from these joints supported by the extremely dense *posterior sacroiliac ligaments* lying posteriorly to the auricular joint surfaces. These support the whole weight of the body, tending to

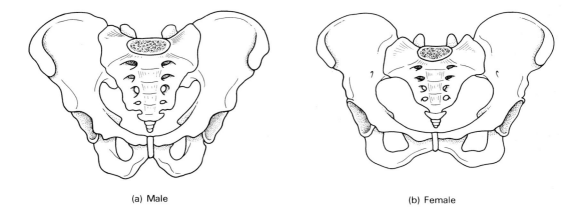

(a) Male   (b) Female

Fig. 94 (a) Male and (b) female pelvis compared.

drag the sacrum forward into the pelvis and, not surprisingly, are the strongest ligaments in the body.

Their action is assisted by an interlocking of the grooves between the auricular surfaces of the sacrum and ilium.

The *sacrotuberous ligament* passes from the ischial tuberosity to the side of the sacrum and coccyx.

The *sacrospinous ligament* passes from the ischial spine to the side of the sacrum and coccyx.

These two ligaments help to define two important exits from the pelvis:

1   the *greater sciatic foramen*—formed by the sacrospinous ligament and the greater sciatic notch;
2   the *lesser sciatic foramen*—formed by the sacrotuberous ligament and the lesser sciatic notch.

Note. There is a useful surface landmark in this region, the dimple constantly seen on each side immediately above the buttock, which defines:

1   the posterior superior iliac spine;
2   the centre of the sacroiliac joint;
3   the level of the second sacral segment;
4   the level of the end of the dural canal of the spinal meninges.

Differences between the male and female pelvis (Fig. 94)

The pelvis demonstrates a large number of sex differences associated principally with two features: first the heavier build and stronger muscles in the male, accounting for the stronger bone structure and better defined muscle markings in this sex; second, the comparatively wider and shallower pelvic cavity in the female, correlated with its role as the bony part of the birth canal.

The sex differences are summarized in Table 2.

When looking at a radiograph of the pelvis, the sex is best determined by three features:

1   the pelvic inlet, heart-shaped in the male, oval in the female;

2   the angle between the inferior pubic rami, which is narrow in the male, wide in the female. In the former, it corresponds almost exactly to the angle between the index and middle fingers when these are held apart; in the latter the angle equals that between the fully extended thumb and the index finger. This is a particularly reliable feature;

3   the soft tissue shadow of the penis and scrotum can usually be seen or, if not, the dense shadow of the lead screen used to shield the testes from harmful radiation.

Obstetrical pelvic measurements (Fig. 95)

The figures for the measurements of the inlet, mid-cavity and outlet of the true pelvis are readily committed to memory in the form shown in Table 3.

The *transverse diameter of the outlet* is assessed clinically by measuring the distance between the ischial tuberosities along a plane

Table 2 Comparison of male and female pelvis.

	Male	Female
General structure	Heavy and thick	Light and thin
Joint surfaces	Large	Small
Muscle attachments	Well marked	Rather indistinct
False pelvis	Deep	Shallow
Pelvic inlet	Heart shaped	Oval
Pelvic canal	'Long segment of a short cone' i.e. long and tapered	'Short segment of a long cone' i.e. short with almost parallel sides
Pelvic outlet	Comparatively small	Comparatively large
1st piece of sacrum	The superior surface of the body occupies nearly half the width of sacrum	Oval superior surface of the body occupies about one-third the width of sacrum
Sacrum	Long, narrow, with smooth concavity	Short, wide, flat, curving forward in lower part
Sacroiliac articular facet (auricular surface)	Extends well down the 3rd piece of the sacrum	Extends down only to upper border of 3rd piece
Subpubic angle (between inferior pubic rami)	'The angle between the middle and index finger'	'The angle between the thumb and index finger'
Inferior pubic ramus	Presents a strong everted surface for attachment of the crus of the penis	This marking is not present
Acetabulum	Large	Small
Ischial tuberosities	Inturned	Everted
Obturator foramen	Round	Oval

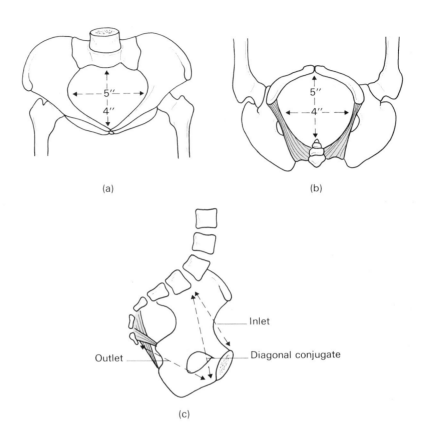

Fig. 95 The measurements of the female pelvis. (a) The inlet, (b) the outlet. (c) Lateral view to show the diagonal conjugate.

passing across the anus; the *anteroposterior outlet diameter* is measured from the pubis to the sacrococcygeal joint. The most useful measurement clinically is, however, the *diagonal conjugate*—from the lower border of the pubic symphysis to the promontory of the sacrum. This normally measures 5 in (12.5 cm); from the practical point of view, it is not possible in the normal pelvis to reach the sacral promontory on vaginal examination either readily or without discomfort to the patient.

Another useful clinical guide is the *subpubic arch*: the examiner's four knuckles (i.e. his clenched fist) should rest comfortably between the ischial tuberosities below the pubic symphysis.

Note that these measurements are all of the bony pelvis; the

Table 3 Obstetrical pelvic measurements.

	Transverse	Oblique	Anteroposterior
Inlet	5 in (12.5 cm)	4.5 (11.5 cm)	4 (10 cm)
Mid-pelvis	4.5 in (11.5 cm)	4.5 (11.5 cm)	4.5 (11.5 cm)
Outlet	4 in (10 cm)	4.5 (11.5 cm)	5 (12.5 cm)

'dynamic pelvis' of the birth-canal, in fact, is narrowed by the pelvic musculature, the rectum and the thickness of the uterine wall. Today accurate radiological techniques enable exact measurements to be made of the bony pelvis.

Variations of the pelvic shape (Fig. 96)

The female pelvic shapes may be subdivided (after Caldwell and Moloy) as follows.

1   The normal and its variants

(a)   *Gynaecoid*—normal.
(b)   *Android*—the masculine type of pelvis.
(c)   *Platypelloid* — shortened in the anteroposterior diameter, increased in the transverse diameter (the 'non-rachitic flat pelvis').
(d)   *Anthropoid*—resembling that of an anthropoid ape with a much lengthened anteroposterior and a shortened transverse diameter.

2   Symmetrically contracted pelvis

That of a small woman but with a symmetrical shape.

3   The Rachitic flat pelvis

The sacrum is rotated so that the sacral promontory projects forward and the coccyx tips backwards. The anteroposterior diameter of the inlet is therefore narrowed, but that of the outlet is increased.

4   The asymmetrical

Asymmetry can be due to a variety of causes such as scoliosis, long-standing hip disease (e.g. congenital dislocation), poliomyelitis, pelvic fracture, congenital abnormality due to thalidomide and the *Naegele pelvis* which is due to the congenital absence of one wing of the sacrum or its destruction by disease.

Clinical features

Fractures of the pelvis

These may be isolated lesions due to a localized blow or may be displacements of part of the pelvic ring due to compression injuries. Lateral compression usually results in fractures through both pubic

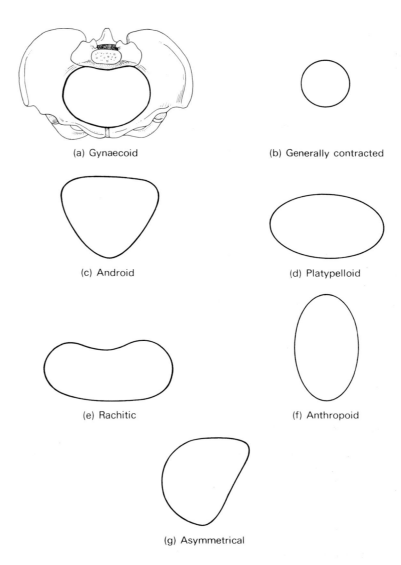

(a) Gynaecoid

(b) Generally contracted

(c) Android

(d) Platypelloid

(e) Rachitic

(f) Anthropoid

(g) Asymmetrical

Fig. 96 Pelvic variations and abnormalities— shown as diagrammatic outlines of the pelvic inlet.

rami on each side, or both rami on one side with dislocation at the symphysis; anteroposterior compression may be followed by dislocation at the symphysis or fractures through the pubic rami accompanied by dislocation at the sacroiliac joint.

Displacement of part of the pelvic ring must, of course, mean that the ring has been broken in two places.

Falls upon the leg may force the head of the femur through the acetabulum, the so-called central dislocation of the hip. Isolated fractures may be produced by local trauma, especially to the iliac wing, sacrum and pubis.

Associated with pelvic fractures one must always consider soft tissue injuries to bladder, urethra and rectum, which may be penetrated by spicules of bone or torn by wide displacements of the pelvic fragments.

Occasionally in these pelvic displacements the iliolumbar branch of the internal iliac artery is ruptured as it crosses above the sacroiliac joint; this may be followed by a severe or even fatal extraperitoneal haemorrhage.

Sacral (caudal) anaesthesia

The sacral hiatus, between the last piece of sacrum and coccyx, can be entered by a needle which pierces skin, fascia and the tough posterior sacrococcygeal ligament to enter the sacral canal. The hiatus can be defined by palpating the sacral cornua on either side (Fig. 93) immediately above the natal left.

Anaesthetic solution injected here will travel extradurally to bathe the spinal roots emerging from the dural sheath, which terminates at the level of the 2nd sacral segment.

The muscles of the pelvic floor and perineum

The canal of the bony and ligamentous pelvis is closed by a diaphragm of muscles and fasciae which the rectum, urethra and, in the female, the vagina, must pierce to reach the exterior. The muscles are divided into (a) the *pelvic diaphragm*, formed by the levator ani and the coccygeus; and (b) the *superficial muscles* of the (a) anterior (urogenital) perineum and the (b) posterior (anal) perineum.

Levator ani (Fig. 97) is the largest and most important muscle of the pelvic floor. It arises from the posterior aspect of the body of the pubic bone, the fascia of the side wall of the pelvis (covering obturator internus) and the spine of the ischium. From this wide origin it sweeps down in a series of loops:

1   to form a sling around the prostate (*levator prostate*) or vagina (*sphincter vaginae*), inserting into the perineal body;
2   to form a sling around the rectum and also insert into, and reinforce the deep part of, the anal sphincter at the anorectal ring (*puborectalis*);
3   the posterior fibres are attached to the sides of the coccyx and to a

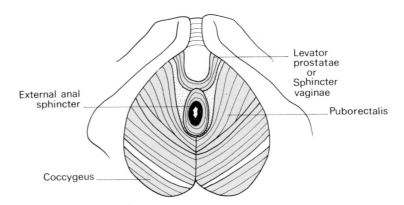

External anal sphincter

Levator prostatae or Sphincter vaginae

Puborectalis

Coccygeus

Fig. 97 Levator ani—inferior aspect.

median fibrous raphe, which stretches between the apex of the coccyx and the anorectal junction.

(Note that the *coccygeus* is in the same tissue plane as levator ani. It corresponds almost exactly with the sacrospinous ligament, which it overlies, and the latter is commonly regarded as a degenerate part of the muscle. The muscle is well developed and the ligament is often missing in those mammals with a mobile tail.)

The muscle acts as the principal support of the pelvic floor, has a sphincter action on the rectum and vagina and assists in increasing intra-abdominal pressure during defaecation, micturition and parturition.

Its deep aspect is related to the pelvic viscera and its perineal aspect forms the inner wall of the ischiorectal fossa (see below).

The anterior (urogenital) perineum (Figs 98, 99)

A line joining the ischial tuberosities passes just in front of the anus. Between this line and the ischiopubic inferior rami lies the urogenital part of the perineum or the *urogenital triangle*.

Attached to the sides of this triangle is a tough fascial sheet termed the *perineal membrane* which is pierced by the urethra in the male and by the urethra and the vagina in the female. Deep to this membrane is the external *sphincter* of the urethra consisting of voluntary muscle fibres surrounding the membranous urethra; these are competent even when the internal sphincter has been completely destroyed. In the female the superficial sphincter is also pierced by the vagina.

Enclosing the deep aspect of the external sphincter is a second fascial sheath (comprising areolar tissue on the deep aspect of levator ani), so that this muscle is, in fact, contained within a fascial capsule

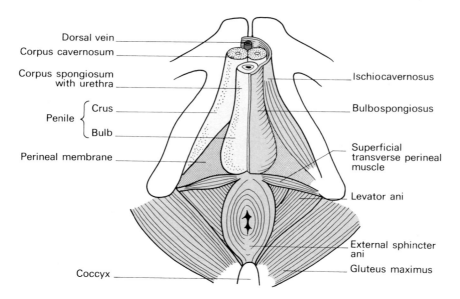

Fig. 98 The male perineum—on the right side the muscles of the anterior perineum have been dissected away.

which is termed the *deep perineal pouch*. This pouch contains, in addition, the deep transverse perineal muscles and, in the male, the two *bulbo-urethral glands of Cowper* whose ducts pass forward to open into the bulbous urethra. Superficial to the perineal membrane is the *superficial perineal pouch* which contains, in the male:

1   the *bulbospongiosus muscle* covering the corpus spongiosum which, in turn, surrounds the urethra (the distal corpus spongiosum expands into the glans penis);

2   the *ischiocavernosus muscle* on each side, arising from the ischial ramus and covering the corpus cavernosum. The urethra is thus enclosed in a spongy sheath supported by a cavernous tube on each side containing thin-walled venous sinuses which become engorged with blood when erection occurs;

3   the superficial transverse perineal muscle, running transversely from the perineal body to the ischial ramus. It is of no functional importance but is seen during perineal excision of the rectum.

In the female the same muscles are present although much less well developed and the bulbospongiosus is pierced by the vagina.

The perineal body

This fibromuscular node lies in the midline at the junction of the anterior and posterior perineum. It is the point of attachment for the anal sphincters, the bulbospongiosus, the transverse perineal muscles and fibres of levator ani.

The posterior (anal) perineum (Figs 99, 100)

This is the triangle lying between the ischial tuberosities on each side

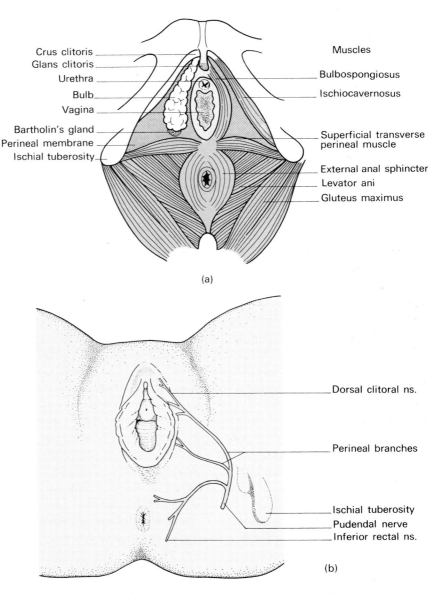

Crus clitoris
Glans clitoris
Urethra
Bulb
Vagina
Bartholin's gland
Perineal membrane
Ischial tuberosity

Muscles
Bulbospongiosus
Ischiocavernosus
Superficial transverse perineal muscle
External anal sphincter
Levator ani
Gluteus maximus

(a)

Dorsal clitoral ns.

Perineal branches

Ischial tuberosity
Pudendal nerve
Inferior rectal ns.

(b)

Fig. 99 (a) The female perineum—on the right side the muscles of the anterior perineum have been dissected away. (b) Distribution of the pudendal nerve to the female perineum.

Fig. 100 The ischiorectal fossa.

and the coccyx. It comprises, in essentials, the anus with its superficial sphincters, levator ani and, at each side, the ischiorectal fossa.

The *ischiorectal fossa* (Fig. 100) is of considerable surgical importance because of its great tendency to become infected. Its boundaries are:

- laterally — the fascia over obturator internus (i.e. the side wall of the pelvis); contained in this wall within a fascial tunnel termed the *pudendal* or *Alcock's canal* are the pudendal vessels and nerve which give off respectively the inferior rectal vessels and nerve, which supply the external sphincter and perianal skin, then pass forward to supply the perineal tissues;
- medially — the fascia over levator ani and the external anal sphincter;
- posteriorly — the sacrotuberous ligament covered posteriorly by gluteus maximus;
- anteriorly — the urogenital perinueum;
- floor — skin and subcutaneous fat.

Clinical features

1   The content of the fossa is coarsely lobulated fat. It is important to note that the ischiorectal fossae communicate with each other behind the anus — infection in one passes readily to the other.

Infection of this space may occur from boils or abrasions of the perianal skin, from lesions within the rectum and anal canal, from pelvic infection bursting through levator ani or, rarely, via the bloodstream. The fossa contains no important structures and can, therefore, be fearlessly incised when infected.

2   The pudendal nerves can be blocked in Alcock's canal on either side to give useful regional anaesthesia in obstetrical forceps delivery (see Fig. 99 and page 273).

The female genital organs

The vulva

The *vulva* (or *pudendum*) is the term applied to the female external genitalia.

The *labia majora* are the prominent hair-bearing folds extending back from the *mons pubis* to meet posteriorly in the midline of the perineum. They are the equivalent of the male scrotum.

The *labia minora* lie between the labia majora as lips of soft skin which meet posteriorly in a sharp fold, the *fourchette*. Anteriorly, they split to enclose the *clitoris*, forming an anterior *prepuce* and posterior *frenulum*.

The *vestibule* is the area enclosed by the labia minora and contains the *urethral orifice* (which lies immediately behind the clitoris) and the *vaginal orifice*.

The vaginal orifice is guarded in the virgin by a thin mucosal fold, the *hymen*, which is perforated to allow the egress of the menses, and may have an annular, semilunar, septate or cribriform appearance. Rarely, it is imperforate and menstrual blood distends the vagina (haematocolpos). At first coitus the hymen tears, usually posteriorly or posterolaterally, and after childbirth nothing is left of it but a few tags termed *carunculae myrtiformes*.

Bartholin's glands (the greater vestibular glands) are a pair of lobulated, pea-sized, mucus-secreting glands lying deep to the posterior parts of the labia majora. They are impalpable when healthy but become obvious when inflamed or distended. Each drains by a duct 1 in long which opens into the groove between the hymen and the posterior part of the labium minus.

Anteriorly, each gland is overlapped by the *bulb of the vestibule*— a mass of cavernous erectile tissue equivalent to the bulbus spongiosum of the male. This tissue passes forwards, under cover of bulbospongiosus, around the sides of the vagina to the roots of the clitoris.

Clinical features

At childbirth the introitus may be enlarged by making an incision in the perineum (*episiotomy*). This starts at the fourchette and extends mediolaterally on the right side for 1.5 in (3 cm). The skin, vaginal epithelium, subcutaneous fat, perineal body and superficial transverse perineal muscle are incised. After delivery the episiotomy is carefully sutured in layers.

The vagina (Fig. 101)

The vagina surrounds the cervix of the uterus, then passes downwards and forwards through the pelvic floor to open into the vestibule.

The cervix projects into the anterior part of the vault of the vagina so that the continuous gutter surrounding the cervix is shallow anteriorly (where the vaginal wall is 3 in (7.5 cm) in length) and is deep posteriorly (where the wall is 4 in (10 cm) long). This continuous gutter is, for convenience of description, divided into the anterior, posterior and lateral *fornices*.

Relations

• Anteriorly — the base of the bladder and the urethra (which is embedded in the anterior vaginal wall).
• Posteriorly — from below upwards, the anal canal (separated by the perineal body), the rectum and then the peritoneum of the pouch of Douglas which covers the *upper quarter of the posterior vaginal wall*.
• Laterally — levator ani, pelvic fascia and the *ureters*, which lie *immediately above the lateral fornices*.

The amateur abortionist (or inexperienced gynaecologist) without a knowledge of anatomy fails to realize that the uterus passes upwards and forwards from the vagina; he pushes the instrument or IUCD (intra-uterine contraceptive device), which he intends to enter the cervix, directly backwards through the posterior fornix. To

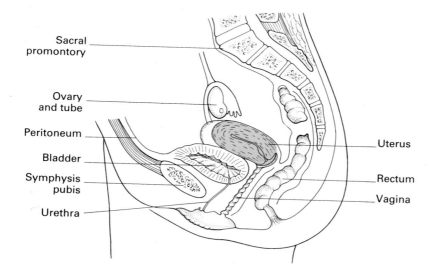

Sacral promontory

Ovary and tube

Peritoneum

Bladder

Symphysis pubis

Urethra

Uterus

Rectum

Vagina

Fig. 101 Sagittal section of the uterus and its relations.

make matters worse, this is the only part of the vagina which is intraperitoneal; the peritoneal cavity is thus entered and peritonitis follows.

Blood supply

Arterial supply is from the internal iliac artery via its vaginal, uterine, internal pudendal and middle rectal branches.

A venous plexus drains via the vaginal vein into the internal iliac vein.

Lymphatic drainage (see Fig. 105)

- Upper third to the external and internal iliac nodes.
- Middle third to the internal iliac nodes.
- Lower third to the superficial inguinal nodes.

Structure

A stratified squamous epithelium lines the vagina and the vaginal cervix; it contains no glands and is lubricated partly by cervical mucus and partly by desquamated vaginal epithelial cells. In nulliparous women the vaginal wall is rugose, but it becomes smoother after childbirth. The rugae of the anterior wall are situated transversely; this allows for filling of the bladder and for intercourse. In contrast, the rugae on the posterior wall run longitudinally. This allows for sideways stretching to accommodate a rectum distended with stool and for the passage of the fetal head.

Beneath the epithelial coat is a thin connective tissue layer separating it from the muscular wall which is made up of a criss-cross arrangement of involuntary muscle fibres. This muscle layer is ensheathed in a fascial capsule which blends with adjacent pelvic connective tissues, so that the vagina is firmly supported in place.

In old age the vagina shrinks in length and diameter. The cervix projects far less into it so that the fornices all but disappear.

The uterus (Figs 101, 102)

The uterus is a pear-shaped organ, 3 in (7.5 cm) in length, made up of the *fundus*, *body* and *cervix*. The *Fallopian (uterine) tubes* enter into each superolateral angle (the *cornu*) above which lies the fundus.

The body of the uterus narrows to a waist termed the *isthmus*, continuing into the *cervix* which is embraced about its middle by the

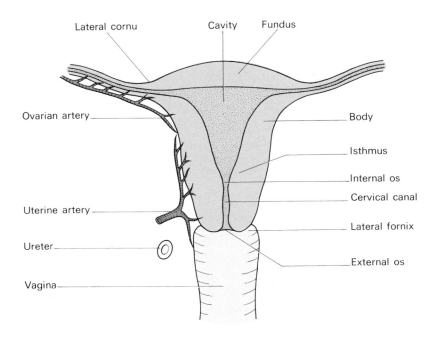

Fig. 102 Coronal section of the uterus and vagina. Note the important relationships of ureter and uterine artery.

vagina; this attachment delimits a supravaginal and vaginal part of the cervix.

The isthmus is 1.5 mm wide. The anatomical *internal os* marks its junction with the uterine body but its mucosa is histologically similar to the endometrium. The isthmus is that part of the uterus which becomes the lower segment in pregnancy.

The cavity of the uterine body is triangular in coronal section, but in the sagittal plane it is no more than a slit. This cavity communicates via the internal os with the *cervical canal* which, in turn, opens into the vagina by the *external os*.

The nulliparous external os is circular but after childbirth it becomes a transverse slit with an anterior and a posterior lip.

The non-pregnant cervix has the firm consistency of the nose; the pregnant cervix has the soft consistency of the lips.

In fetal life the cervix is considerably larger than the body; in childhood (the *infantile uterus*) the cervix is still twice the size of the body but, during puberty, the uterus enlarges to its adult size and proportions by relative overgrowth of the body. The adult uterus is bent forward on itself at about the level of the internal os to form an angle of 170°; this is termed *anteflexion of the uterus*. Moreover, the axis of the cervix forms an angle of 90° with the axis of the vagina—*anteversion of the uterus*. The uterus thus lies in an almost horizontal plane.

In *retroversion of the uterus*, the axis of the cervix is directed upwards and backwards. Normally on vaginal examination the lowermost part of the cervix to be felt is its anterior lip; in retroversion either the os or the posterior lip becomes the presenting part.

In *retroflexion* the axis of the body of the uterus passes upwards and backwards in relation to the axis of the cervix.

Frequently these two conditions co-exist. They may be mobile and symptomless — as a result of distension of the bladder or purely as a development anomaly. Indeed, mobile retroversion is found in a quarter of the female population and may be regarded as a normal variant. Less commonly, they are fixed, the result of adhesions, previous pelvic infection, endometriosis or the pressure of a tumour in front of the uterus (Fig. 103).

Relations

• Anteriorly — the body is related to the uterovesical pouch of peritoneum and lies either on the superior surface of the bladder or on coils of intestine. The supravaginal cervix is related directly to bladder, separated only by connective tissue. The infravaginal cervix has the anterior fornix immediately in front of it.

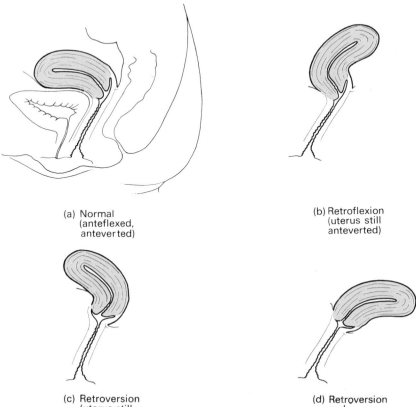

(a) Normal
(anteflexed,
anteverted)

(b) Retroflexion
(uterus still
anteverted)

(c) Retroversion
(uterus still
anteflexed)

(d) Retroversion
and
retroflexion

Fig. 103 Variations in uterine position and their terminology.

- Posteriorly — lies the pouch of Douglas, with coils of intestine within it.
- Laterally — the broad ligament and its contents (see below); *the ureter lies 12 mm lateral to the supravaginal cervix.*

Clinical features

The most important single practical relationship in this region is that of the ureter to the supravaginal cervix. At this point, the ureter lies just above the level of the lateral fornix, below the uterine vessels as these pass across within the broad ligament (Fig. 104). In performing a hysterectomy, the ureter may be accidentally divided in clamping the uterine vessels, especially when the pelvic anatomy has been distorted by a previous operation, a mass of fibroids, infection or malignant infiltration.

The ureter is readily infiltrated by lateral extension of a carcinoma of the uterus; bilateral hydronephrosis with uraemia is a frequent mode of termination of this disease.

The close relationship of ureter to the lateral fornix is best appreciated by realizing that a ureteric stone at this site can be palpated on vaginal examination. (This is the answer to the examination question: 'When can a stone in the ureter be felt?')

Blood supply

The *uterine artery* (from the internal iliac) runs in the base of the broad ligament and crosses above and at right angles to the ureter to reach the uterus at the level of the internal os. The artery then ascends

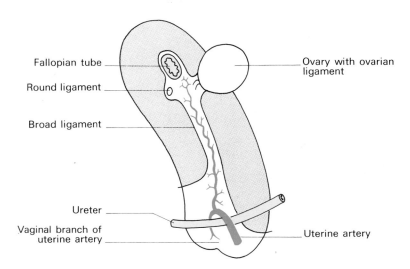

Fallopian tube

Round ligament

Broad ligament

Ureter

Vaginal branch of uterine artery

Ovary with ovarian ligament

Uterine artery

Fig. 104 Lateral view of the uterus (schematic) to show composition of the broad ligament, the relations of ureter and uterine artery, and the peritoneal covering of the uterus (pink stippled).

in a tortuous manner alongside the uterus, supplying the corpus, and then anastomoses with the *ovarian artery*. The uterine artery also gives off a descending branch to the cervix and branches to the upper vagina. The veins accompany the arteries and drain into the internal iliac veins, but they also communicate via the pelvic plexus with the veins of the vagina and bladder.

Lymph drainage (Fig. 105)

1   The fundus (together with the ovary and Fallopian tube) drains along the ovarian vessels to the aortic nodes, apart from some lymphatics which pass along the round ligament to the inguinal nodes.
2   The body drains via the broad ligament to nodes lying alongside the external iliac vessels.
3   The cervix drains in three directions—laterally, in the broad ligament, to the external iliac nodes; posterolaterally along the uterine vessels to the internal iliac nodes; and posteriorly along the recto-uterine folds to the sacral nodes.

Always examine the inguinal nodes in a suspected carcinoma of the corpus uteri—they may be involved by lymphatic spread along the round ligament.

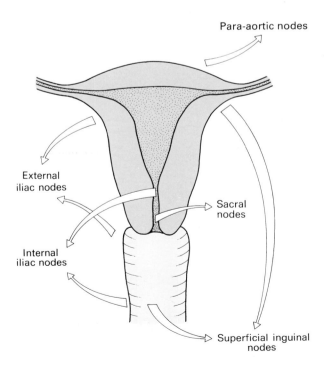

Fig. 105 Lymph drainage of the uterus and vagina.

Structure

The body of the uterus is covered with peritoneum except where this is reflected off at two sites, anteriorly on to the bladder at the uterine isthmus and laterally at the broad ligaments. Anteriorly, the peritoneum is only loosely adherent to the supravaginal cervix; this allows for bladder distension. The muscle wall is thick and made up of a criss-cross of involuntary fibres mixed with fibroelastic connective tissue.

The *mucosa* is applied directly to muscle with no submucosa intervening. The mucosa of the body of the uterus is the *endometrium*, made up of a single layer of cuboidal ciliated cells forming simple tubular glands which dip down to the underlying muscular wall. Below this epithelium is a stroma of connective tissue containing blood vessels and round cells.

The cervical canal epithelium is made up of tall columnar cells which form a series of complicated branching glands; these secrete an alkaline mucus which forms a protective '*cervical plug*' filling the canal.

The vaginal aspect of the cervix is covered with a stratified squamous epithelium continuous with that of the vagina.

The mucosa of the corpus undergoes extensive changes during the menstrual cycle which may be briefly summarized thus:
1   first 4 days — desquamation of its superficial two-thirds with bleeding;
2   subsequent 2–3 days — rapid reconstitution of the raw mucosal surface by growth from the remaining epithelial cells in the depths of the glands;
3   by the 14th day the endometrium has reformed; this is the end of the *proliferative phase*;
4   from the 14th day until the menstrual flow commences is the *secretory phase*; the endometrium thickens, the glands lengthen and distend with fluid and the stroma becomes oedematous and stuffed with white cells.

At the end of this phase three layers can be defined:
1   a compact superficial zone;
2   a spongy middle zone — with dilated glands and oedematous stroma;
3   a basal zone of inactive non-secreting tubules.

With degeneration of the corpus luteum there is shrinkage of the endometrium, the arteries retract and coil, producing ischaemia of the middle and superficial zones, which then desquamate. It is probable that spasm of the vessels in the basal layer (which remains non-desquamated) prevents the woman bleeding to death.

Only very slight desquamation and bleeding takes place in the mucosa of the cervical canal.

The Fallopian tubes (Fig. 106)

The Fallopian, or uterine, tubes are about 4 in (10 cm) long; they lie in the free edge of the broad ligaments and open into the cornu of the uterus. Each comprises four parts.

1　The *infundibulum* — the bugle-shaped extremity extending beyond the broad ligament and opening into the peritoneal cavity by the *ostium*. Its mouth is fimbriated and overlies the ovary, to which one long fimbria actually adheres (*fimbria ovarica*).

2　The *ampulla*—wide, thin-walled and tortuous.

3　The *isthmus*—narrow, straight and thick-walled.

4　The *interstitial part*—which pierces the uterine wall.

Structure

Apart from the interstitial part, the tube is clothed in peritoneum. Beneath this is a muscle of outer longitudinal and inner circular fibres.

The mucosa is formed of columnar, mainly ciliated cells and lies in longitudinal ridges, each of which is thrown into numerous folds.

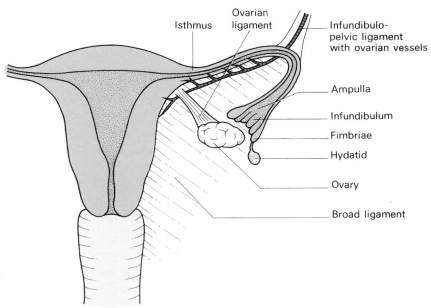

Fig. 106 The Fallopian tube, ovary and broad ligament.

The ova are propelled to the uterus along this tube, partly by peristalsis and partly by cilial action.

Clinical features

1   Note that the genital canal in the female is the only direct communication into the peritoneum from the exterior and is a potential pathway for infection (for example, in gonorrhoea).
2   The fertilized ovum may implant ectopically, i.e. in a site other than the endometrium of the corpus uteri. When this occurs in the Fallopian tube it is called, according to the exact site, fimbrial, ampullary, isthmic or interstitial, of which the ampullary is the commonest and interstitial the rarest.

As the ectopic embryo enlarges, it may abort into the peritoneal cavity (where rarely it continues to grow as a secondary abdominal pregnancy), or else ruptures the tube. This second fate is particularly likely to occur in the narrow and relatively non-distensible isthmus; rupture is usually into the peritoneal cavity but may rarely occur into the broad ligament.

The ovary (Fig. 106)

The ovary is an almond-shaped organ, 1.5 in (4 cm) long, attached to the back of the broad ligament by the *mesovarium*. The ovary has two other attachments, the *infundibulopelvic ligament*, along which pass the ovarian vessels and lymphatics from the side wall of the pelvis, and the *ovarian ligament*, which passes to the cornu of the uterus.

Relations

The ovary is usually described as lying on the side wall of the pelvis opposite the *ovarian fossa*, which is a depression bounded by the external iliac vessels in front and the ureter and internal iliac vessels behind and which contains the obturator nerve. However, the ovary is extremely variable in its position and is frequently found prolapsed into the pouch of Douglas in perfectly normal women.

The ovary, like the testis, develops from the genital ridge and then descends into the pelvis. In the same way as the testis, it therefore drags its blood supply and lymphatic drainage downwards with it from the posterior abdominal wall.

Blood supply, lymph drainage and nerve supply

Blood supply is from the ovarian artery which arises from the aorta at the level of the renal arteries. The ovarian vein drains, on the right side, to the inferior vena cava, on the left, to the left renal vein, exactly comparable to the venous drainage of the testis.

Lymphatics pass to the aortic nodes at the level of the renal vessels, following the general rule that lymphatic drainage accompanies the venous drainage of an organ.

Nerve supply is from the aortic plexus (T10).

All these structures pass to the ovary in the infundibulopelvic ligament.

Structure

The ovary has no peritoneal covering; the serosa ends at the mesovarian attachment. It consists of a connective tissue stroma containing *Graafian follicles* at various stages of development, *corpora lutea* and *corpora albicantia* (hyalinized, regressing corpora lutea, which take several months to absorb completely).

The surface of the ovary in young children is covered with a so-called 'germinal epithelium' of cuboidal cells. It is now known, however, that the primordial follicles develop in the ovary in early fetal life and do not differentiate from these cells. In adult life, in fact, the epithelial covering of the ovary disappears, leaving only a fibrous capsule termed the *tunica albuginea*.

After the menopause the ovary becomes small and shrivelled; in old age the follicles disappear completely.

The endopelvic fascia and the pelvic ligaments (Fig. 107)

Pelvic fascia is the term applied to the connective tissue floor of the pelvis covering levator ani and obturator internus. The *endopelvic fascia* is the extraperitoneal cellular tissue of the uterus (the *parametrium*), vagina, bladder and rectum. Within this endopelvic fascia are three important condensations of connective tissue which sling the pelvic viscera from the pelvic walls.

1   The *cardinal ligaments* (transverse cervical, or Mackenrodt's ligaments), which pass laterally from the cervix and upper vagina to the side walls of the pelvis along the lines of attachment of levator ani, are composed of white fibrous connective tissue with some involuntary muscle fibres and are pierced in their upper part by the ureters.

2   The *uterosacral ligaments*, which pass backwards from the posterolateral aspect of the cervix at the level of the isthmus and from the lateral vaginal fornices deep to the uterosacral folds of peritoneum in

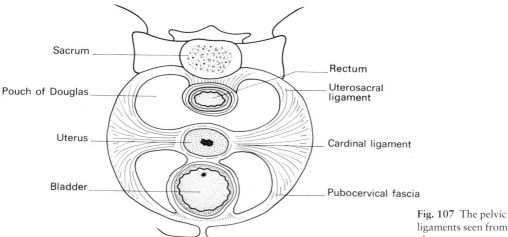

Sacrum

Pouch of Douglas

Uterus

Bladder

Rectum

Uterosacral
ligament

Cardinal ligament

Pubocervical fascia

Fig. 107 The pelvic
ligaments seen from
above.

the lateral boundaries of the pouch of Douglas, are attached to the periosteum in front of the sacroiliac joints and the lateral part of the third piece of the sacrum.

3   The *pubocervical fascia* extends forward from the cardinal ligament to the pubis on either side of the bladder, to which it acts as a sling.

These three ligaments act as supports to the cervix of the uterus and the vault of the vagina, in conjunction with the important elastic muscular foundation provided by levator ani. In prolapse these ligaments lengthen (in procidentia they may be 6 in (15 cm) long) and any repair operation must include their reconstitution.

Two other pairs of ligaments take attachments from the uterus.

1   The *broad ligament* is a fold of peritoneum connecting the lateral margin of the uterus with the side wall of the pelvis on each side. The uterus and its broad ligaments, therefore, form a partition across the pelvic floor dividing off an anterior compartment, containing bladder (the *uterovesical pouch*), from a posterior compartment, containing rectum (the pouch of Douglas or *recto-uterine pouch*).

The broad ligament contains or carries (Figs 104, 106):

• the Fallopian (uterine) tube in its free edge;
• the ovary, attached by the mesovarium to its posterior aspect;
• the round ligament;
• the ovarian ligament, crossing from the ovary to the uterine cornu (see ovary);
• the uterine vessels and branches of the ovarian vessels;
• lymphatics and nerve fibres.

The ureter passes forwards to the bladder deep to this ligament and lateral to and immediately above the lateral fornix of the vagina.

2 The *round ligament* — a fibromuscular cord — passes from the lateral angle of the uterus in the anterior layer of the broad ligament to the internal inguinal ring; thence it traverses the inguinal canal to the labium majus. Taken together with the ovarian ligament, it is equivalent to the male gubernaculum testis and can be thought of as the pathway along which the female gonad might have, but in fact did not, descend to the labium majus (the female homologue of the scrotum).

Vaginal examination

The relations of the vagina to the other pelvic organs must be constantly borne in mind when carrying out a vaginal examination.

Inspection (by means of a speculum) enables the vaginal walls and cervix to be examined and a biopsy or cytological smear to be taken. Inspection of the introitus while straining detects prolapse and the presence of stress incontinence.

- Anteriorly—the urethra, bladder and symphysis pubis are felt.
- Posteriorly — the rectum (invasion of the vagina by a rectal neoplasm must always be sought after in this disease). Collections of fluid, malignant deposits, prolapsed uterine tubes and ovaries or coils of distended bowel may be felt in the pouch of Douglas.
- Laterally—the ovary and tube, and the side wall of pelvis. Rarely, a stone in the ureter may be felt through the lateral fornix. The strength of the perineal muscles can be assessed by asking the patient to tighten up her perineum.
- Apex—the cervix is felt projecting back from the anterior wall of the vagina. In the normal anteverted uterus the anterior lip of the cervix presents; in retroversion either the cervical os or the posterior lip are first to be felt.

Pathological cervical conditions—for example, neoplasm—can be felt, as can the softening of the cervix in pregnancy and its dilatation during labour.

Bimanual examination assesses the pelvic size and position of the uterus, enlargements of ovary or uterine tubes and the presence of other pelvic masses.

The obstetrician can assess the pelvic size both in the transverse and anteroposterior diameter. Particularly important is the distance from the lower border of the symphysis pubis to the sacral promontory, which is termed the *diagonal conjugate*. If the pelvis is of normal size, the examiner's fingers should fail to reach the promontory of the sacrum. If it is readily palpable, pelvic narrowing is present (see 'obstetrical pelvic measurements', page 139).

Embryology of the Fallopian tubes, uterus and vagina (Fig. 108)

The *paramesonephric* (or *Müllerian*) *ducts* develop, one on each side, adjacent to the *mesonephric* (*Wolffian*) *ducts* in the posterior abdominal wall—they are mesodermal in origin. All these four tubes lie close together caudally, projecting into the anterior (urogenital) compartment of the cloaca.

One system disappears in the male, the other in the female, each leaving behind congenital remnants of some interest to the clinician.

In the male, the paramesonephric duct disappears, apart from the appendix testis and the prostatic utricle. In the female, the mesonephric system (which in the male develops into the vas deferens and epididymal ducts) persist as remnants in the broad ligament termed the *epöophoron*, *paröophoron* and *ducts of Gärtner*.

The paramesonephric ducts in the female form the Fallopian tubes cranially. More caudally, they come together and fuse in the midline

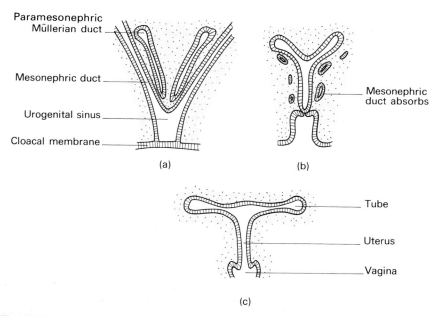

Fig. 108 Diagrams of the development of the Fallopian tubes, uterus and vagina from the paramesonephric (Müllerian) ducts and the urogenital sinus (after Hollinshead).

(dragging, as they do so, a peritoneal fold from the side wall of the pelvis which becomes the broad ligament). The median structure so formed differentiates into the epithelium of the uterine body (endometrium), cervical canal and upper one-third of the vagina, which are first solid and later become canalized. The rest of the vaginal epithelium develops by canalization of the solid sinuvaginal node at the back of the urogenital sinus. This accounts for the differences in lymphatic drainage of the upper and lower vagina (Fig. 105). The muscle of the Fallopian tubes, uterine body, cervix and vagina develops from surrounding mesoderm, so that remnants of the mesonephric duct system of the female are found in the myometrium, cervix and vaginal wall.

Developmental abnormalities of this system can easily be deduced. All stages of division of the original double tube may persist from a bicornuate uterus to a complete reduplication of the uterus and vagina. Alternatively, there may be absence, hypoplasia or atresia of the duct system on one or both sides.

Failure of canalization of the originally solid caudal end of the duct results, after puberty, in the accumulation of menstrual blood above the obstruction. First the vagina may distend with blood, then the uterus and then the tubes (haematocolpos, haematometra and haematosalpinx, respectively).

The posterior abdominal wall

The bed of the posterior abdominal wall is made up of three bony and four muscular structures.

The bones are:
- the bodies of the lumbar vertebrae;
- the sacrum;
- the wings of the ilium.

The muscles are:
- the diaphragm—posterior part;
- the quadratus lumborum;
- the psoas major;
- the iliacus.

The diaphragm has been considered in the section on thorax.

The psoas must be dealt with in more detail because of the involvement of its sheath in the formation of a psoas abscess.

The *psoas major* arises from the transverse processes of all the lumbar vertebrae and from the sides of the bodies and the intervening discs of T12 to L5 vertebrae. It passes downwards and laterally at the margin of the brim of the pelvis, narrowing down to a tendon which crosses the front of the hip joint beneath the inguinal ligament to be inserted, with iliacus, into the lesser trochanter of the femur (Fig. 109).

The psoas major, together with iliacus, flexes the hip on the trunk, or, alternatively, the trunk on the hips (e.g. in sitting up from the lying position). In addition, psoas and iliacus internally rotate the hip. *Psoas minor*, absent in 40% of subjects, lies on psoas major and attaches to the iliopubic eminence.

Clinical features

1   The femoral artery lies on the psoas tendon, and it is this firm posterior relation of the femoral artery at the groin which enables it here to be identified and compressed easily by the finger.
2   The psoas is enclosed in the psoas sheath which is a compartment of the lumbar fascia. Pus from a tuberculous infection of the lumbar vertebrae is limited in its anterior spread by the anterior longitudinal vertebral ligament, and therefore passes laterally into its sheath (*psoas abscess*), which may also be entered by pus tracking down from the posterior mediastinum in disease of the thoracic vertebrae. Pus may then spread under the inguinal ligament into the femoral tri-

Fig. 109 Psoas sheath and psoas abscess. On the right is a normal psoas sheath; on the left it is shown distended with pus, which tracks under the inguinal ligament to present in the groin.

angle where it produces a soft swelling (Fig. 109). Occasionally, in completely neglected cases, pus tracks along the femoral vessels, along the subsartorial canal and eventually appears in the popliteal fossa.

The retroperitoneal organs are: the pancreas, kidneys and ureters (which have already been considered), the suprarenals, the aorta and inferior vena cava and their main branches, the para-aortic lymph nodes and the lumbar sympathetic chain.

The suprarenal glands (Fig. 81)

The suprarenal glands cap the upper poles of the kidneys and lie against the diaphragm. The left is related anteriorly to the stomach across the lesser sac, the right lies behind the right lobe of the liver and tucks medially behind the inferior vena cava.

Each gland, although weighing only 3 to 4 g, has three arteries supplying it:
1   a direct branch from the aorta;
2   a branch from the phrenic artery;
3   a branch from the renal artery.

The single main vein drains from the hilum of the gland into the nearest available vessel—the inferior vena cava on the right, the renal vein on the left. The stubby right suprarenal vein, coming directly from the inferior vena cava, presents the most dangerous feature in performing an adrenalectomy — the tiro should always choose the easier left side and leave the right to his chief.

The suprarenal gland comprises a cortex and medulla, which represent two developmentally and functionally independent endocrine glands within the same anatomical structure. The medulla is derived from the neural crest (ectoderm) whose cells also give rise to the sympathetic ganglia. The cortex, on the other hand, is derived from the mesoderm. The suprarenal medulla receives preganglionic sympathetic fibres from the greater splanchnic nerve and secretes adrenaline and noradrenaline. The cortex secretes the adrenocortical hormones.

Abdominal aorta (Fig. 110)

The aorta enters the abdomen via the aortic hiatus in the diaphragm at the level of the 12th thoracic vertebra and ends at L4. It lies throughout this course against the vertebral bodies.

Anteriorly, from above down, it is related to the pancreas (separating it from the stomach), the third part of the duodenum and coils of small intestine. It is crossed by the left renal vein. A large tumour of

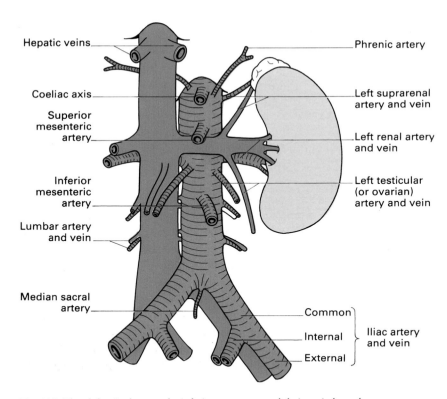

Hepatic veins

Coeliac axis

Superior
mesenteric
artery

Inferior
mesenteric
artery

Lumbar artery
and vein

Median sacral
artery

Phrenic artery

Left suprarenal
artery and vein

Left renal artery
and vein

Left testicular
(or ovarian)
artery and vein

Common
Internal
External

Iliac artery
and vein

Fig. 110 The abdominal aorta, the inferior vena cava and their main branches.

pancreas or stomach, a mass of enlarged para-aortic nodes, or a large
ovarian cyst may transmit the pulsations of the aorta and be mistaken
for an aneurysm.

The branches of the aorta are:
1   three anterior unpaired branches passing to the viscera:
  (a)   the *coeliac axis*—giving off the
    • hepatic artery
    • splenic artery
    • left gastric artery
  (b)   the *superior mesenteric artery*
  (c)   the *inferior mesenteric artery*
2   three lateral paired branches passing to viscera:
  (a)   the *suprarenal artery*
  (b)   the *renal artery*
  (c)   the *testicular* or *ovarian artery*
3   five lateral paired branches to the parietes:
  (a)   the *inferior phrenic artery*
  (b)   four *lumbar* branches

4   terminal branches:
    (a)   the *common iliacs*
    (b)   the *median sacral artery*

The *common iliac arteries* pass, one on each side, downwards and outwards to bifurcate into the internal and external iliacs in front of the sacroiliac joint, at the level of the sacral promontory. They give no other branches.

At the bifurcation, the common iliac artery is crossed superficially by the ureter — a convenient site to identify this latter structure in pelvic operations.

The *external iliac artery* runs along the brim of the pelvis on the medial side of psoas major. The artery passes below the inguinal ligament to form the femoral artery, giving off, immediately before its termination, the *inferior epigastric artery*, which demarcates the medial edge of the internal inguinal ring (Fig. 45) and also the *deep circumflex iliac artery*.

The *internal iliac artery* passes backwards and downwards into the pelvis, sandwiched between the ureter anteriorly and the internal iliac vein posteriorly. At the upper border of the greater sciatic notch it divides into an anterior and posterior division, which give off numerous branches to supply the pelvic organs, perineum, buttock and sacral canal.

Inferior vena cava (Fig. 110)

The inferior vena cava commences at L5 by the junction of the common iliac veins *behind* the right common iliac artery (unlike the usual arrangement of a vein being superficial to its corresponding artery). It lies to the right of the aorta as it ascends until separated from it by the right crus of the diaphragm when the aorta pierces this muscle. The inferior vena cava itself passes through the diaphragm at T8 (Fig. 11), traverses the pericardium and drains into the right atrium.

As the inferior vena cava ascends, it is related anteriorly to coils of small intestine, the third part of the duodenum, the head of the pancreas with the common bile duct, and the first part of duodenum. It then passes behind the foramen of Winslow, in front of which lies the portal vein, separating it from the common bile duct and hepatic artery. Finally, the inferior vena cava lies in a deep groove in the liver before piercing the diaphragm. Within the liver it receives the right and left hepatic veins. Occasionally these veins fuse into a single trunk which opens directly into the inferior vena cava; on other occasions the central hepatic vein (which usually enters the left hepatic near its termination) drains directly into the inferior vena cava (Fig. 74). These variations are now of importance because of the possibility of carrying out resection of one or other lobe of the liver.

Lumbar sympathetic chain

The lumbar part of the sympathetic trunk commences deep to the medial arcuate ligament of the diaphragm as a continuation of the thoracic sympathetic chain (see Fig. 276). On each side it lies against the bodies of the lumbar vertebrae overlapped, on the right side, by the inferior vena cava and on the left by the aorta.

The lumbar arteries lie deep to the chain but the lumbar veins may cross superficial to it and are of importance because they may be damaged in performing a sympathectomy.

Below, the lumbar trunk passes deep to the iliac vessels to continue as the sacral trunk in front of the sacrum. Inferiorly, the chains converge and unite in front of the coccyx as the small ganglion impar.

Usually the lumbar trunk carries four ganglia, although sometimes these are condensed to three. All four send grey rami communicantes to the lumbar spinal nerves; in addition, the upper two ganglia receive white rami.

Branches from the chain pass to plexuses around the abdominal

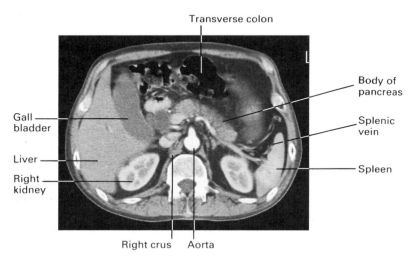

Fig. 111 CT scan at the level of the first lumbar vertebra. This demonstrates the liver, gall-bladder, aorta with the commencement of the superior mesenteric artery, the inferior vena cava, the crura of the diaphragm, the kidneys, pancreas and the spleen. The splenic vein can be seen as it passes to the splenic hilum posterior to the body of the pancreas. The vena cava lies on the right crus. The vessels have been enhanced by an intravenous injection of contrast.

Superior mesenteric vessels

Inferior vena cava

Liver

Right kidney

Left renal vein and artery

Aorta   Psoas   Quadratus lumborum

Fig. 112 CT scan at the level of the second lumbar vertebra demonstrating the kidneys, aorta, i.v.c., the liver, renal and superior mesenteric vessels, and the muscles of the abdominal wall. The blood vessels, again, have been enhanced by an intravenous injection of contrast.

aorta and its branches, which also receive fibres from the splanchnic nerves and the vagus. Other branches pass in front of the common iliac vessels as the hypogastric plexus ('presacral nerves') to supply the pelvic viscera via plexuses of nerves distributed along the branches of the internal iliac artery.

The parasympathetic supply to the pelvic viscera arises from the anterior primary rami of S2, 3 and 4 and is distributed with the pelvic plexuses (see page 432).

Clinical features

Lumbar sympathectomy is carried out via an extraperitoneal approach. A paramedian or transverse midabdominal incision is used, the peritoneum exposed and peeled medially from the posterior abdominal wall. The ureter, which adheres to the peritoneum like a fly to fly-paper, is seen and carefully preserved. Psoas major comes into view with the genitofemoral nerve upon it, then the lumbar vertebrae, against which the sympathetic chain can be felt.

Usually the 2nd, 3rd and 4th ganglia are excised with the intermediate chain; this effects an adequate sympathectomy of the lower limb, the skin of which then becomes warm, pink and dry.

Computerized axial tomography (CT scanning) has revealed a fresh dimension to the importance of topographical anatomy. It is now necessary for clinicians to possess a detailed knowledge of the cross-sectional relationships of the body in health so that pathological abnormalities can be appreciated. Figures 111 and 112 demonstrate CT scan cuts of the abdomen at the levels of the 1st and 2nd lumbar vertebral bodies respectively. Clinical students should take every opportunity of studying normal scans with the help of a skilled radiologist.

Part 3
The Upper Limb

The female breast

The female breast overlies the 2nd to the 6th rib; two-thirds of it rests on pectoralis major, one-third on serratus anterior, while its lower medial edge just overlaps the upper part of the rectus sheath.

Structure

The breast is made up of 15–20 lobules of glandular tissue embedded in fat; the latter accounts for its smooth contour and most of its bulk. These lobules are separated by fibrous septa running from the subcutaneous tissues to the fascia of the chest wall (the *ligaments of Cooper*).

Each lobule drains by its lactiferous duct on to the *nipple*, which is surrounded by the pigmented *areola*. This area is lubricated by the *areolar glands of Montgomery*; these are large, modified sebaceous glands which may form sebaceous cysts which may, in turn, become infected.

The male breast is rudimentary, comprising small ducts without alveoli and supported by fibrous tissue and fat. Insignificant it may be, but it is still prone to the major diseases that affect the female organ.

Blood supply

1   From the axillary artery via its lateral thoracic and acromiothoracic branches.
2   From the internal thoracic (internal mammary) artery via its perforating branches; these pierce the first to the fourth intercostal spaces, then traverse pectoralis major to reach the breast along its medial edge. The first and second perforators are the largest of these branches.
3   From the intercostal arteries via their lateral perforating branches; a relatively unimportant source.

The venous drainage is to the corresponding veins.

Lymphatic drainage

This is of considerable importance in the spread of breast tumours.

The lymph drainage of the breast, as with any other organ, follows the pathway of its blood supply and therefore travels:

1   along tributaries of the axillary vessels to axillary lymph nodes;

2   along the tributaries of the internal thoracic vessels, piercing pectoralis major to traverse each intercostal space to lymph nodes along the internal mammary chain; these also receive lymphatics penetrating along the lateral perforating branches of the intercostal vessels.

Although the lymph vessels lying between the lobules of the breast freely communicate, there is a tendency for the lateral part of the breast to drain towards the axilla and the medial part to the internal mammary chain (Fig. 113).

A subareolar plexus of lymphatics below the nipple (the *plexus of Sappey*) and another deep plexus on the pectoral fascia have, in the past, been considered to be the central points to which, respectively, the superficial and deep parts of the breast drain before communicating with main efferent lymphatics. These plexuses appear, however, to be relatively unimportant, the vessels, in the main, passing directly to the regional lymph nodes.

The axillary lymph nodes (some 20–30 in number) drain not only the lymphatics of the breast, but also those of the pectoral region, upper abdominal wall and the upper limb, and are arranged in five groups (Fig. 114):

1   anterior—lying deep to pectoralis major along the lower border of pectoralis minor;

2   posterior—along the subscapular vessels;

3   lateral—along the axillary vein;

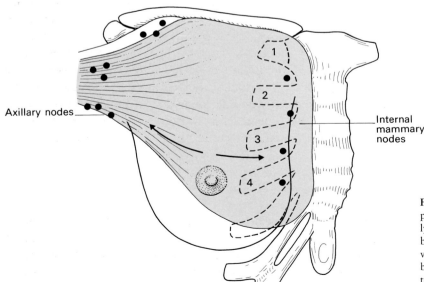

Axillary nodes

Internal mammary nodes

Fig. 113 Diagram of the principal pathways of lymphatic drainage of the breast. These follow the venous drainage of the breast—to the axilla and to the internal mammary chain.

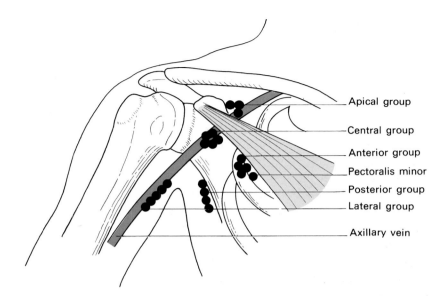

Apical group
Central group
Anterior group
Pectoralis minor
Posterior group
Lateral group
Axillary vein

Fig. 114 The lymph nodes of the axilla.

4   central—in the axillary fat;

5   apical (through which all the other axillary nodes drain)—immediately behind the clavicle at the apex of the axilla above pectoralis minor and along the medial side of the axillary vein.

From the apical nodes emerges the subclavian lymph trunk. On the right, this either drains directly into the subclavian vein or else joins the right jugular trunk; on the left it usually drains directly into the thoracic duct.

Lymphatic spread of a growth of the breast may occur further afield when these normal pathways have become interrupted by malignant deposits, surgery or radiotherapy. Secondaries may then be found in the lymphatics of the opposite breast or in the opposite axillary lymph nodes, the groin lymph nodes (via lymph vessels in the trunk wall), the cervical nodes (as a result of retrograde extension from the blocked thoracic duct or jugular trunk), or in peritoneal lymphatics spreading there in a retrograde manner from the lower internal mammary nodes: this in addition, of course, to spread via the blood stream.

Development

The breasts develop as an invagination of chest wall ectoderm, which forms a series of branching ducts. Shortly before birth this site of invagination everts to form the nipple. At puberty, alveoli sprout from the ducts and considerable fatty infiltration of the breast tissue takes place. With pregnancy there is tremendous development of the alveoli which, in lactation, secrete the fatty droplets of milk. At the menopause the gland tissue atrophies.

Clinical features

1   Developmental abnormalities are not uncommon. The nipple may fail to evert and it is important to find out from the patient whether or not an inverted nipple is a recent event or has been present since birth. Supernumerary nipples or even breasts may occur along a vertical 'milk line'—a reminder of the line of mammary glands in more primitive mammals; on the other hand, the breast on one or both sides may be small or even absent (amazia).

2   An abscess of the breast should be opened by a radial incision to avoid cutting across a number of lactiferous ducts. Such an abscess may rupture from one fascial compartment into its neighbours, and it is important at operation to break down any loculi which thus form in order to provide ample drainage.

3   Dimpling of the skin over a carcinoma of the breast results from malignant infiltration and fibrous contraction of Cooper's ligaments —as these pass from breast to skin, their shortening results in tethering of the skin to the underlying tumour. This may also occur, however, in chronic infection, after trauma and, very rarely, in fibroadenosis, so that skin fixation to a breast lump is not necessarily diagnostic of malignancy.

4   Retraction of the nipple, if of recent origin, is suggestive of involvement of the milk ducts in the fibrous contraction of a scirrhous tumour.

5   The excision of a breast carcinoma by *radical mastectomy* involves the removal of a wide area of skin around the tumour, all the breast tissue, the pectoralis major (through which lymphatics pass to the internal mammary chain), the pectoralis minor (which lies as a gateway to the axilla), and the whole axillary contents of fatty tissue and contained lymph nodes. This excision also removes the bulk of the lymphatics from the arm which pass along the anterior and medial aspects of the axillary vein. A few lymph vessels from the upper limb pass above the axillary vein and are therefore saved.

Most surgeons today perform less extensive surgery for breast cancer; for example, a *simple mastectomy*, in which the breast alone is removed, or an *extended simple mastectomy*, which combines this with clearance of the axillary fat and its contained nodes.

Oedema of the arm after mastectomy usually only occurs if further damage is done to this precarious lymph drainage by infection, malignant infiltration or heavy irradiation, or if additional strain is put on the evacuation of fluid from the limb by ligation or thrombosis of the axillary vein.

Surface anatomy and surface markings of the upper limb

Much of the anatomy of the limbs can be revised on oneself; otherwise choose a thin colleague.

Bones and joints

The subcutaneous border of the *clavicle* can be palpated throughout its length; the supraclavicular nerves crossing it can be rolled against the bone.

The *acromion process* forms a sharp bony edge at the lateral extremity of the *scapular spine*. It lies immediately above the smooth bulge of the *deltoid muscle* which itself covers the *greater tubercle of the humerus*. Less easily identified is the *coracoid process* of the scapula, lying immediately below the clavicle at the junction of the middle and outer thirds, and covered by the anterior fibres of the deltoid.

The medial border of the scapula can be both seen and felt. Abduction of the arm is a complex affair made up of abduction at the shoulder joint, depression at the sternoclavicular joint and rotation of the scapula; the last two are readily confirmed on self-palpation.

With the shoulder abducted, the *head of the humerus* can be felt in the axilla; note its movement with rotation of the arm.

At the elbow, the three bony landmarks are the *olecranon process* and the *medial* and *lateral epicondyles*. A supracondylar fracture lies above these points, which therefore remain in their triangular relationship to each other; in dislocation of the elbow, however, the olecranon comes more or less in line with the epicondyles (Fig. 115).

Note a hollow in the posterolateral aspect of the extended elbow distal to the lateral epicondyle; this lies over the *head of the radius* which can be felt to rotate during pronation and supination.

The posterior border of the *ulna* is completely subcutaneous and

(a) Normal   (b) Dislocation   (c) Supracondylar fracture

Fig. 115 The relationship of the medial and lateral epicondyles to the olecranon process (a) is disturbed in a dislocation of the elbow (b) but maintained in a supracondylar fracture (c).

crossed by no named vessels or nerve; it can therefore be exposed surgically from end to end without danger.

At the wrist, the *styloid processes* of the radius and ulna can be felt; the former extends more distally. The *dorsal tubercle of Lister* is palpable on the posterior aspect of the distal end of the radius.

In the palm of the hand, palpate the *pisiform* at the base of the hypothenar eminence. *Flexor carpi ulnaris* is inserted into it and when this tendon is relaxed by flexing the wrist the pisiform can be moved a little from side to side. The *hook of the hamate* can be felt by deep palpation just distal to the pisiform. The *scaphoid* is felt at the base of the thenar eminence and also within the anatomical snuff-box, where there is characteristic tenderness when this bone is fractured.

Muscles and tendons

The anterior fold of the axilla is formed by the *pectoralis major*, and its posterior fold by the *teres major* and *latissimus dorsi*. The digitations of *serratus anterior* can be seen in a muscular subject on the medial axillary wall.

In the upper arm the *deltoid* forms the smooth contour of the shoulder. The *biceps* and *brachialis* constitute the bulk of the anterior aspect of the arm, and the *triceps* its posterior aspect.

When the forearm is flexed against resistance, the *brachioradialis* presents prominently along its radial border.

At the wrist (Figs 116–18) it is convenient to commence at the radial pulse. The tendon medial to this is that of the *flexor carpi radialis*, then *palmaris longus* (which may be absent), then the cluster of tendons of *flexor digitorum superficialis*. The tendon of *flexor carpi ulnaris* lies most medially, inserting into the pisiform; the ulnar pulse can be felt just lateral to this tendon.

Flexor carpi radialis
Radial artery
Median nerve

Palmaris longus
Flexor digitorum superficialis
Flexor carpi ulnaris
Ulnar artery and nerve
Flexor retinaculum

Fig. 116 The structures on the anterior aspect of the wrist.

On the dorsal aspect of the wrist (Figs 117, 118) the anatomical snuff-box is formed by the tendons of *abductor pollicis longus* and of *extensor pollicis brevis* laterally and that of *extensor pollicis longus* medially (i.e. towards the ulnar border) — the latter can be traced

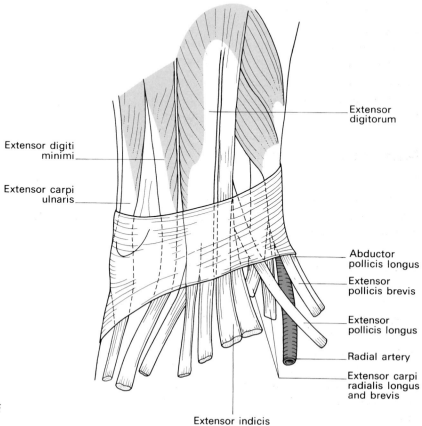

Extensor digitorum

Extensor digiti minimi

Extensor carpi ulnaris

Abductor pollicis longus
Extensor pollicis brevis
Extensor pollicis longus
Radial artery
Extensor carpi radialis longus and brevis

Extensor indicis

Fig. 117 The structures on the posterior aspect of the right wrist.

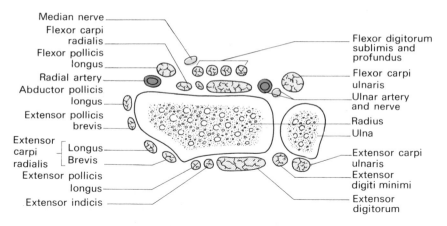

Median nerve

Flexor carpi radialis

Flexor pollicis longus

Radial artery

Abductor pollicis longus

Extensor pollicis brevis

Extensor carpi radialis — Longus — Brevis

Extensor pollicis longus

Extensor indicis

Flexor digitorum sublimis and profundus

Flexor carpi ulnaris

Ulnar artery and nerve

Radius

Ulna

Extensor carpi ulnaris

Extensor digiti minimi

Extensor digitorum

Fig. 118 Schematic section immediately above the wrist joint.

to the base of the terminal phalanx of the extended thumb. The tendons of extensor digitorum are seen in the extended hand passing to be inserted into the bases of the proximal phalanges of the fingers.

Vessels

Feel the pulsations of the *subclavian artery* against the first rib, the *brachial artery* against the humerus, the *radial* and *ulnar* arteries at the wrist and the *radial artery* again in the anatomical snuff-box.

The brachial artery bifurcates into its radial and ulnar branches at the level of the neck of the radius and the line of the radial artery then corresponds to the slight groove which can be seen along the ulnar border of the tensed brachioradialis.

The veins of the upper limb (Fig. 119) comprise the *deep venae comitantes*, which accompany all the main arteries, usually in pairs, and the much more important superficial veins — more important both in size and in practical value because of their use for venepuncture and transfusion.

These superficial veins can be seen as a dorsal venous network on the back of the hand which drains into a lateral *cephalic* and medial *basilic* vein.

The cephalic vein at its origin lies fairly constantly in the superficial fascia just posterior to the radial styloid; even if not visible it can be cut down upon confidently at this site. It then runs up the anterior aspect of the forearm to lie in a groove along the lateral border of the biceps and then, after piercing the deep fascia, in the groove between pectoralis major and the deltoid, where again it can readily be

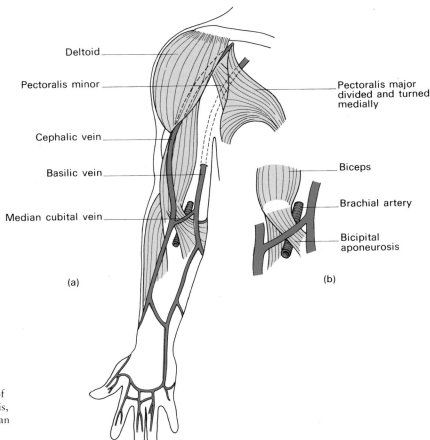

Deltoid

Pectoralis minor

Cephalic vein

Basilic vein

Median cubital vein

Pectoralis major
divided and turned
medially

Biceps

Brachial artery

Bicipital
aponeurosis

(a)

(b)

Fig. 119 (a) The
superficial veins of the
upper limb. (b) Detail of
the bicipital aponeurosis,
lying between the median
cubital vein and the
brachial artery.

exposed for an emergency cut-down. It finally penetrates the clavipectoral fascia to enter the axillary vein.

The basilic vein runs along the posteromedial aspect of the forearm, passes on to the anterior aspect just below the elbow and pierces the deep fascia at about the middle of the upper arm. At the edge of the posterior axillary fold it is joined by the venae comitantes of the brachial artery to form the axillary vein.

Linking the cephalic and basilic veins just distal to the front of the elbow is the *median cubital* vein, usually the most prominent superficial vein in the body and visible or palpable when all others are hidden in fat or collapsed in shock.

It was this vein that was favoured for the operation of bleeding, or phlebotomy, in former days; the underlying brachial artery was protected from the barber-surgeon's knife by the *bicipital aponeurosis*, a condensation of deep fascia passing across from the biceps tendon,

which was therefore termed the 'grâce à Dieu' (praise be to God)
fascia.

In more modern times one tries to avoid using this vein for injection of intravenous barbiturates and other irritating drugs because of
the slight risk of entering the brachial artery and also because of the
danger of piercing a superficially placed abnormal ulnar artery in
occasional instances of high brachial bifurcation.

Nerves

A number of nerves in the upper limb can be palpated, particularly in
a thin subject; these are the *supraclavicular nerves*, as they pass over
the clavicle, the cords of the *brachial plexus* against the humeral head
(with the arm abducted), the *median nerve* in the mid-upper arm,
crossing over the brachial artery, the *ulnar nerve* in the groove of the
medial epicondyle and the *superficial radial nerve* fibres as it passes
over the tendon of extensor pollicis longus at the wrist.

The median nerve lies first lateral then medial to the brachial
artery, crossing it at the mid-upper arm, usually superficially but
occasionally deeply. This close relationship is of historical interest:
Nelson had his median nerve accidentally incorporated in the ligature
around the artery when his arm was amputated above the elbow.

Useful surface markings of other, impalpable, nerves may be listed
as follows.

1   The *axillary nerve* is related closely to the surgical neck of the
humerus a hand's breadth below the acromion process.

2   The *radial nerve* crosses the posterior aspect of the humeral shaft
at its mid-point.

3   The *posterior interosseous branch* of the radial nerve is located by
Henry's method as it winds round the radius. Place three fingers along
the radius, the uppermost lying just distal to the radial head; the 3rd
finger then lies over this nerve.

4   The *median nerve* (Figs 116, 118) in the forearm lies, as its name
suggests, in the median plane; its area of distribution in the hand is
thus anaesthetized if local anaesthetic be injected exactly in the midline at the wrist.

5   The *ulnar nerve* at the wrist lies immediately medial to the ulnar
pulse (Figs 116, 118). In the hand, it passes on the radial side of the
pisiform and then lies on the hook of the hamate. If you press with
your fingernail just lateral to the pisiform bone, you will experience
tingling in your ulnar two fingers.

The bones and joints of the upper limb

The scapula (Fig. 120)

This triangular bone bears three prominent features: the *glenoid fossa* laterally (which is the scapula's contribution to the shoulder joint), the *spine* on its posterior aspect, projecting laterally as the *acromion process*, and the *coracoid process* on its anterior aspect.

Its strong muscular coverings protect the scapula and it is rarely fractured, only by direct and severe violence.

The clavicle (Fig. 120)

This long bone has a number of unusual features.
1 It has no medullary cavity.
2 It is the first to ossify in the fetus (5th–6th week).
3 Although a long bone, it develops in membrane and not in cartilage.
4 It is the most commonly fractured bone in the body.

The clavicle is made up of a medial two-thirds which is circular in section and convex anteriorly, and a lateral one-third which is flattened in section and convex posteriorly.

Medially it articulates with the manubrium at the sternoclavicular joint (this joint containing an articular disc), and is also attached to the first costal cartilage by the costoclavicular ligament.

Fig. 120 The left scapula and clavicle.

Laterally it articulates with the acromion at the acromioclavicular joint (the joint containing an incomplete articular disc) and, in addition, is attached to the coracoid process by the tough coracoclavicular ligament.

The third parts of the subclavian vessels and the trunks of the brachial plexus pass behind the medial third of the shaft of the clavicle, separated only by the thin subclavius muscle. Rarely, these vessels are torn by the fragments of a fractured clavicle; this was the cause of death of Sir Robert Peel following a fall from his horse.

The sternal end of the clavicle has important posteror relations; behind the sternoclavicular joints lie the common carotid artery on the left and the bifurcation of the brachiocephalic artery on the right. The internal jugular vein lies a little more laterally on either side. These vessels are separated from bone by the strap muscles—the sternohyoid and sternothyroid.

Clinical features

The clavicle has three functions:
1   to transmit forces from the upper limb to the axial skeleton;
2   to act as a strut holding the arm free from the trunk, to hang supported principally by trapezius;
3   to provide attachment for muscles.

The weakest point along the clavicle is the junction of the middle and outer third. Transmission of forces to the axial skeleton in falls on the shoulder or hand may prove greater than the strength of the bone at this site and this indirect force is the usual cause of fracture.

When fracture occurs, the trapezius is unable to support the weight of the arm so that the characteristic picture of the patient with a fractured clavicle is that of a man supporting his sagging upper limb with his opposite hand. The lateral fragment is not only depressed but also drawn medially by the shoulder adductors, principally the teres major, latissimus dorsi and pectoralis major (Fig. 121).

The humerus (Fig. 122)

The upper end of the humerus consists of a *head* (one-third of a sphere) facing medially, upwards and backwards, separated from the *greater* and *lesser tubercles* by the *anatomical neck*. The tubercles, in turn, are separated by the *bicipital groove* along which emerges the long head of biceps from the shoulder joint.

Where the upper end and the shaft of the humerus meet there is the narrow *surgical neck* against which lie the axiallary nerve and cir-

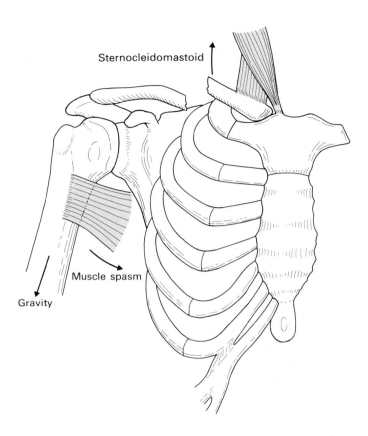

Sternocleidomastoid

Muscle spasm

Gravity

Fig. 121 The deformity of
a fractured clavicle—
downward displacement
and adduction of the
outer fragment by gravity
and muscle spasm
respectively; slight
elevation of the inner
fragment by the
sternocleidomastoid.

cumflex humeral vessels. The shaft itself is circular in section above
and flattened in its lower part. The posterior aspect of the shaft bears
the faint *spiral groove*, demarcating the origins of the medial and
lateral heads of the triceps between which wind the radial nerve and
the profunda vessels.

The lower end of the humerus bears the rounded *capitulum* laterally, for articulation with the radial head, and the spool-shaped
trochlea medially, articulating with the trochlear notch of the ulna.

The *medial* and *lateral epicondyles*, on either side, are extracapsular; the medial is the larger of the two, extends more distally and
bears a groove on its posterior aspect for the ulnar nerve.

Three important nerves thus come into close contact with the
humerus — the axillary, the radial and the ulnar; they may be
damaged, respectively, in fractures of the humeral neck, midshaft and
lower end (Fig. 122).

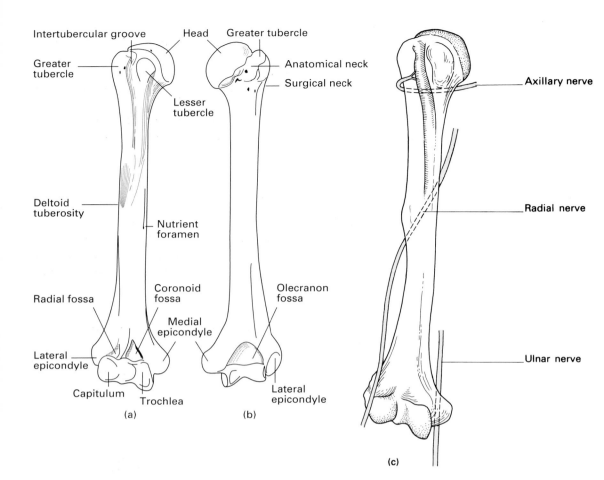

Intertubercular groove   Head   Greater tubercle

Greater tubercle

Anatomical neck

Surgical neck

Lesser tubercle

Deltoid tuberosity

Nutrient foramen

Coronoid fossa

Radial fossa

Olecranon fossa

Medial epicondyle

Lateral epicondyle

Capitulum   Trochlea

Lateral epicondyle

(a)   (b)

Axillary nerve

Radial nerve

Ulnar nerve

(c)

It is an important practical point to note that the lower end of the humerus is angulated forward 45° on the shaft. This is easily confirmed by examining a lateral radiograph of the elbow, when it will be seen that a vertical line continued downwards along the front of the shaft bisects the capitulum. Any decrease of this angulation indicates backward displacement of the distal end of the humerus and is good radiographic evidence of a supracondylar fracture.

Fig. 122 The (a) anterior and (b) posterior view of the humerus. (c) The humerus with its three major related nerves—axillary, radial and ulnar —all of which are in danger of injury in humeral fractures.

The radius and ulna (Fig. 123)

The radius consists of the *head*, *neck*, *shaft* (with its *radial tuberosity*) and expanded distal end. The ulna comprises *olecranon*, *trochlear fossa*, *coronoid process* (with its *radial notch* for articulation with the radial head), *shaft* and small distal *head*, which articulates with the medial side of the distal end of the radius at the inferior radio-ulnar joint.

In pronation and supination, the head of the radius rotates against the radial notch of the ulna, the shaft of the radius swings round the relatively fixed ulnar shaft (the two bones being connected by a fibrous interosseous ligament) and the distal end of the radius rotates against the head of the ulna. This axis of rotation passes from the radial head proximally to the ulnar head distally.

Clinical features

1   The pronator teres is inserted midway along the radial shaft. If the radius is fractured proximal to this, the proximal fragment is supinated (by the action of the biceps) and the distal fragment is pronated by pronator teres. The fracture must, therefore, be splinted

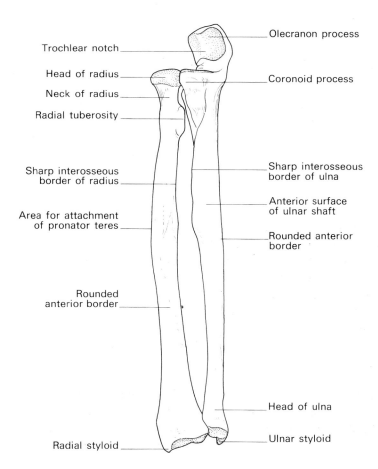

Trochlear notch

Head of radius

Neck of radius

Radial tuberosity

Sharp interosseous border of radius

Area for attachment of pronator teres

Rounded anterior border

Radial styloid

Olecranon process

Coronoid process

Sharp interosseous border of ulna

Anterior surface of ulnar shaft

Rounded anterior border

Head of ulna

Ulnar styloid

Fig. 123 The right radius and ulna—anterior aspect.

with the forearm supinated so that the distal fragment is aligned with the supinated proximal end. If the fracture is distal to the midshaft, the actions of biceps and the pronator muscles more or less balance and the fracture is, therefore, immobilized with the forearm in the neural position (Fig. 124).

2   The force of a fall on the hand produces different effects in different age groups; in a child it may cause a posterior displacement of the distal radial epiphysis, in the young adult the shafts of the radius and ulna may fracture, or the scaphoid may fracture (see page 212), whereas, in the elderly, the most likely result will be a *Colles' fracture*. In the last injury, the radius fractures about 1 in (2.5 cm)

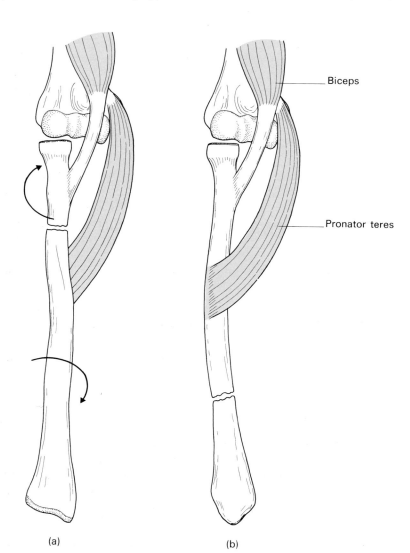

(a)   (b)

Fig. 124 The important role of pronator teres in radial fractures. (a) In proximal fractures, above the insertion of pronator teres, the distal fragment is pronated. Such a fracture must be splinted in the supinated position. (b) When the fracture is distal to pronator teres insertion, the action of this muscle on the proximal fragment is cancelled by the supinator action of biceps. This fracture is, therefore, held reduced in the neutral position, midway between pronation and supination.

proximal to the wrist joint; the distal fragment is displaced posteriorly and usually becomes impacted. The shortening which results brings the styloid processes of the radius and ulna more or less in line with each other.

Another forearm injury resulting from a fall on the outstretched hand is fracture of the head of the radius, due to its being crushed against the capitulum of the humerus.

3   The olecranon process may be fractured by direct violence but more often it is avulsed by forcible contraction of the triceps, which is inserted into its upper aspect. In these circumstances the bone ends are widely displaced and operative repair, to reconstruct the integrity of the elbow joint, becomes essential.

4   A subcutaneous bursa is constantly present over the olecranon and is likely to become inflamed when exposed to repeated trauma. Students and coal miners share this hazard so that olecranon bursitis goes by the nicknames of 'student's elbow' and 'miner's elbow'. Although I have seen many miners with this lesion, I have yet to see a medical student thus disabled.

The bones of the hand (Fig. 125)

The *carpus* is made up of two rows each containing four bones. In the proximal row, from the lateral to the medial side, are the *scaphoid*, *lunate* and *triquetral*, the last bearing the *pisiform* on its anterior surface, into which sesamoid bone the flexor carpi ulnaris is inserted.

In the distal row, from the lateral to the medial side, are the *trapezium*, *trapezoid*, *capitate* and *hamate*.

The carpus as a whole is arched transversely, the palmar aspect being concave. This is maintained by:

1   the shapes of the individual bones, which are broader posteriorly than anteriorly (except for the lunate, which is broader anteriorly);

2   the tough *flexor retinaculum* passing from the scaphoid and the ridge of the trapezium laterally to the pisiform and the hook of the hamate medially (Fig. 126).

Clinical features

1   A fall on the hand may dislocate the rest of the carpal arch backwards from the lunate which, as commented on above, is wide-based anteriorly (perilunate dislocation of the carpus). The dislocated carpus may then reduce spontaneously, only to push the lunate forward and tilt it over so that its distal articular surface faces forward (dislocation of the lunate).

2   The scaphoid may be fractured by a fall on the palm with the hand

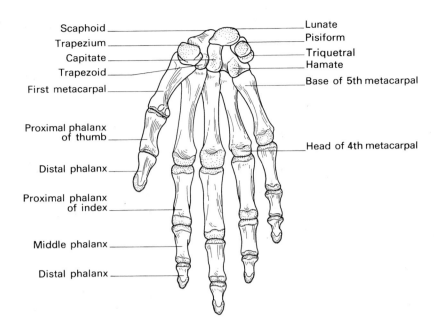

Scaphoid
Trapezium
Capitate
Trapezoid
First metacarpal

Proximal phalanx of thumb

Distal phalanx

Proximal phalanx of index

Middle phalanx

Distal phalanx

Lunate
Pisiform
Triquetral
Hamate
Base of 5th metacarpal

Head of 4th metacarpal

Fig. 125 The right carpus, metacarpus and phalanges.

abducted, in which position the scaphoid lies directly facing the radius.

The blood supply of the scaphoid in one-third of cases enters distally along its waist so that, if the fracture is proximal, the blood supply to this small proximal fragment may be completely cut off with resultant aseptic necrosis of this portion of bone (Fig. 127).

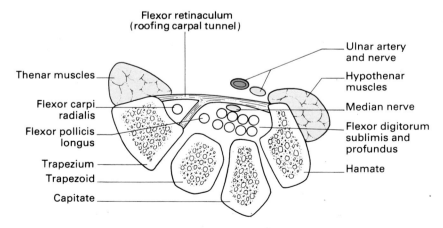

Flexor retinaculum
(roofing carpal tunnel)

Thenar muscles

Flexor carpi radialis

Flexor pollicis longus

Trapezium
Trapezoid
Capitate

Ulnar artery and nerve

Hypothenar muscles

Median nerve

Flexor digitorum sublimis and profundus

Hamate

Fig. 126 Transverse section through the carpus, showing the attachments of the flexor retinaculum. Note the separate osseofascial compartment for the tendon of flexor carpi radialis.

3 '*The carpal tunnel syndrome.*' The flexor retinaculum forms the roof of a tunnel the floor and walls of which are made up of the concavity of the carpus. Packed within this tunnel are the long flexor tendons of the fingers and thumb together with the median nerve. Any lesion diminishing the size of the compartment—for example, an old fracture or arthritic change — may result in compression of the median nerve, resulting in paraesthesiae, numbness and motor weakness in its distribution. Since the superficial palmar branch of the nerve is given off proximal to the retinaculum, there is usually no sensory impairment in the palm.

It is interesting that this syndrome also often occurs without any very obvious cause, although symptoms are relieved by dividing the retinaculum longitudinally.

The shoulder (Figs 128, 129)

The shoulder is a ball-and-socket joint between the relatively large *head of humerus* and relatively small and shallow *glenoid fossa*, although the latter is deepened somewhat by the cartilaginous ring of the *labrum glenoidale*.

The joint *capsule* is lax and is attached around the epiphyseal lines of both the glenoid and the humeral head. However, it does extend down on to the diaphysis on the medial aspect of the neck of the humerus, so that an osteomyelitis of the upper end of the humeral shaft may involve the joint by direct spread.

The capsule is lined by synovial membrane which is prolonged

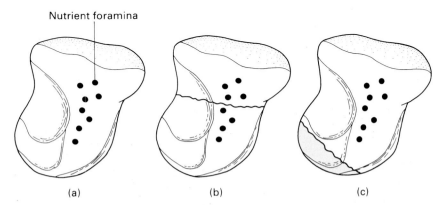

Fig. 127 Blood supply of the scaphoid. (a) Blood vessels enter the bone principally in its distal half. (b) A fracture through the waist of the scaphoid—vessels to the proximal fragment are preserved. (c) A fracture near the proximal pole of the scaphoid —in this case there are no vessels supplying the proximal fragment and aseptic necrosis of bone is therefore inevitable.

Coraco-
clavicular
ligament

Coracoid
process

Weak
gleno-humeral
ligaments
in capsule

Thin capsule

Acromio-
clavicular joint

Coracoacromial
ligament

Acromion

Long head
of biceps

Glenoid fossa

Opening into
subscapularis
bursa

Labrum glenoidale

Fig. 128 The left
shoulder joint—its
ligaments are shown
after removal of the
humerus.

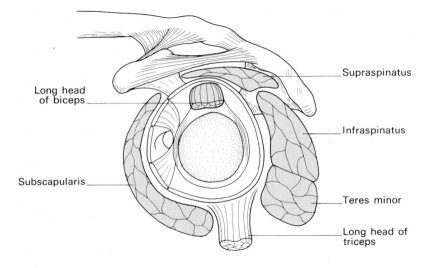

Long head
of biceps

Subscapularis

Supraspinatus

Infraspinatus

Teres minor

Long head of
triceps

Fig. 129 The shoulder
joint—the same view as in
Fig. 128 but now with the
addition of the
surrounding muscles.

along the tendon of the long head of the biceps as this traverses the
joint. The synovium also communicates with the *subscapular bursa*
beneath the tendon of subscapularis.

The stability of the shoulder joint depends almost entirely on
the strength of the surrounding muscles, which may be grouped
into:

1   the closely related short muscles of the 'rotator cuff' (see
below);

2   the long head of biceps, arising from the supraglenoid tubercle
and crossing over the head of the humerus, thus lying actually within
the joint, although enclosed in a tube of synovium;

3   the more distantly related long muscles of the shoulder; the
deltoid, long head of triceps, pectoralis major, latissimus dorsi and
teres major.

Movements of the shoulder girdle

The movements of the shoulder joint itself cannot be divorced from those of the whole shoulder girdle. Even if the shoulder joint is fused, a wide range of movement is still possible by elevation, depression, rotation and protraction of the scapula, leverage occurring at the sternoclavicular joint, the pivot being the costoclavicular ligament.

Abduction of the shoulder is initiated by the supraspinatus; the deltoid can then abduct to 90°. Further movement to 180° (elevation) is brought about by rotation of the scapula upwards by the trapezius and serratus anterior. Shoulder and shoulder girdle movements combine into one smooth action. As soon as abduction commences at the shoulder joint, so rotation of the scapula begins. Test this on yourself or on a colleague by palpating the lower pole of the scapula. This will be felt to swing outwards on initiation of shoulder abduction. Movements of the scapula occur with reciprocal movements at the sternoclavicular joint. Place a finger on this joint; elevate the shoulder and the joint will be felt to depress, swing the shoulder forwards and it will be felt to move backwards, and so on.

Rotator cuff (Fig. 130) is the name given to the sheath of tendons of the short muscles of the shoulder which covers and blends with all but the inferior aspect of that joint. The muscles are the supraspinatus, infraspinatus and teres minor, which are inserted from above down into the humeral greater tubercle, and the subscapularis, which is inserted into the lesser tubercle. All originate from the scapula.

Of these muscles, the supraspinatus is of the greatest practical importance. It passes over the apex of the shoulder beneath the acromion process and coracoacromial ligament, from which it is separated by the *subacromial bursa*. This bursa is continued beneath the deltoid as the *subdeltoid bursa* forming, together, the largest bursa in the body.

The supraspinatus initiates the abduction of the humerus on the scapula; if the tendon is torn as a result of injury, active initiation of abduction becomes impossible and the patient has to develop the trick movement of tilting his body towards the injured side so that gravity passively swings the arm from his trunk. Once this occurs, the deltoid and the scapular rotators can then come into play.

Inflammation of the supraspinatus tendon ('supraspinatus tendinitis') is characterized by a painful arc of shoulder movement between 60° and 120°; in this range, the tendon impinges against the overlying acromion and the coracoacromial ligament. The investiga-

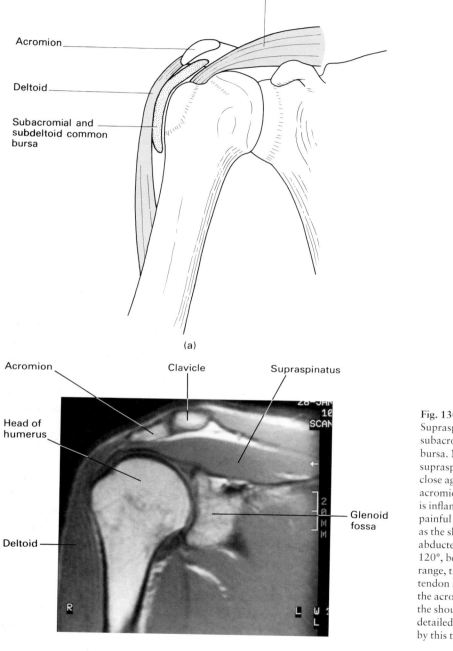

(a)

(b)

Fig. 130 (a) Supraspinatus and the subacromial-subdeltoid bursa. Note that the supraspinatus tendon lies close against the acromion—if this tendon is inflamed, there is a painful arc of movement as the shoulder is abducted from 60° to 120°, because, in this range, the inflamed tendon impinges against the acromion. (b) MRI of the shoulder showing the detailed anatomy revealed by this technique.

tion of soft tissue lesions around the shoulder has been greatly facilitated by magnetic resonance imaging (MRI) which reveals the anatomical structures in exquisite detail (Fig. 130b).

Principal muscles acting on the shoulder joint

Abductors
  supraspinatus
  deltoid
Flexors
  pectoralis major
  coracobrachialis
  deltoid (anterior fibres)
Medial rotators
  pectoralis major
  latissimus dorsi
  teres major
  deltoid (anterior fibres)
  subscapularis

Adductors
  pectoralis major
  latissimus dorsi
Extensors
  teres major
  latissimus dorsi
  deltoid (posterior fibres)
Lateral rotators
  infraspinatus
  teres minor
  deltoid (posterior fibres)

Clinical features

Dislocation of the shoulder

The wide range of movement possible at the shoulder is achieved only at the cost of stability, and for this reason it is the most commonly dislocated major joint. Its inferior aspect is completely unprotected by muscles and it is here that, in violent abduction, the humeral head may slip away from the glenoid to lie in the subglenoid region, whence it usually passes anteriorly into a subcoracoid position (Fig. 131).

The axillary nerve, lying in relation to the surgical neck of the humerus, may be torn in this injury.

The head of the humerus is drawn medially by the powerful adductors of the shoulder; its greater tubercle, therefore, no longer remains the most lateral bony projection of the shoulder region, being replaced for this honour by the acromion process. The normal bulge of the deltoid over the greater tubercle is lost; instead there is the characteristic flattening of this muscle.

In reducing the dislocation by *Kocher's method* the elbow is flexed and the forearm rotated outwards; this stretches the subscapularis which is holding the humeral head internally rotated. The elbow is then swung medially across the trunk, thus levering the head of the humerus laterally so that it slips back into place.

In the *Hippocratic method*, the foot is used as a fulcrum in the axilla, traction and adduction being applied to the forearm; in this way the humeral head is levered outwards into its normal position.

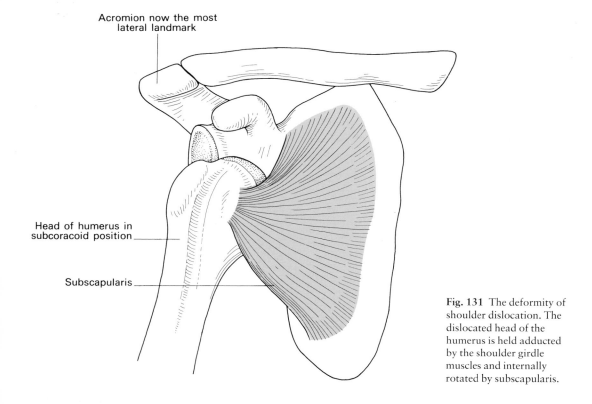

Acromion now the most
lateral landmark

Head of humerus in
subcoracoid position

Subscapularis

Fig. 131 The deformity of
shoulder dislocation. The
dislocated head of the
humerus is held adducted
by the shoulder girdle
muscles and internally
rotated by subscapularis.

The elbow joints (Figs 132, 133)

The elbow joint, although a single synovial cavity, is made up of three
distinct articulations, which are:
1   the *humero-ulnar*, between the trochlea of the humerus and the
trochlear notch of the ulna (a hinge-joint);
2   the *humeroradial*, between the capitulum and the upper concave
surface of the radial head (a ball and socket joint);
3   the *superior radio-ulnar*, between the head of the radius and the
radial notch of the ulna, the head being held in place by the tough
annular ligament (a pivot joint).

The *capsule* of the elbow joint is closely applied around this
complex articular arrangement; the non-articular medial and lateral
epicondyles are extracapsular. The capsule is thin and loose anteriorly
and posteriorly to allow flexion and extension, whereas it is strongly
thickened on either side to form the *medial* and *lateral collateral* ligaments. The lateral ligament is attached distally to the *annular ligament* around the radial head. In order to allow rotation of the radius,
the lower margin of the annular ligament is free and, beneath it, the

synovium of the elbow bulges downwards on to the neck of the radius.

Two sets of movements take place at the elbow:

1 flexion and extension at the humero-ulnar and humeroradial joints;

2 pronation and supination at the proximal radio-ulnar (in conjunction with associated movements of the distal radio-ulnar joint).

Fig. 132 The bony components of the elbow joint. Note the three sets of articular surfaces.

Fig. 133 The joint capsule of the right elbow —lateral aspect.

Muscles acting on the elbow

Flexors
  biceps
  brachialis
  brachioradialis
  the forearm flexor
  muscles

Extensors
  triceps
  anconeus

Pronators
  pronator teres
  pronator quadratus
  flexor carpi radialis

Supinators
  biceps
  supinator
  extensor pollicis longus
  extensor pollicis brevis
  abductor pollicis longus

The supinator action of the biceps is due to its insertion on to the *posterior* aspect of the tuberosity of the radius. When the biceps contracts, not only is the forearm flexed, but the radius 'unwinds' as its tuberosity is rotated anteriorly, i.e. the forearm supinates (Fig. 134). Biceps is a powerful muscle, hence supination is more powerful than pronation—try it on yourself.

Clinical features

1 The elbow joint is safely approached by a vertical posterior incision which divides the triceps expansion.
2 As the capsule is relatively weak anteriorly and posteriorly it will be distended at these sites by an effusion, particularly posteriorly, since the anterior aspect is covered by muscles and dense deep fascia.

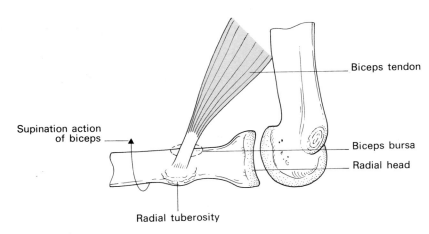

Supination action of biceps

Biceps tendon

Biceps bursa

Radial head

Radial tuberosity

Fig. 134 The supination action of biceps.

Aspiration of such an effusion is readily performed posteriorly on one or other side of the olecranon.

3 The annular ligament is funnel-shaped in adults, but its sides are vertical in young children. A sudden jerk on the arm of a child under the age of 8 years may subluxate the radial head through this ligament ('*pulled elbow*'). Reduction is easily affected by firm supination of the elbow which 'screws' the radial head back into place.

4 *Posterior dislocation of the elbow* may occur as a result of the indirect violence of a fall on the hand. Occasionlly the coronoid process of the ulna is fractured in this injury, being snapped off against the trochlea of the humerus. Characteristically, the triangular relationship between the olecranon and the two humeral epicondyles is lost (Fig. 115).

Reduction is effected by traction to overcome the protective spasm of the muscles acting on the joint, together with flexion of the elbow, which levers the humero-ulnar joint back into place.

The wrist joint (Fig. 135)

The articular disc of the inferior radio-ulnar joint covers the head of the ulna and is attached to the base of the ulnar styloid process. This disc, together with the distal end of the radius, form the proximal face of the wrist joint, the distal surface being the proximal articular surfaces of the scaphoid, lunate and triquetral.

The wrist is a condyloid joint — that is to say, it allows flexion, extension, abduction, adduction and circumduction, the last being a

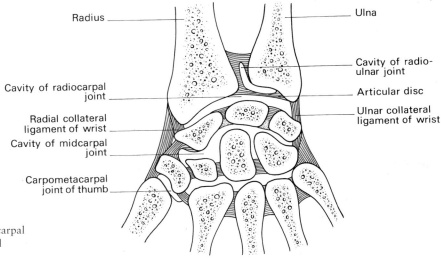

Radius

Ulna

Cavity of radio-ulnar joint

Cavity of radiocarpal joint

Articular disc

Radial collateral ligament of wrist

Ulnar collateral ligament of wrist

Cavity of midcarpal joint

Carpometacarpal joint of thumb

Fig. 135 The wrist, carpal and carpometacarpal joints in section.

combination of the previous four. Flexion and extension are increased
by associated sliding movements of the intercarpal joints; although
the range of flexion at the wrist is actually less than that of extension,
these associated movements make it apparently greater.

Because of the greater distal projection of the radial styloid, the
range of abduction at the wrist is considerably less than that of
adduction.

Muscles acting on the wrist

Flexors—all the long muscles crossing the anterior aspect of the wrist
joint.
Extensors—all the long muscles crossing the posterior aspect of the
joint.
Adductors—flexor carpi ulnaris acting in concord with extensor carpi
ulnaris.
Abductors—flexor carpi radialis and extensores carpi radialis longus
and brevis together with the long abductor and short extensor of the
thumb.

The joints of the hand (Fig. 135)

The joints between the individual carpal bones allow gliding movements to occur which increase the range of extension and, more particularly, flexion permitted at the wrist joint.

The carpometacarpal joint of the thumb is saddle-shaped and
permits flexion and extension (in a plane parallel to the palm of the
hand), abduction and adduction (in a plane at a right angle to the
palm) and opposition, in which the thumb is brought across in
contact with the 5th finger. This joint's range contrasts with the
limited movements of the other carpometacarpal joints which allow
a few degrees of gliding movement of the 2nd and 3rd metacarpals
and a small range of flexion and extension of the 4th and 5th
metacarpals.

The opposite state of affairs holds at the *metacarpophalangeal
(m/p) joints*; only a 60° range of flexion and extension is possible
at the m/p joint of the thumb, whereas a 90° range of flexion and
extension, together with abduction, adduction and circumduction,
are possible at the four other m/p joint's, which are condyloid in
shape.

Note that when the m/p joints of the fingers are flexed, abduction
and adduction become impossible. This is because each metacarpal
head, although rounded at its distal extremity, is flattened anteriorly;
when the base of the proximal phalanx moves on to this flattened

surface, side movements become impossible. Moreover, the collateral ligaments on either side of the m/p joints become taut in flexion and thus prevent abduction and adduction.

The m/p joints of the fingers, but not the thumb, are linked by the tough *deep transverse ligaments*, which prevent any spreading of the palm when a firm grip is taken.

All the *interphalangeal* (*i/p*) *joints* have pulley-shaped opposing surfaces and are therefore hinge-joints allowing flexion and extension only. At all the m/p and i/p joints the ligamentous arrangements are the same.

1   Posteriorly—the joint capsule is replaced by the expansion of the extensor tendon of the digit concerned.

2   Anteriorly—the capsule is formed by a dense plate of fibrocartilage. This *palmar ligament* is the response to the friction of the adjacent flexor tendons.

3   On either side the joints are reinforced by the *collateral ligaments*, which are lax in extension and taut in flexion of the joint.

The muscles acting on the hand

The long flexors of the fingers are:

1   *flexor digitorum profundus*, inserted into the base of the four distal phalanges;

2   *flexor digitorum superficialis*, inserted into the sides of the four middle phalanges.

The profundus tendon pierces that of superficialis over the proximal phalanx.

The profundus flexes the distal phalanx, superficialis the middle phalanx; acting together they flex the fingers and the wrist (Fig. 136).

The long extensors of the fingers are:

- *extensor digitorum longus*, reinforced by
- *extensor indicis*   ⎫   which join the appropriate tendons
- *extensor digiti minimi*   ⎬   of extensor digitorum longus on their
                              ⎭   medial sides.

The tendons of extensor digitorum terminate in each finger by an aponeurotic extensor expansion which covers the dorsum of the proximal phalanx and the sides of its base. It then attaches by a central slip into the base of the middle phalanx and by two lateral slips to the distal phalanx (Fig. 136).

The margins of the extensor expansion are reinforced by the tendons of the intrinsic muscles of the fingers:

1   the *dorsal* and *palmar interossei*, arising from the sides and the fronts of the metacarpals respectively;

Fig. 136 The tendons of a finger. (a) Lateral view. (b) Posterior view.

2   the *lumbricals*, which arise from the four profundus tendons and run on the radial side of the m/p joints to join the extensor expansion.

These intrinsic muscles, arising from the palmar aspect of the hand and inserting along the dorsal aspects of the fingers, have the unique action in that they *flex* the m/p joints and *extend* the i/p joints.

The interossei, together with *abductor digiti minimi*, are responsible for abduction and adduction of the fingers. A weak abduction movement accompanies the action of extensor digitorum and the long flexors adduct the fingers in the movement of full flexion. However, if these movements of extension and flexion are eliminated by laying the hand flat on the table, abduction and adduction become purely the actions of the intrinsic muscles. A card gripped between the fingers in this position of the hand is kept there entirely by intrinsic muscle action.

The 5th finger receives two further intrinsic muscles, *opponens digiti minimi* and *flexor digiti minimi*, from the *hypothenar eminence*.

The eight muscles acting ȯn the thumb may be divided into the long (proceeding from the forearm), and the short or intrinsic muscles.

- Long
  flexor pollicis longus—inserted into the distal phalanx
  extensor pollicis longus—into the distal phalanx
  extensor pollicis brevis—into the proximal phalanx
  abductor pollicis longus—into the metacarpal
- Short

adductor pollicis
flexor pollicis brevis } into the base of the proximal phalanx
abductor pollicis brevis
opponens pollicis—along the metacarpal

The flexors and extensors of the wrist play an important synergic role in movements of the hand. Notice how weak the grip becomes when the wrist is fully flexed; it must be held firmly in the extended or neutral position by balanced muscle action in order to allow the long flexors of the fingers and thumb to work at their full stretch and, therefore, at their maximum efficiency.

The arteries of the upper limb

The axillary artery

The axillary artery commences at the lateral border of the first rib, as a continuation of the subclavian, and ends at the lower border of the axilla (i.e. the lower border of teres major) to become the brachial artery. It is divided into three parts by pectoralis minor and, apart from its distal extremity, it lies covered by pectoralis major.

Above pectoralis minor, the brachial plexus lies above and behind the artery, but, distal to this, the cords of the plexus take up their positions around the artery according to their names, i.e. lateral, medial and posterior.

The branches of the axillary artery supply the chest wall and shoulder; conveniently, the 1st, 2nd and 3rd parts give off one, two and three branches respectively:
- 1st part: 1, superior thoracic artery
- 2nd part: 1, acromiothoracic trunk
             2, lateral thoracic artery
- 3rd part: 1, subscapular artery
             2, anterior circumflex humeral artery
             3, posterior circumflex humeral artery

All but the circumflex humeral vessels are encountered in the axillary dissection of a radical mastectomy.

The brachial artery

The brachial artery continues on from the axillary and ends at the level of the neck of the radius by dividing into the radial and ulnar arteries. It is superficial (immediately below the deep fascia) along its whole course, except where it is crossed, at the level of the middle of the humerus, by the median nerve which passes superficially from its lateral to medial side; occasionally the nerve crosses deep to the artery. Fairly frequently the artery divides into its two terminal branches in the upper arm.

The named branches of the artery are:

- the *profunda* (accompanying the radial nerve);
- *superior ulnar collateral* (accompanying the ulnar nerve);
- *nutrient* (to the humerus);
- *inferior ulnar collateral*.

The radial artery

The radial artery (Fig. 137) commences at the level of the radial neck by lying on the tendon of biceps. In its upper half it lies overlapped by brachioradialis, the surface marking of the artery being the groove which can be seen on the medial side of this tensed muscle in the mus-

Brachialis

Biceps

Lateral cutaneous nerve of forearm

Brachioradialis

Supinator

Superficial radial nerve

Extensor carpi radialis longus

Pronator teres deep and superficial heads

Anterior interosseus branch of median nerve

Radial artery

Flexor pollicis longus

Median nerve

Medial cutaneous nerve of forearm

Common flexor origin

Ulnar nerve

Ulnar artery

Flexor digitorum profundus

Flexor carpi ulnaris

Dorsal branch of ulnar nerve

Fig. 137 Dissection of the forearm to show principal vessels and nerves.

cular subject. Distally in the forearm the artery lies superficially between brachioradialis and flexor carpi radialis, and it is between these two tendons that it is palpated at the wrist (Fig. 116).

In the middle third of the forearm the radial nerve lies along the lateral side of the artery; the nerve may here be incorporated in a carelessly placed ligature.

Distal to the radial pulse, the artery gives off a branch to assist in forming the superficial palmar arch. It then passes deep to the tendons of abductor pollicis longus and extensor pollicis brevis to enter the anatomical snuff-box (in which it can be felt), pierces the first dorsal interosseous muscle and adductor pollicis, between the 1st and 2nd metacarpals, and goes on to form the deep palmar arch with the deep branch of the ulnar artery.

The ulnar artery

The ulnar artery (Fig. 137) is the larger of the two terminal branches of the brachial artery. From its commencement it passes beneath the muscles arising from the common flexor origin, lies upon flexor digitorum profundus and is overlapped by flexor carpi ulnaris. The median nerve crosses superficially to the ulnar artery, separated from it by only part of one muscle, the deep head of pronator teres.

In the distal half of the forearm the artery becomes superficial between the tendons of flexor carpi ulnaris and flexor digitorum sublimis; it then crosses the flexor retinaculum to form the superficial palmar arch with the superficial branch of the radial artery.

The ulnar nerve accompanies the artery on its medial side in the distal two-thirds of its course in the forearm and across the flexor retinaculum (Fig. 116).

Note. There is a rich anastomosis of arteries around all major joints. Apart from remembering this fact, the clinical student need not commit to memory the numerous named branches involved.

The brachial plexus

The brachial plexus is of great practical importance to the surgeon. It may be damaged in open, closed or obstetrical injuries, be pressed upon by a cervical rib or be involved in tumour. It is encountered, and hence put in danger, in operations upon the root of the neck.

C5

C6

C7

Musculocutaneous
nerve

Axillary nerve

Radial nerve

Median nerve

C8

T1

Ulnar nerve

Roots

Upper, middle and
lower trunks

Lateral, posterior and
medial cords

Anterior and
posterior divisions

Fig. 138 Scheme of the brachial plexus.

The plexus is formed as follows (Fig. 138):

1   five *roots* derived from the anterior primary rami of C5, 6, 7, 8 and T1; link up into:

2   three *trunks* formed by the union of
  • C5 and 6 (upper);
  • C7 alone (middle);
  • C8 and T1 (lower)
  which split into:

3   six *divisions* formed by each trunk dividing into an anterior and posterior division; which link up again into:

4   three *cords*
  • a lateral, from the fused anterior divisions of the upper and middle trunks;
  • a medial, from the anterior division of the lower trunk;
  • a posterior, from the union of all three posterior divisions.

The *roots* lie between the anterior and middle scalene muscles. The *trunks* traverse the posterior triangle of the neck. The *divisions* lie behind the clavicle. The *cords* lie in the axilla.

The cords continue distally to form the main nerve trunks of the upper limb thus:

1   the lateral cord continues as the *musculocutaneous nerve*;

2   the medial cord, as the *ulnar nerve*;

3   the posterior cord, as the *radial nerve* and the *axillary nerve*;

4 a cross-communication between the lateral and medial cords forms the *median nerve*.

For reference purposes, the derivatives of the various components of the brachial plexus are given below (Fig. 139).

From the roots
- nerve to rhomboids;
- nerve to subclavius;
- nerve to serratus anterior (C5, 6, 7).

From the trunk
- suprascapular nerve—from the upper trunk (supplies supraspinatus and infraspinatus).

From the lateral cord
- musculocutaneous nerve;
- lateral pectoral nerve;
- lateral root of median nerve.

From the medial cord
- medial pectoral nerve;
- medial cutaneous nerves of arm and forearm;

Fig. 139 The derivatives of the brachial plexus. The lightly coloured areas show the posterior divisions.

- ulnar nerve;
- medial root of median nerve.

From the posterior cord

- subscapular nerves;
- nerve to latissimus dorsi (thoracodorsal nerve);
- axillary nerve;
- radial nerve.

Note that the posterior cord supplies the skin and muscles of the

Fig. 140 The segmental cutaneous innervation of the body.

posterior aspect of the limb whereas the anteriorly placed lateral and medial cords supply the anterior compartment structures.

The segmental cutaneous supply of the upper limb (Fig. 140)

In spite of this complex interlacing of the nerve roots in the brachial plexus, the skin of the upper limb, as with the skin of the rest of the body, has a perfectly regular segmental nerve supply. This is derived from C4 to T2 which is arranged approximately as follows:
- C4—supplies skin over the shoulder tip;
- C5—radial side of upper arm;
- C6—radial side of forearm;
- C7—the skin of the hand;
- C8—ulnar side of forearm;
- T1—ulnar side of upper arm;
- T2—skin of the axilla.

The course and distribution of the principal nerves of the upper limb

The nerves of the upper limb are derived from the brachial plexus.

The axillary nerve

The axillary (circumflex) nerve (C5, 6) arises from the posterior cord of the plexus and winds round the surgical neck of the humerus in company with the circumflex humeral vessels (Figs 122, 141). Its branches are:
- muscular—to deltoid and teres minor;
- cutaneous—to a palm-sized area of skin over the deltoid.

The axillary nerve may be injured in fractures of the humeral neck or in dislocations of the shoulder. This will be followed by weakness of shoulder abduction, wasting of the deltoid and a small patch of anaesthesia over this muscle.

The radial nerve

The radial nerve (C5, 6, 7, 8, T1) is the main branch of the posterior cord. Lying first behind the axillary artery, it then passes backwards

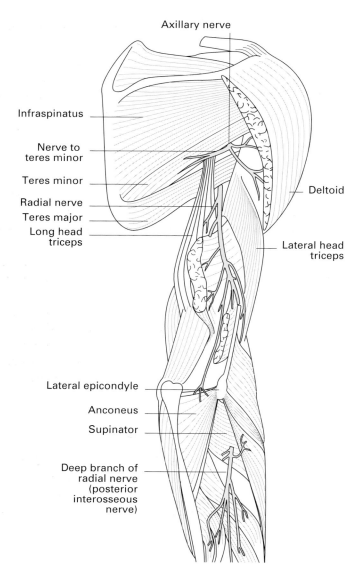

Axillary nerve

Infraspinatus

Nerve to
teres minor

Teres minor

Radial nerve

Teres major

Long head
triceps

Deltoid

Lateral head
triceps

Lateral epicondyle

Anconeus

Supinator

Deep branch of
radial nerve
(posterior
interosseous
nerve)

Fig. 141 The distribution
of the radial nerve.

between the long and medial heads of the triceps to lie in the
spiral groove on the back of the humerus between the medial and
lateral heads of triceps (Fig. 141). The profunda branch of the
brachial artery accompanies the nerve in this part of its course (Fig.
122).

At the lower third of the humerus, the radial nerve pierces the
lateral intermuscular septum to re-enter the anterior compartment of
the arm between brachialis and brachioradialis (a convenient site for
surgical exposure, Fig. 137). At the level of the lateral epicondyle its

important *posterior interosseous nerve* is given off, which winds round the radius within the supinator muscle then sprays out to be distributed to the extensor muscles of the forearm.

The radial nerve itself continues as the superficial radial nerve, lying deep to brachioradialis (Fig. 137). Above the wrist, it emerges posteriorly from beneath this muscle to end by dividing into cutaneous nerves to the posterior aspects of the radial $3\frac{1}{2}$ digits.

Branches

The radial nerve is the nerve of supply to the extensor aspect of the upper limb. The main trunk itself innervates: triceps anconeus brachioradialis and extensor carpi radialis longus. It also gives a twig to the lateral part of brachialis.

The posterior interosseous branch supplies all the remaining extensor muscles of the forearm together with the supinator and abductor pollicis longus.

Cutaneous branches are distributed to the back of the arm, forearm and radial side of the dorsum of the hand. So great is the overlap from adjacent nerves, however, that division of the radial nerve results, surprisingly, in only a small area of anaesthesia over the dorsum of the hand, in the web between the thumb and index finger.

The musculocutaneous nerve

The musculocutaneous nerve (C5, 6, 7) continues on from the lateral cord of the plexus. It pierces coracobrachialis then runs between biceps and brachialis (supplying all these three muscles) to innervate, by its terminal cutaneous branch, the skin of the lateral forearm.

The ulnar nerve

The ulnar nerve (C7, 8, T1; Fig. 142) is formed from the medial cord of the plexus. It lies medial to the axillary and brachial artery as far as the middle of the humerus, then pierces the medial intermuscular septum (in company with the superior ulnar collateral artery) to descend on the anterior face of triceps. It passes behind the medial epicondyle (where it can readily be rolled against the bone), to enter the forearm (Fig. 122). Here it descends beneath flexor carpi ulnaris until this muscle thins out into its tendon, leaving the nerve to lie superficially on its radial side. In the distal two-thirds of the forearm the nerve is accompanied by the ulnar artery which lies on the nerve's radial side. About 2 in (5 cm) above the wrist, a dorsal cutaneous

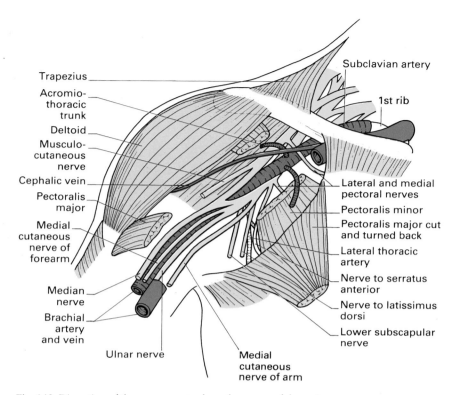

Trapezius
Acromio-thoracic trunk
Deltoid
Musculo-cutaneous nerve
Cephalic vein
Pectoralis major
Medial cutaneous nerve of forearm
Median nerve
Brachial artery and vein
Ulnar nerve
Medial cutaneous nerve of arm
Subclavian artery
1st rib
Lateral and medial pectoral nerves
Pectoralis minor
Pectoralis major cut and turned back
Lateral thoracic artery
Nerve to serratus anterior
Nerve to latissimus dorsi
Lower subscapular nerve

Fig. 142 Dissection of the upper arm to show the course of the major nerves.

branch passes deep to flexor carpi ulnaris to supply the dorsal aspects of the ulnar $1\frac{1}{2}$ fingers.

The ulnar nerve crosses the flexor retinaculum superficially (Fig. 137) to break up into a superficial terminal branch, supplying the ulnar $1\frac{1}{2}$ fingers, and a deep terminal branch which supplies the hypothenar muscles and the intrinsic muscles of the hand.

Its branches are:
• muscular—to flexor carpi ulnaris, medial half of flexor digitorum profundus, the hypothenar muscles, the interossei, 3rd and 4th lumbricals and the adductor pollicis (i.e. it supplies all the intrinsic muscles of the hand apart from those of the thenar eminence and the 1st and 2nd lumbricals, which are innervated by the median nerve);
• cutaneous—to the ulnar side of both aspects of the hand and both surfaces of the ulnar $1\frac{1}{2}$ fingers.

The median nerve

The median nerve (C6, 7, 8, T1; Fig. 142) arises by the junction of a branch from the medial and another from the lateral cord of the plexus, which unite anterior to the third part of the axillary artery. Continuing along the lateral aspect of the brachial artery, the nerve then crosses superficially (occasionally deep) to the artery at the mid-humerus to lie on its medial side. The nerve enters the forearm between the heads of pronator teres, the deeper of which separates it from the ulnar artery (Fig. 137). Here the nerve gives off its *anterior interosseous branch* (which supplies flexor pollicis longus, flexor digitorum profundus to the index and middle fingers, and pronator quadratus), and then lies on the deep aspect of flexor digitorum superficialis, to which it adheres.

At the wrist, the median nerve becomes superficial on the ulnar side of flexor carpi radialis, exactly in the midline (Fig. 116). Here it gives off a *palmar cutaneous branch*, which supplies the skin of the midpalm. It then passes deep to the flexor retinaculum, giving off an important branch to the thenar muscles beyond the distal skin crease,

(a)

(c)

(b)

Fig. 143 The usual cutaneous distribution (shown in pale blue) of the (a) median, (b) ulnar and (c) radial nerves in the hand (considerable variations and overlap occur).

twigs to the radial two lumbricals and cutaneous branches to the palmar aspects of the radial $3\frac{1}{2}$ digits.

Its branches are:
- muscular—to all the muscles of the flexor aspects of the forearm, apart from the flexor carpi ulnaris and the ulnar half of flexor digitorum profundus, and to the thenar eminence muscles and the radial two lumbricals;
- cutaneous—to the skin of the radial side of the palm, the palmar, and a variable degree of the dorsal, aspect of the radial $3\frac{1}{2}$ digits.

Note that there is considerable variation in the exact cutaneous distribution of the nerves in the hand; for example, the ulnar nerve may encroach on median territory and supply the whole of the 4th and 5th digits (Fig. 143).

The anatomy of upper limb deformities

Many deformities of the upper limb, particularly those resulting from nerve injuries, are readily interpreted anatomically.

Brachial plexus injuries may occur from traction on the arm during birth. The force of downward traction falls upon roots C5 and 6, resulting in paralysis of the deltoid and short muscles of the shoulder, and of brachialis and biceps which flex and supinate at the elbow. The arm, therefore, hangs limply by the side with the forearm pronated and the palm facing backwards, like a porter hinting for a tip (*Erb-Duchenne paralysis*). In adults this lesion is seen in violent falls on the side of the head and shoulder forcing the two apart and thus putting a tearing strain on the upper roots of the plexus.

Upward traction on the arm (e.g. in a forcible breech delivery) may tear the lowest root, T1, which is the segmental supply of the intrinsic hand muscles. The hand assumes a clawed appearance because of the unopposed action of the long flexors and extensors of the fingers; the extensors, inserting into the bases of the proximal phalanges, extend the m/p joints while the flexor profundus and sublimis, inserting into the distal and middle phalanges, flex the i/p joints (*Klumpke's paralysis*). There is often an associated *Horner's syndrome* (ptosis and constriction of the pupil), due to traction on the cervical sympathetic chain.

A mass of malignant supraclavicular lymph nodes or the direct invasion of a pulmonary carcinoma (*Pancoast's syndrome*) may produce a similar neurological picture by involvement of the lowest root of the plexus.

Not infrequently, the lower trunk of the plexus (C8, T1) is pressed upon by a cervical rib, or by the fibrous strand running from the extremity of such a rib, resulting in paraesthesiae along the ulnar border of the arm and weakness and wasting of the small muscles of the hand.

The *radial nerve* may be injured in the axilla by the pressure of a crutch ('crutch palsy') or may be compressed when a drunkard falls into an intoxicated sleep with the arm hanging over the back of a chair ('Saturday night palsy'). Fractures of the humeral shaft may damage the main radial nerve, whereas its posterior interosseous branch, to the extensor muscles of the forearm, may be injured in fractures or dislocations of the radial head. An ill-placed incision to expose the head of the radius taken more than three fingers' breadth below the head will divide the nerve as it lies in the supinator muscle.

Damage to the main trunk of the radial nerve results in a *wrist drop* due to paralysis of all the wrist extensors (Fig. 144). Damage to the posterior interosseous nerve, however, leaves extensor carpi radialis longus intact, as it is supplied from the radial nerve above its division; this muscle alone is sufficiently powerful to maintain extension of the wrist.

The disability produced by a wrist drop is inability to grip firmly, since, unless the flexor muscles are stretched by extending the wrist, they act at a mechanical disadvantage. Try yourself to grip strongly with the wrist flexed and realize how, by operative fusion of the wrist joint in extension, the weakness produced by a radial nerve paralysis would be overcome.

Nerve overlap means that division of the radial nerve produces only a small area of anaesthesia of the dorsum of the hand between the 1st and 2nd metacarpals.

The *ulnar nerve*, in its vulnerable position behind the medial epicondyle of the humerus, may be damaged in fractures or dislocations of the elbow; it is also frequently divided in lacerations of the wrist. In the latter case, all the intrinsic muscles of the fingers (apart from the radial two lumbricals) are paralysed so that the hand assumes the clawed position already described under Klumpke's palsy (Fig. 144). The clawing is slightly less intense in the 2nd and 3rd digits because of their intact lumbricals, supplied by the median nerve. In late cases, wasting of the interossei is readily seen on inspecting the dorsum of the hand. Sensory loss over the ulnar $1\frac{1}{2}$ fingers is present.

If the nerve is injured at the elbow, the flexor digitorum profundus to the 4th and 5th fingers is paralysed so that the clawing of these two

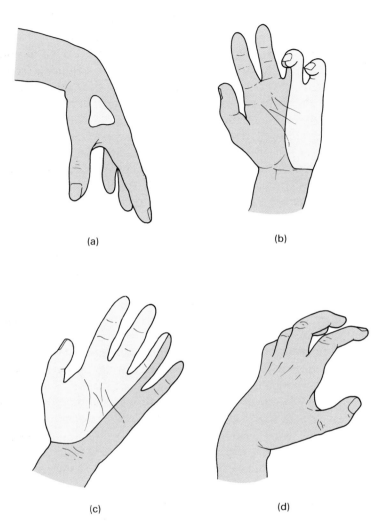

(a)

(b)

(c)

(d)

Fig. 144 Deformities of the hand. (a) Radial palsy —wrist drop. (b) Ulnar nerve palsy—'Main en griffe' or claw hand. (c) Median nerve palsy— 'Monkey's hand'. (d) Volkmann's contracture— another claw hand deformity. The pale blue areas represent the usual distribution of anaesthesia.

fingers is less intense than in division at the wrist. Paralysis of the flexor carpi ulnaris results in a tendency to radial deviation of the wrist.

Division of the ulnar nerve leaves a surprisingly efficient hand. The long flexors enable a good grip to be taken; the thumb, apart from loss of adductor pollicis, is intact and sensation over the palm of the hand is largely maintained. Indeed, it may be difficult to determine clinically with certainty that the nerve is injured; a reliable test is loss of ability to adduct and abduct the fingers with the hand laid flat, palm downwards on the table; this eliminates 'trick' movements of adduction and abduction of the fingers brought about as part of their flexion and extension respectively.

The *median nerve* is occasionally damaged in supracondylar fractures but it is in greatest danger in lacerations of the wrist.

If divided at the wrist, only the thenar muscles (excluding adductor pollicis) and the radial two lumbricals are paralysed and wasting of the thenar muscles occurs. It might be thought that such a lesion is relatively trivial since the only motor defect is loss of accurate opposition movement of the thumb to other fingers. In point of fact this injury is a serious disability because of the loss of sensation over the thumb, adjacent $2\frac{1}{2}$ fingers and the radial two-thirds of the palm of the hand, which prevents the accurate and delicate adjustments the hand makes in response to tactile stimuli (Fig. 144).

If the median nerve is divided at the elbow, there is serious muscle impairment. Pronation of the forearm is lost and is replaced by a trick movement of rotation of the upper arm. Wrist flexion is weak and accompanied by ulnar deviation, since this now depends on the flexor carpi ulnaris and the ulnar half of flexor digitorum profundus.

Volkmann's contracture of the hand follows ischaemia and subsequent fibrosis and contraction of the long flexor and extensor muscles of the forearm (Fig. 144).

The deformities are readily explained as follows.

1 Since the flexors of the wrist are bulkier than the extensors, their fibrous contraction is greater and the wrist is therefore flexed.

2 The long extensors of the fingers are inserted into the proximal phalanges; their contracture extends the m/p joints.

3 The long flexors are inserted into the distal and middle phalanges and therefore flex the i/p joints.

There is, therefore, flexion at the wrist, extension at the m/p and flexion at the i/p joints.

If the wrist is passively further flexed by the examiner, the tight flexor tendons are somewhat relaxed and therefore the fingers become a little less clawed.

Dupuytren's contracture results from a fibrous contraction of the palmar aponeurosis, particularly of the 4th and 5th fingers.

The palmar aponeurosis is merely part of the deep fascial sheath of the upper limb; it passes from the palm along either side of each finger, blends with the fibrous flexor sheath of the fingers and is attached to the sides of the proximal and middle phalanges. Contracture of this fascia results in a longitudinal thickening in the palm together with flexion of the m/p and proximal i/p joints. However, the distal i/p joints are not involved and, in fact, in an advanced case, are actually *extended* by the distal phalanx being pushed backwards against the palm of the hand.

The spaces of the hand

The spaces of the hand are of practical significance because they may become infected and, in consequence, become distended with pus. The important spaces are:

1 the superficial pulp spaces of the fingers;
2 the synovial tendon sheaths of the 2nd, 3rd and 4th fingers;
3 the ulnar bursa;
4 the radial bursa;
5 the midpalmar space;
6 the thenar space.

The superficial pulp space of the fingers
(Fig. 145)

The tips of the fingers and thumb are composed entirely of subcutaneous fat broken up and packed between fibrous septa, which pass from the skin down to the periosteum of the terminal phalanx. The tight packing of this compartment is responsible for the severe pain of a 'septic finger'—there is little room for the expansion of inflamed and oedematous tissues.

The blood vessels to the shaft of the distal phalanx must traverse this space and may become thrombosed in a severe pulp infection with resulting necrosis of the diaphysis of the bone. The base of the distal phalanx receives its blood supply more proximally from a branch of the digital artery in the middle segment of the finger and therefore survives. At each of the skin creases of the fingers, the skin is bound down to the underlying flexor sheath so that the pulp over each phalanx is in a separate compartment cut off from its neighbours. Infection may, however, track from one space to another along the neurovascular digital bundles.

Fig. 145 The distal pulp space of the finger—note the distribution of the arterial supply to the distal phalanx.

Over the palm of the hand there is very little subcutaneous tissue, the skin adhering to the underlying palmar aponeurosis; in contrast, the skin of the dorsum of the fingers and hand is loose and fluid can, therefore, readily collect beneath it. Unless this is remembered, the marked dorsal oedema which may accompany sepsis of the palmar aspect of the fingers or hand may result in the primary site of the infection being overlooked.

The ulnar and radial bursae and the synovial tendon sheaths of the fingers (Fig. 146)

The flexor tendons traverse a fibro-osseous tunnel in each digit. This tunnel is made up posteriorly by the metacarpal head, the phalanges and the fronts of the intervening joints. The anterior fibrous part consists of condensed deep fascia attached to the sharp anterolateral margin of each phalanx and termed the *fibrous flexor sheath*. This is particularly tough over the phalanges but loose over the front of each joint; it therefore holds the flexor tendons in place without 'bow-stringing' during flexion of the fingers, but does not impede movement of the joints.

Distally, the fibrous sheath ends at the insertion of the profundus

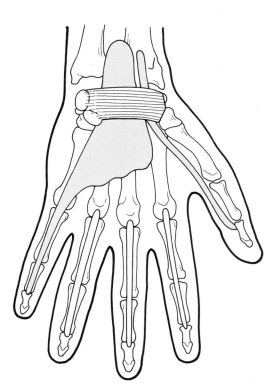

Fig. 146 The synovial sheaths of the flexor tendons of the hand—the radial and ulnar bursae track proximally deep to the flexor retinaculum and provide a potential pathway of infection into the forearm. In many cases these bursae communicate.

tendon (or flexor pollicis longus tendon in the case of the thumb) at the base of the distal phalanx.

These fibrous sheaths are lined by synovial membrane which is reflected around each tendon. The tendons of the 2nd, 3rd and 4th fingers have synovial sheaths which are closed off proximally at the metacarpal head, but the synovial sheaths of the thumb and little finger extend proximally into the palm.

That of the long flexor tendon of the thumb extends through the palm, deep to the flexor retinaculum, to about 1 in (2.5 cm) proximal to the wrist and is termed the *radial bursa*. The synovial sheath of the 5th finger continues as the *ulnar bursa*, an expanded synovial sheath which encloses all the finger tendons in the palm and which also extends proximally below the flexor retinaculum for 1 in (2.5 cm) above the wrist. In about 50% of cases the radial and ulnar bursae communicate. These synovial sheaths may become infected either

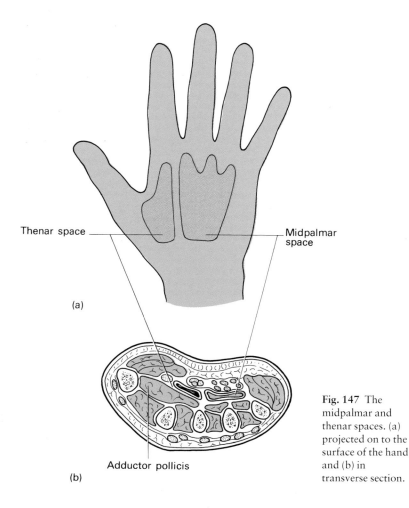

Thenar space

Midpalmar space

(a)

Adductor pollicis

(b)

Fig. 147 The midpalmar and thenar spaces. (a) projected on to the surface of the hand and (b) in transverse section.

directly — for example, following the entry of a splinter — or may be secondarily involved from a neglected pulp-space infection. Infection of the 2nd, 3rd and 4th sheaths are confined to the finger concerned, but sepsis in the 1st and 5th sheaths may spread proximally into the palm through the radial and ulnar bursa respectively, and may pass from one bursa to the other via the frequent cross-communication between the two.

Since these bursae both extend proximally beyond the wrist, infection may, on occasion, spread into the forearm.

Two spaces deep in the palm of the hand may rarely become distended with pus; these are the midpalmar and thenar spaces (Fig. 147).

The *midpalmar space* lies behind the flexor tendons and ulnar bursa in the palm and in front of the 3rd, 4th and 5th metacarpals with their attached interossei. The 1st and 2nd metacarpals are curtained off from this space by the adductor pollicis, which arises from the shaft of the 3rd metacarpal and passes as a triangular sheet to the base of the proximal phalanx of the thumb.

The *thenar space* is the space superficial to the 2nd and 3rd metacarpals and the adductor pollicis. It is separated from the midpalmar space by a fibrous partition.

Infection of these two spaces sometimes results from penetrating wounds or may be due to secondary involvement from a long-neglected tendon sheath infection. Nowadays they are fortunately extremely rare.

Part 4
The Lower Limb

The anatomy and surface markings of the lower limb

Bones and joints

The tip of the *anterior superior spine of the ilium* is easily felt and may be visible in the thin subject. The *greater trochanter* of the femur lies a hand's breadth below the iliac crest; it is best palpated with the hip abducted so that the overlying hip abductors (tensor fasciae latae and gluteus medius and minimus) are relaxed. In the very thin, wasted patient the greater trochanter may be seen as a prominent bulge and its overlying skin is a common site for a pressure sore to form in such a case.

The *ischial tuberosity* is covered by gluteus maximus when one stands. In the sitting position, however, the muscle slips away laterally so that weight is taken directly on the bone. To palpate this bony point, therefore, feel for it uncovered by gluteus maximus in the flexed position of the hip.

At the knee, the *patella* forms a prominent landmark. When quadriceps femoris is relaxed, this bone is freely mobile from side to side; note that this is so when you stand erect. The *condyles* of the femur and tibia, the *head of the fibula* and the joint line of the knee are all readily palpable; less so is the *adductor tubercle* of the femur, best identified by running the fingers down the medial side of the thigh until they are halted by it, the first bony prominence so to be encountered.

The *tibia* can be felt throughout its course along its anterior sub-cutaneous border. The *fibula* is subcutaneous for its terminal 3 in (7 cm) above the *lateral malleolus*, which extends more distally than the stumpier *medial malleolus* of the tibia.

Immediately in front of the malleoli can be felt a block of bone which is the *head of the talus*.

The *tuberosity of the navicular* stands out as a bony prominence 1 in (2.5 cm) in front of the medial malleolus; it is the principal point of insertion of tibialis posterior. The base of the 5th metatarsal is easily felt on the lateral side of the foot and is the site of insertion of peroneus brevis.

If the *calcaneus* (*os calcis*) is carefully palpated, the *peroneal tubercle* can be felt 1 in (2.5 cm) below the tip of the lateral malleolus and the *sustentaculum tali* 1 in (2.5 cm) below the medial malleolus; these represent pulleys respectively for peroneus longus and for flexor hallucis longus.

Bursae of the lower limb

A number of these bony prominences are associated with overlying
bursae which may become distended and inflamed: the one over the
ischial tuberosity may enlarge with too much sitting ('weaver's
bottom'); that in front of the patella is affected by prolonged kneeling
forwards, as in scrubbing floors or hewing coal ('housemaid's knee',
the 'beat knee' of north-country miners, or prepatellar bursitis);
whereas the bursa over the ligamentum patellae is involved by years
of kneeling in a more erect position—as in praying ('clergyman's knee'
or infrapatellar bursitis). Young women who affect fashionable but
tight shoes are prone to bursitis over the insertion of the tendo
Achillis into the calcaneus and may also develop bursae over the navicular tuberosity and dorsal aspects of the phalanges.

A 'bunion' is a thickened bursa on the inner aspect of the first
metatarsal head, usually associated with hallux valgus deformity.

Mensuration in the lower limb

Measurement is an important part of the clinical examination of the
lower limb. Unfortunately, students find difficulty in carrying this out
accurately and still greater difficulty in explaining the results they
obtain, yet this is nothing more or less than a simple exercise in
applied anatomy.

First note the differences between *real* and *apparent* shortening of
the lower limbs. Real shortening is due to actual loss of bone length—
for example, where a femoral fracture has united with a good deal of
overriding of the two fragments. Apparent shortening is due to a *fixed*
deformity of the limb (Fig. 148). Stand up and flex your knee and hip
on one side, imagine these are both ankylosed at 90° and note that,
although there is no loss of tissue in this leg, it is apparently some 2 ft
(60 cm) shorter than its partner.

If there is a fixed pelvic tilt or fixed joint deformity in one limb,
there may be this apparent difference between the lengths of the two
legs. By experimenting on yourself you will find that adduction
apparently shortens the leg, whereas it is apparently lengthened in
abduction.

To measure the real length of the limbs (Fig. 149), overcome
any disparity due to fixed deformity by putting both legs into exactly
the same position; where there is no joint fixation, this means that
the patient lies with his pelvis 'square', his legs abducted symmetrically and both lying flat on the couch. If, however, one hip is in 60° of
fixed flexion, for example, the other hip must first be put into this
identical position. The length of each limb is then measured from

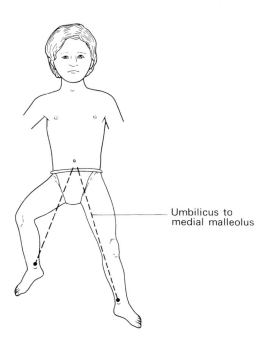

Fig. 148 Apparent shortening—one limb may be apparently shorter than the other because of fixed deformity; the legs in this illustration are actually equal in length but the right is *apparently* considerably shorter because of a gross flexion contracture at the hip. Apparent shortening is measured by comparing the distance from the umbilicus to the medial malleolus on each side.

Umbilicus to medial malleolus

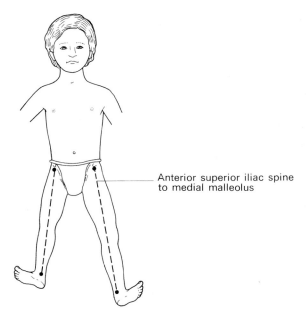

Fig. 149 Measuring real shortening—the patient lies with the pelvis 'square' and the legs placed symmetrically. Measurement is made from the anterior superior spine to the medial malleolus on each side.

Anterior superior iliac spine to medial malleolus

the anterior superior iliac spine to the medial malleolus. In order to obtain identical points on each side, side the finger upwards along Poupart's inguinal ligament and mark the bony point first encountered by the finger. Similarly, slide the finger upwards from just distal to the malleolus to determine the apex of this landmark on each side.

To determine apparent shortening, the patient lies with his legs parallel (as they would be when he stands erect) and the distance from umbilicus to each medial malleolus is measured (Fig. 148).

Now suppose we find 4 in (10 cm) of apparent shortening and 2 in (5 cm) of real shortening of the limb; we interpret this as meaning that 2 in (5 cm) of the shortening is due to true loss of limb length and another 2 in (5 cm) is due to fixed postural deformity.

If the apparent shortening is *less* than the real, this can only mean that the hip has ankylosed in the abducted, and hence apparently elongated, position.

Note this important point: one reason why the orthopaedic surgeon immobilizes a tuberculous hip in the abducted position is that, when the hip becomes ankylosed, shortening due to actual destruction at the hip (i.e. true shortening) will be compensated, to a considerable extent, by the apparent lengthening produced by the fixed abduction.

Having established that there is real shortening present, the examiner must then determine whether this is at the hip, the femur or the tibia, or at a combination of these sites.

At the hip

Place the thumb on the anterior superior spine and the index finger on the greater trochanter on each side; a glance is sufficient to tell if there is any difference between the two sides.

Examiners may still ask about Nelaton's line and Bryant's triangle (Fig. 150).

Nelaton's line joins the anterior superior iliac spine to the ischial tuberosity and should normally lie above the greater trochanter; if the line passes through or below the trochanter, there is shortening at the head or neck of the femur.

Bryant's triangle might better be called 'Bryant's T' because it is not necessary to construct all of its three sides. With the patient supine, a perpendicular is dropped from each anterior superior spine and the distance between this line and the greater trochanter compared on each side. (The third side of the triangle, joining the trochanter to the anterior spine, need never be completed.)

At the femur

Measure the distance from the anterior superior spine (if hip disease has been excluded) or from the greater trochanter to the line of the

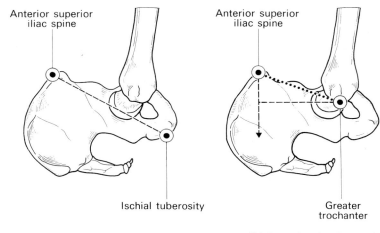

Fig. 150 (a) Nelaton's line joins the anterior superior iliac spine to the ischial *tuberosity*—normally this passes above the greater trochanter. (b) Bryant's triangle— drop a vertical from each superior spine; compare the perpendicular distance from this line to the greater trochanter on either side. (There is no need to complete the third side of the triangle.)

knee joint (not to the patella, whose height can be varied by contraction of the quadriceps).

At the tibia

Compare the distance from the line of the knee joint to the medial malleolus on each side.

Muscles and tendons

Quadriceps femoris forms the prominent muscle mass on the anterior aspect of the thigh; its insertion into the medial aspect of the patella can be seen to extend more distally than on the lateral side. In the well-developed subject, *sartorius* can be defined when the hip is flexed and externally rotated against resistance. It extends from the anterior superior iliac spine to the medial side of the upper end of the tibia and, as the lateral border of the *femoral triangle*; it is an important landmark.

Gluteus maximus forms the bulk of the buttock and can be felt to contract in extension of the hip.

Gluteus medius and *minimus* and the *adductors* can be felt to tighten respectively in resisted abduction and adduction of the hip.

Define the tendons around the knee with this joint comfortably flexed in the sitting position:

• laterally—the *biceps* tendon passes to the head of the fibula, the *iliotibial tract* lies about 0.5 in (12 mm) in front of this tendon and passes to the lateral condyle of the tibia;

• medially — the bulge which one feels is the *semimembranosus insertion* on which two tendons, *semitendinosus* laterally and *gracilis* medially and more anteriorly, are readily palpable.

Between the tendons of biceps and semitendinosus can be felt the heads of origin of *gastrocnemius*. This muscle, with soleus, forms the bulk of the posterior bulge of the calf; the two end distally in the *tendo Achillis* (calcaneal tendon).

At the front of the ankle (Fig. 151) the tendon of *tibialis anterior* lies most medially, passing to its insertion at the base of the first metatarsal and the medial cuneiform. More laterally, the tendons of *extensor hallucis longus* and *extensor digitorum longus* are readily visible in the dorsiflexed foot. *Peroneus longus* and *brevis* tendons

Peroneus brevis

Perforating branch of peroneal artery

Extensor digitorum longus and brevis

Peroneus tertius

Anterior tibial artery

Superior and inferior extensor retinacula

Dorsalis pedis artery

Tibialis anterior

Extensor hallucis longus and brevis

Fig. 151 The structures passing over the dorsum of the ankle.

pass behind the lateral malleolus. Behind the medial malleolus, from the medial to the lateral side, pass the tendons of *tibialis posterior* and *flexor digitorum longus*, the *posterior tibial artery* with its venae comitantes, the *tibial nerve* and, finally, *flexor hallucis longus* (Fig. 152).

Vessels

The *femoral artery* (Fig. 153) can be felt pulsating at the mid-inguinal point, half-way between the anterior superior iliac spine and the pubic symphysis. The upper two-thirds of a line joining this point to the adductor tubercle, with the hip somewhat flexed and externally rotated, accurately defines the surface markings of this vessel. A finger on the femoral pulse lies directly over the head of the femur, immediately lateral to the femoral vein and a finger's breadth medial to the femoral nerve.

The pulse of the *popliteal artery* is often not easy to detect. It is most readily felt with the patient prone, his knee flexed and his muscles relaxed by resting the leg on the examiner's arm. The pulse is sought by firm pressure downwards against the popliteal fossa of the femur.

The pulse of *dorsalis pedis* (Fig. 151) is felt between the tendons of extensor hallucis longus and extensor digitorum on the dorsum of the foot—it is absent in about 2% of normal subjects. The *posterior tibial artery* (Fig. 152) may be felt a finger's breadth below and behind the medial malleolus. In about 1% of healthy subjects this artery is replaced by the peroneal artery.

Fig. 152 The structures passing behind the medial malleolus.

Anterior superior iliac spine

Inguinal ligament

Midline

Femoral artery

Adductor tubercle

Fig. 153 The surface markings of the femoral artery; the upper two-thirds of a line joining the mid-inguinal point (halfway between the anterior superior iliac spine and the symphysis pubis), to the adductor tubercle.

The absence of one or both pulses at the ankle is not, therefore, in itself diagnostic of vascular disease.

The *small (or short) saphenous vein* commences as a continuation of the veins on the lateral side of the dorsum of the foot, runs proximally behind the lateral malleolus, and terminates by draining into the popliteal vein behind the knee. The *great (or long) saphenous vein* arises from the medial side of the dorsal network of veins, passes upwards in front of the medial malleolus, with the saphenous nerve anterior to it, to enter the femoral vein in the groin, one inch below the inguinal ligament and immediately medial to the femoral pulse.

These veins are readily studied in any patient with extensive varicose veins and are usually visible, in their lower part, in the thin normal subject on standing. (The word 'saphenous' is derived from the Greek for 'clear'.)

From the practical point of view, the position of the long saphenous vein immediately in front of the medial malleolus is perhaps the most important single anatomical relationship; no matter how collapsed or obese, or how young and tiny the patient, the vein can be relied upon to be available at this site when urgently required for transfusion purposes (Fig. 154).

Nerves

Only one nerve can be felt in the lower limb; this is the *common peroneal nerve* which can be rolled against the bone as it winds round the neck of the fibula (Fig. 155). Not unnaturally, it may be injured at this site in adduction injuries to the knee or compressed by a tight plaster cast or firm bandage, with a resultant foot drop.

The *femoral nerve* emerges from under the inguinal ligament 0.5 in (12 mm) lateral to the femoral pulse. After a course of only about 2 in (5 cm) the nerve breaks up into its terminal branches.

The surface markings of the *sciatic nerve* (Fig. 156) can be represented by a line which commences at a point midway between the posterior superior iliac spine and the ischial tuberosity, curves outwards and downwards through a point midway between the greater trochanter and ischial tuberosity and then continues vertically downwards in the midline of the posterior aspect of the thigh. The nerve

Great saphenous vein

Medial malleolus

Fig. 154 The relationship of the great (long) saphenous vein to the medial malleolus.

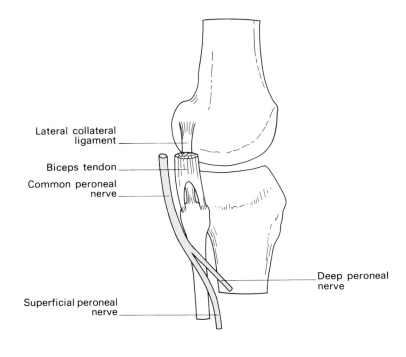

Lateral collateral ligament

Biceps tendon

Common peroneal nerve

Deep peroneal nerve

Superficial peroneal nerve

Fig. 155 The close relationship of the common peroneal nerve to the neck of the fibula; at this site it may be compressed by a tight bandage or plaster cast.

Posterior superior iliac spine

Sciatic nerve

Greater trochanter

Ischial tuberosity

Fig. 156 The surface markings of the sciatic nerve. Join the midpoint between the ischial tuberosity and posterior superior iliac spine to the midpoint between the ischial tuberosity and the greater trochanter by a curved line; continue this line vertically down the leg—it represents the course of the sciatic nerve.

ends at a variable point above the popliteal fossa by dividing into the medial and lateral popliteal nerves, now renamed the *tibial* and *common peroneal nerves* respectively.

It would seem inconceivable that a nerve with such constant and well-defined landmarks could be damaged by intramuscular injections, yet this has happened so frequently that it has seriously been proposed that this site should be prohibited. The explanation is, I believe, a psychological one. The standard advice is to employ the upper outer quadrant of the buttock for these injections, and when

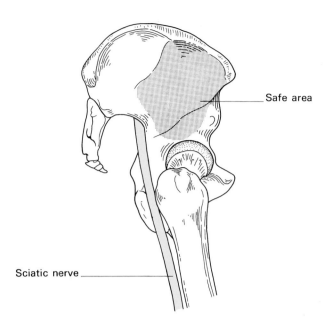

Safe area

Sciatic nerve

Fig. 157 The 'safe area' for injections in the buttock.

the full anatomical extent of the buttock—extending upwards to the iliac crest and outwards to the greater trochanter—is implied, perfectly sound and safe advice this is. Many nurses, however, have an entirely different mental picture of the buttock; a much smaller and more aesthetic affair comprising merely the hillock of the natus. An injection into the upper outer quadrant of this diminutive structure lies in the immediate area of the sciatic nerve!

A better surface marking for the 'safe area' of buttock injections can be defined as that area which lies under the outstretched hand when the thumb and thenar eminence are placed along the iliac crest with the tip of the thumb touching the anterior superior iliac spine (Fig. 157).

The bones and joints of the lower limb

The os innominatum (see 'The pelvis', pages 133–42)

The femur (Figs 158 and 159)

The femur is the largest bone in the body. It is 18 in (45 cm) in length, a measurement it shares with the vas, the spinal cord and the thoracic duct and which is also the distance from the teeth to the cardia of the stomach.

The femoral *head* is two-thirds of a sphere and faces upwards, medially and forwards. It is covered with cartilage except for its central *fovea* where the ligamentum teres is attached.

The *neck* is 2 in (5 cm) long and is set at an angle of 125° to the shaft. In the female, with her wider pelvis, the angle is smaller.

The junction between the neck and the shaft is marked anteriorly by the *trochanteric line*, laterally by the *greater trochanter*, medially and somewhat posteriorly by the *lesser trochanter* and posteriorly by the prominent *trochanteric crest*, which unites the two trochanters.

The blood supply to the femoral head is derived from vessels travelling up from the diaphysis along the cancellous bone, from vessels in the hip capsule, where this is reflected on to the neck in longitudinal bands or retinacula, and from the artery in the ligamentum teres; this third source is negligible (Fig. 160).

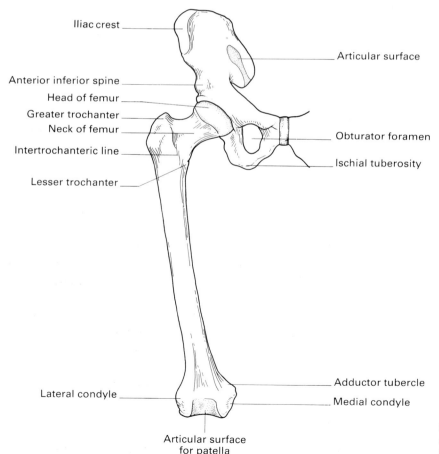

Iliac crest

Articular surface

Anterior inferior spine

Head of femur

Greater trochanter

Neck of femur

Intertrochanteric line

Obturator foramen

Ischial tuberosity

Lesser trochanter

Lateral condyle

Adductor tubercle

Medial condyle

Articular surface
for patella

Fig. 158 The anterior aspect of the right femur.

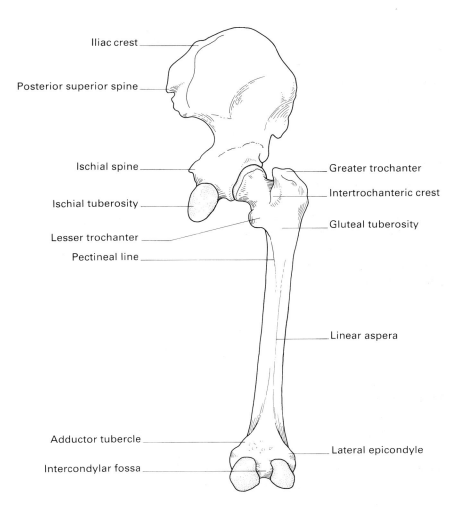

Iliac crest

Posterior superior spine

Ischial spine

Ischial tuberosity

Lesser trochanter

Pectineal line

Greater trochanter

Intertrochanteric crest

Gluteal tuberosity

Linear aspera

Adductor tubercle

Intercondylar fossa

Lateral epicondyle

Fig. 159 The posterior aspect of the right femur.

The femoral *shaft* is roughly circular in section at its middle but is flattened posteriorly at each extremity. Posteriorly also it is marked by a strong crest, the *linea aspera*. Inferiorly, this crest splits into the medial and lateral *supracondylar lines*, leaving a flat *popliteal surface* between them. The medial supracondylar line ends distally in the *adductor tubercle*.

The lower end of the femur bears the prominent *condyles* which are separated by a deep *intercondylar notch* posteriorly but which blend anteriorly to form an articular surface for the patella. The lateral condyle is the more prominent of the two and acts as a buttress to assist in preventing lateral displacement of the patella.

Clinical features

1   The upper end of the femur is a common site for fracture in the

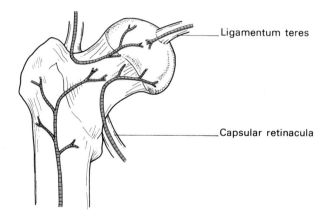

Ligamentum teres

Capsular retinacula

Fig. 160 The sources of
blood supply of the
femoral head—along the
ligamentum teres,
through the diaphysis and
via the retinacula.

elderly. The neck may break immediately beneath the head (subcapital), near its midpoint (cervical) or adjacent to the trochanters (basal),
or the fracture line may pass between, along or just below the
trochanters (Fig. 161).

Fractures of the femoral neck will interrupt completely the blood
supply from the diaphysis and, should the retinacula also be torn,
avascular necrosis of the head will be inevitable. The nearer the fracture to the femoral head, the more tenuous the retinacular blood
supply and the more likely it is to be disrupted.

In contrast, pertrochanteric fractures, being outside the joint
capsule, leave the retinacula undisturbed; avascular necrosis, therefore, never follows such injuries (Fig. 162).

There is a curious age pattern of hip injuries; children may sustain
greenstick fractures of the femoral neck, schoolboys may displace the
epiphysis of the femoral head, in adult life the hip dislocates and, in

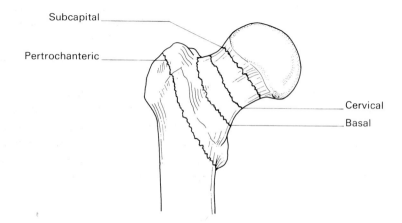

Subcapital

Pertrochanteric

Cervical

Basal

Fig. 161 The head and
neck of the femur,
showing the terminology
of the common fracture
sites.

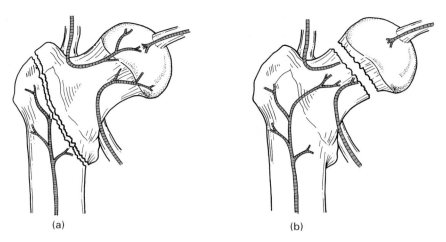

(a)   (b)

Fig. 162 (a) A pertrochanteric fracture does not damage the retinacular blood supply
—aseptic bone necrosis does not occur. (b) A subcapital fracture cuts off most of the
retinacular supply to the head—aseptic bone necrosis is common.

old age, fracture of the neck of the femur again becomes the usual
lesion.

2   Fractures of the femoral shaft are accompanied by considerable
shortening due to the longitudinal contraction of the extremely
strong surrounding muscles.

The proximal segment is flexed by iliacus and psoas and abducted
by gluteus medius and minimus, whereas the distal segment is pulled
medially by the adductor muscles. Reduction requires powerful traction, to overcome the shortening, and then manipulation of the distal
fragment into line with the proximal segment; the limb must therefore
be abducted and also pushed forwards by using a large pad behind the
knee.

Fractures of the lower end of the shaft, immediately above the
condyles, are relatively rare; fortunately so, because they may be
extremely difficult to treat since the small distal fragment is tilted
backwards by gastrocnemius, the only muscle which is attached to it.
The sharp proximal edge of this distal fragment may also tear the
popliteal artery, which lies directly behind it (Fig. 163).

3   The angle subtended by the femoral neck to the shaft may be
decreased, producing a *coxa vara* deformity. This may result from
adduction fractures, slipping of the femoral epiphysis or bone-
softening diseases. *Coxa valga*, where the angle is increased, is much
rarer but occurs in impacted abduction fractures. Note, however, that
in children the normal angle between the neck and shaft is about
160°.

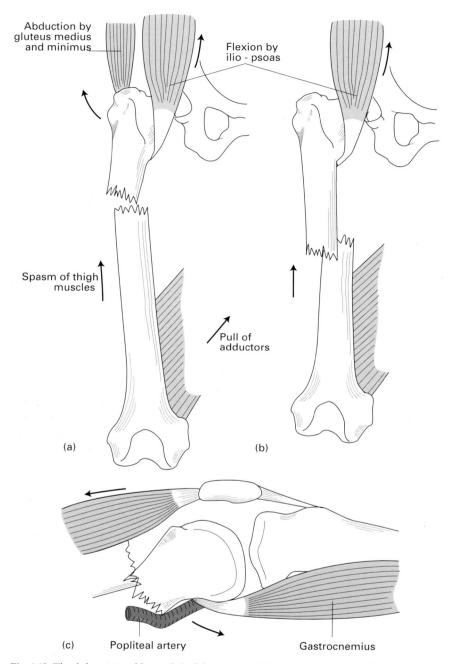

Abduction by
gluteus medius
and minimus

Flexion by
ilio - psoas

Spasm of thigh
muscles

Pull of
adductors

(a)   (b)

(c)   Popliteal artery   Gastrocnemius

Fig. 163 The deformities of femoral shaft fractures. (a) Fracture of the proximal
shaft—the proximal fragment is flexed by iliacus and psoas and abducted by gluteus
medius and minimus. (b) Fracture of the mid-shaft—flexion of the proximal fragment
by iliacus and psoas. (c) Fracture of the distal shaft—the distal fragment is angulated
backwards by gastrocnemius—the popliteal artery may be torn in this injury. (In all
these fractures overriding of the bone ends is produced by muscle spasm.)

The patella

The patella is a sesamoid bone, the largest in the body, in the expansion of the quadriceps tendon, which continues from the apex of the bone as the ligamentum patellae.

The posterior surface of the patella is covered with cartilage and articulates with the two femoral condyles by means of a larger lateral and smaller medial facet.

Clinical features

1 Lateral dislocation of the patella is resisted by the prominent articular surface of the lateral femoral condyle and by the medial pull of the lowermost fibres of vastus medialis which insert almost horizontally along the medial margin of the patella. If the lateral condyle of the femur is underdeveloped, or if there is a considerable *genu valgum* (knock-knee deformity), recurrent dislocations of the patella may occur (Fig. 164).

2 A direct blow on the patella may split or shatter it but the fragments are not avulsed because the quadriceps expansion remains intact.

The patella may also be fractured transversely by violent contraction of the quadriceps—for example, in trying to stop a backward fall. In this case, the tear extends outwards into the quadriceps expansion, allowing the upper bone fragment to be pulled proximally; there may be a gap of over 2 in (5 cm) between the bone ends. Reduction is impossible by closed manipulation and operative repair of the extensor expansion is imperative.

Occasionally this same mechanism of sudden forcible quadriceps contraction tears the quadriceps expansion above the patella, ruptures the ligamentum patellae or avulses the tibial tubercle.

It is interesting that following complete excision of the patella for a comminuted fracture, knee function and movement may return to 100% efficiency; it is difficult, then, to ascribe any particular function to this bone.

The tibia (Fig. 165)

The upper end of the tibia is expanded into the *medial* and *lateral condyles*, the former having the greater surface area of the two. Between the condyles is the *intercondylar area* which bears, at its waist, the *intercondylar eminence*, projecting upwards slightly on either side as the *medial* and *lateral intercondylar tubercles*.

The *tuberosity* of the tibia is at the upper end of the anterior border of the shaft and gives attachment to the ligamentum patellae.

The anterior aspect of this tuberosity is subcutaneous, only excepting the infrapatellar bursa immediately in front of it.

The shaft of the tibia is triangular in cross-section, its anterior border and anteromedial surface being subcutaneous throughout their whole extent.

The posterior surface of the shaft bears a prominent oblique line at its upper end termed the *soleal line*, which not only marks the tibial origin of the soleus but also delimits an area above into which is inserted the popliteus.

The lower end of the tibia is expanded and quadrilateral in section, bearing an additional surface, the *fibular notch*, for the lower tibiofibular joint.

The *medial malleolus* projects from the medial extremity of the bone and is grooved posteriorly by the tendon of tibialis posterior.

The inferior surface of the lower end of the tibia is smooth,

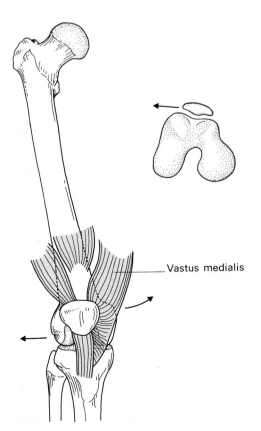

Vastus medialis

Fig. 164 Factors in the stability of the patella: (i) the medial pull of vastus medialis and (ii) the high patellar articular surface of the lateral femoral condyle. These resist the tendency for lateral displacement of the patella which results from the valgus angulation between the femur and the tibia.

Intercondylar eminence

Lateral condyle

Styloid process

Head of fibula

Tuberosity of tibia

Medial condyle

Neck of fibula

Interosseous membrane

Soleal line

Lateral surface

Medial surface

Posterior surface

Medial malleolus

Medial malleolus

Lateral malleolus

Lateral malleolus

Calcaneus

Talus

Fig. 165 The tibia and fibula.

cartilage-covered and forms, with the malleoli, the upper articular surface of the ankle joint.

Clinical features

1   The upper end of the tibial shaft is one of the most common sites for acute osteomyelitis. Fortunately, the capsule of the knee joint is attached closely around the articular surfaces so that the upper

extremity of the tibial diaphysis is extracapsular; involvement of the knee joint therefore only occurs in the late and neglected case.

2 The shaft of the tibia is subcutaneous and unprotected anteromedially throughout its course and is particularly slender in its lower third. It is not surprising that the tibia is the commonest long bone to be fractured and to suffer compound injury.

3 The extensive subcutaneous surface of the tibia makes it a delightfully accessible donor site for bone-grafts.

The fibula (Fig. 166)

The fibula serves three functions. It is:
1 an origin for muscles;
2 a part of the ankle joint;
3 a pulley for the tendons of peroneus longus and brevis.

It comprises the *head* with a *styloid process* (into which is inserted the tendon of biceps), the *neck* (around which passes the common peroneal nerve; Fig. 155), the *shaft* and the *lower end* or *lateral malleolus*. The latter bears a medial roughened surface for the lower tibiofibular joint, below which is the articular facet for the talus. A groove on the posterior aspect of the malleolus lodges the tendons of peroneus longus and brevis.

A note on growing ends and nutrient foramina in the long bones

The shaft of every long bone bears one or more *nutrient foramina*

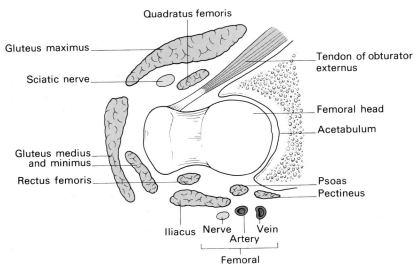

Fig. 166 The immediate relations of the hip joint (in diagrammatic horizontal section).

which are obliquely placed; this obliquity is due to unequal growth at the upper and lower epiphyses. The artery is obviously dragged in the direction of more rapid growth and the direction of slope of entry of the nutrient foramen therefore points *away* from the more rapid growing end of the bone.

The direction of growth of the long bones can be remembered by a little jingle which runs:

'From the knee, I flee

To the elbow, I grow.'

With one exception, the epiphysis of the growing end of a long bone is the first to appear and last to fuse with its diaphysis; the exception is the epiphysis of the upper end of the fibula which, although at the growing end, appears *after* the distal epiphysis and fuses after the latter has blended with the shaft.

The site of the growing end is of considerable practical significance; for example, if a child has to undergo and above-elbow amputation, the humeral upper epiphyseal line continues to grow and the elongating bone may well push its way through the stump end, requiring reamputation.

The bones of the foot

These are best considered as a functional unit and are therefore dealt with together under 'the arches of the foot' (see page 253).

The hip (Figs 166, 167)

The hip is the largest joint in the body. To the surgeon, the examiner and, therefore, the student it is also the most important.

It is a perfect example of a ball-and-socket joint. Its articular surfaces are the femoral head and the horse-shoe shaped articular surface of the acetabulum, which is deepened by the fibrocartilaginous *labrum acetabulare*. The non-articular lower part of the acetabulum, the *acetabular notch*, is closed off below by the *transverse acetabular ligament*. From this notch is given off the *ligamentum teres*, passing to the fovea on the femoral head.

The *capsule* of the hip is attached proximally to the margins of the acetabulum and to the transverse acetabular ligament. Distally, it is attached along the trochanteric line, the bases of the greater and lesser trochanters and, posteriorly, to the femoral neck about 0.5 in (12 mm) from the trochanteric crest. From this distal attachment, capsular fibres are reflected on to the femoral neck as *retinacula* and provide one pathway for the blood supply to the femoral head (see 'The femur', page 233; Fig. 160).

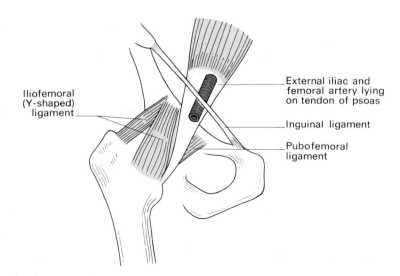

Iliofemoral (Y-shaped) ligament

External iliac and femoral artery lying on tendon of psoas

Inguinal ligament

Pubofemoral ligament

Fig. 167 The anterior aspect of the hip. Note that the psoas tendon and the femoral artery are intimate anterior relations of the joint.

Note that acute osteomyelitis of the upper femoral metaphysis will involve the neck which is intracapsular and which will therefore rapidly produce a secondary pyogenic arthritis of the hip joint.

Three ligaments reinforce the capsule:

1 the *iliofemoral* (Y-shaped ligament of Bigelow) — which arises from the anterior inferior iliac spine, bifurcates, and is inserted at each end of the trochanteric line (Fig. 167);

2 the pubofemoral — arising from the iliopubic junction to blend with the medial aspect of the capsule;

3 the *ischiofemoral*—arising from the ischium to be inserted into the base of the greater trochanter.

Of these, the iliofemoral is by far the strongest and resists hyper-extension strains on the hip. In posterior dislocation it usually remains intact.

The *synovium* of the hip covers the non-articular surfaces of the joint and occasionally bulges out anteriorly to form a bursa beneath the psoas tendon where this crosses the front of the joint.

Movements

The hip is capable of a wide range of movements—flexion, extension, abduction, adduction, rotation and circumduction.

The principal muscles acting on the joint are:

• flexors—iliacus and psoas major assisted by rectus femoris, sartorius, pectineus;

• extensors—gluteus maximus, the hamstrings;

• adductors—adductor longus, brevis and magnus assisted by gracilis and pectineus;
• abductors—gluteus medius and minimus, tensor fasciae latae;
• lateral rotators — principally gluteus maximus assisted by the obturators, gemelli and quadratus femoris;
• medial rotators—tensor fasciae latae and anterior fibres of gluteus medius and minimus.

Relations (Fig. 166)

The hip joint is surrounded by muscles:
• anteriorly—iliacus, psoas and pectineus, together with the femoral artery and vein;
• laterally—tensor fasciae latae, gluteus medius and minimus;
• posteriorly — the tendon of obturator internus with the gemelli, quadratus femoris, the sciatic nerve and, more superficially, gluteus maximus;
• superiorly — the reflected head of rectus femoris lying in contact with the joint capsule;
• inferiorly — the obturator externus, passing back to be inserted into the trochanteric fossa.

Surgical exposure of the hip joint therefore inevitably involves considerable and deep dissection.

The *lateral approach* comprises splitting down through the fibres of tensor fasciae latae, gluteus medius and minimus on to the femoral neck. Further access may be obtained by detaching the greater trochanter with the gluteal insertions.

The *anterior approach* passes between gluteus medius and minimus laterally and sartorius medially, then dividing the reflected head of rectus femoris to expose the anterior aspect of the hip joint. More room may be obtained by detaching these glutei from the external aspect of the ilium.

The *posterior approach* is through an angled incision commencing at the posterior superior iliac spine, passing to the greater trochanter and then dropping vertically downwards from this point.

Gluteus maximus is split in the line of its fibres and then incised along its tendinous insertion. Gluteus medius and minimus are detached from their insertions into the greater trochanter (or the trochanter is detached and subsequently wired back in place), and an excellent view of the hip joint is thus obtained.

Nerve supply

Hilton's law states that the nerves crossing a joint supply the muscles acting on it and the joint itself. The hip is no exception and receives fibres from the femoral, sciatic and obturator nerves. It is important to note that these nerves also supply the knee joint and, for this reason, it is not uncommon for a patient, particularly a child, to complain bitterly of pain in the knee and for the cause of the mischief, the diseased hip, to be overlooked.

Clinical features

Trendelenburg's test

The stability of the hip in the standing position depends on two factors, the strength of the surrounding muscles and the integrity of the lever system of the femoral neck and head within the intact hip joint. When standing on one leg, the abductors of the hip on this side (gluteus medius and minimus and tensor fasciae latae) come into powerful action to maintain fixation at the hip joint, so much so that the pelvis actually rises slightly on the opposite side. If, however, there is any defect in these muscles or lever mechanism of the hip joint, the weight of the body in these circumstances forces the pelvis to tilt downwards on the opposite side.

This positive Trendelenburg test is seen if the hip abductors are paralysed (e.g. poliomyelitis), if there is an old unreduced or congenital dislocation of the hip, if the head of the femur has been destroyed by disease or removed operatively (pseudarthrosis), if there is an ununited fracture of the femoral neck or if there is a very severe degree of coxa vara.

The test may be said to indicate 'a defect in the osseo-muscular stability of the hip joint'.

A patient with any of the conditions enumerated above walks with a characteristic 'dipping gait'.

Dislocation of the hip (Fig. 168)

The hip is usually dislocated backwards and this is produced by a force applied along the femoral shaft with the hip in the flexed position (e.g. the knee striking against the opposite seat when a train runs into the buffers). If the hip is also in the adducted position, the head of the femur is unsupported posteriorly by the acetabulum and dislocation can occur without an associated acetabular fracture. If the hip is abducted, dislocation must be accompanied by a fracture of the posterior acetabular lip.

The sciatic nerve, a close posterior relation of the hip, is in danger

(a)

(b)

(c)

(d)

Fig. 168 Dislocation of the hip. If the hip is forced into posterior dislocation while adducted (a), there is no associated fracture of the posterior acetabular lip (b). Dislocation in the abducted position (c) can only occur with a concomitant acetabular fracture (d). (The inset figure indicates the plane of these diagrams.)

of damage in these injuries, as will be appreciated by a glance at Fig. 156.

Reduction of a dislocated hip is quite simple providing that a deep anaesthetic is used to relax the surrounding muscles; the hip is flexed, rotated into the neutral position and lifted back into the acetabulum. Occasionally, forcible abduction of the hip will dislocate the hip forwards. Violent force along the shaft (e.g. a fall from a height) may thrust the femoral head through the floor of the acetabulum, producing a central dislocation of the hip.

The knee joint (Figs 169, 170)

The knee is a hinge joint made up of the articulations between the

Lateral collateral ligament

Lateral semilunar cartilage

Popliteus tendon

Biceps tendon

Posterior
Anterior } Cruciate ligaments

Medial semilunar cartilage

Medial collateral ligament

Patellar ligament

Patella

Fig. 169 The knee—anterior view; the knee is flexed and the patella has been turned downwards.

Transverse ligament of knee

Medial meniscus

Posterior cruciate

Tuberosity of tibia

Anterior cruciate

Lateral meniscus

Fig. 170 The right knee in transverse section.

femoral and tibial condyles and between the patella and the patellar surface of the femur.

The *capsule* is attached to the margins of these articular surfaces but communicates above with the *suprapatellar bursa* (between the lower femoral shaft and the quadriceps), posteriorly with the bursa under the medial head of gastrocnemius and often, through it, with the bursa under semimembranosus. It may also communicate with the bursa under the lateral head of gastrocnemius. The capsule is also perforated posteriorly by popliteus, which emerges from it in much the same way that the long head of biceps bursts out of the shoulder joint.

The capsule of the knee joint is reinforced on each side by the *medial* and *lateral collateral ligaments*, the latter passing to the head of the fibula and lying free from the capsule.

Anteriorly, the capsule is considerably strengthened by the *ligamentum patellae*, and, on each side of the patella, by the *medial* and *lateral patellar retinacula*, which are expansions from vastus medialis and lateralis.

Posteriorly, the tough *oblique ligament* arises as an expansion from the insertion of semimembranosus and blends with the joint capsule.

Internal structures

Within the joint are a number of important structures.

The *cruciate ligaments* are extremely strong connections between the tibia and femur. They arise from the anterior and posterior intercondylar areas of the superior aspect of the tibia, taking their names from their tibial origins, and pass obliquely upwards to attach to the intercondylar notch of the femur.

The anterior ligament resists forward displacement of the tibia or the femur and becomes taut in hyperextension of the knee, the posterior resists backward displacement of the tibia and becomes taut in hyperflexion.

The *semilunar cartilages* (*menisci*) are crescent-shaped and are triangular in cross-section, the medial being larger and less curved than the lateral. They are attached by their extremities to the tibial intercondylar area and by their periphery to the capsule of the joint, although the lateral cartilage is only loosely adherent and the popliteus tendon intervenes between it and the lateral collateral ligament.

They deepen, although to only a negligible extent, the articulations between the tibial and femoral condyles and probably act as shock absorbers. If both menisci are removed, the knee can regain complete functional efficiency, although it is interesting that, following surgery, a rim of fibrocartilage regenerates from the connective tissue margin of the excised menisci.

An *infrapatellar pad of fat* fills the space between the ligamentum patellae and the femoral intercondylar notch. The synovium covering this pad projects into the joint as two folds termed the *alar folds*.

Movements of the knee

The principal knee movements are flexion and extension, but note on yourself that rotation of the knee is possible when this joint is in the flexed position. In full extension, i.e. in the standing position, the knee is quite rigid because the medial condyle of the tibia, being rather larger than the lateral condyle, rides forward on the medial femoral condyle, thus 'screwing' the joint firmly together. The first step in flexion of the fully extended knee is 'unscrewing' or internal rotation. This is brought about by *popliteus*, which arises from the lateral side of the lateral condyle of the femur, emerges from the joint capsule posteriorly and is inserted into the back of the upper end of the tibia.

The principal muscles acting on the knee are:

- extensor—quadriceps femoris;
- flexors — hamstrings assisted by gracilis, gastrocnemius and sartorius;
- medial rotator—popliteus.

Clinical features

1   The stability of the knee depends upon the strength of its surrounding muscles and of its ligaments. Of the two, the muscles are by far the more important. Providing quadriceps femoris is powerfully developed, the knee will function satisfactorily even in the face of considerable ligamentous damage. Conversely, the most skilful surgical repair of torn ligaments is doomed to failure unless the muscles are functioning strongly; without their support, reconstructed ligaments will merely stretch once more.

2   When considering soft tissue injuries of the knee joint, think of three Cs that may be damaged—the Collateral ligaments, the Cruciates and the Cartilages.

The *collateral ligaments* are taut in full extension of the knee and are, therefore, only liable to injury in this position. The medial ligament may be partly or completely torn when a violent abduction strain is applied, whereas an adduction force may damage the lateral ligament. If one or other collateral ligament is completely torn, the extended knee can be rocked away from the affected side.

The *cruciate ligaments* may both be torn (along with the collateral ligaments) in severe abduction or adduction injuries. The anterior cruciate, which is taut in extension, may be torn by violent hyperextension of the knee or in anterior dislocation of the tibia on the femur. The posterior cruciate tears in a posterior dislocation.

If both the cruciate ligaments are torn, unnatural anteroposterior mobility of the knee can be demonstrated.

If there is only increased forward mobility, the anterior cruciate ligament has been divided or is lax. Increased backward mobility implies a lesion of the posterior cruciate.

The *semilunar cartilages* can only tear when the knee is flexed and is thus able to rotate. If you place a finger on either side of the ligamentum patellae on the joint line and then rotate your flexed knee first internally and then externally, you will note how the lateral and medial cartilages are respectively sucked into the knee joint. If the flexed knee is forcibly abducted and externally rotated, the medial cartilage will be drawn between, and then split by, the grinding surfaces of the medial condyles of the femur and tibia. This occurs when a footballer twists his flexed knee while running or when a miner topples over in the crouched position while hewing coal in a narrow seam. A severe adduction and internal rotation strain may similarly tear the lateral cartilage, but this injury is less common.

The knee 'locks' in this type of injury because the torn and displaced segment of cartilage lodges between the condyles and prevents full extension of the knee.

The tibiofibular joints

The tibia and fibula are connected by:
1 the *superior tibiofibular joint*, a synovial joint between the head of the fibula and the lateral condyle of the tibia;
2 the *interosseous membrane*, which is crossed by the anterior tibial vessels above and pierced by the perforating branch of the peroneal artery below;
3 the *inferior tibiofibular joint*, a fibrous joint, the only one in the limbs, between the triangular areas of each bone immediately above the ankle joint.

The ankle (Fig. 171)

The ankle is a hinge joint between a mortice formed by the malleoli and lower end of the tibia and the body of the talus.

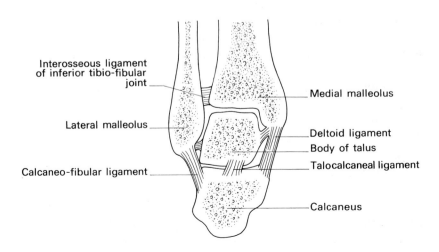

Fig. 171 The ankle in coronal section.

The *capsule* of the joint fits closely around its articular surfaces, and, as in every hinge joint, it is weak anteriorly and posteriorly but reinforced laterally and medially by *collateral ligaments*.

Movements of the ankle

The ankle joint is capable of being flexed and extended (plantar- and dorsiflexion).

The body of the talus is slightly wider anteriorly and, in full extension, becomes firmly wedged between the malleoli. Conversely, in flexion, there is slight laxity at the joint and some degree of side to side tilting is possible: test this fact on yourself.

The principal muscles acting on the ankle are:
•   dorsiflexors — tibialis anterior assisted by extensor digitorum longus and extensor hallucis longus, peroneus tertius;
•   plantarflexors—gastrocnemius and soleus assisted by tibialis posterior, flexor hallucis longus and flexor digitorum longus.

Clinical features

1   The collateral ligaments of the ankle can be sprained or completely torn by forcible abduction or adduction, the lateral ligament being far the more frequently affected. If the ligament is completely disrupted the talus can be tilted in its mortice; this is difficult to demonstrate clinically and is best confirmed by taking an anteroposterior radiograph of the ankle while forcibly inverting the foot.

2   The most usual ankle fracture is that produced by an abduction-external rotation injury; the patient catches his foot in a rabbit hole, his body and his tibia internally rotate while the foot is rigidly held. First there is a torsional spinal fracture of the lateral malleolus, then avulsion of the medial collateral ligament, with or without avulsion of a flake of the medial malleolus and, finally, as the tibia is carried forwards, the posterior margin of the lower end of the tibia shears off against the talus. These stages are termed *1st*, *2nd* and *3rd degree Pott's fractures*. Notice that, with widening of the joint, there is forward dislocation of the tibia on the talus, producing characteristic prominence of the heel in this injury.

The joints of the foot

Inversion and eversion of the foot take place at the talocalcaneal articulations and at the *mid-tarsal joints* between the calcaneum and the cuboid and between the talus and the navicular.

Loss of these rotatory movements of the foot, e.g. after injury or because of arthritis, results in quite severe disability because the foot cannot adapt itself to walking on rough or sloping ground.

Inversion is brought about by tibialis anterior and posterior assisted by the long extensor and flexor tendons of the hallux; eversion is the duty of the peronei.

The other tarsal joints allow slight gliding movements only, and individually, are not of clinical importance. The arrangement of the metacarpophalangeal and interphalangeal joints is on the same basic plan as in the upper limb.

The arches of the foot (Fig. 172)

On standing, the heel and the metatarsal heads are the principal weight-bearing points, but a moment's study of footprints on the wet bathroom floor will show that the lateral margin of the foot and the tips of the phalanges also touch the ground.

The bones of the foot are arranged in the form of two longitudinal

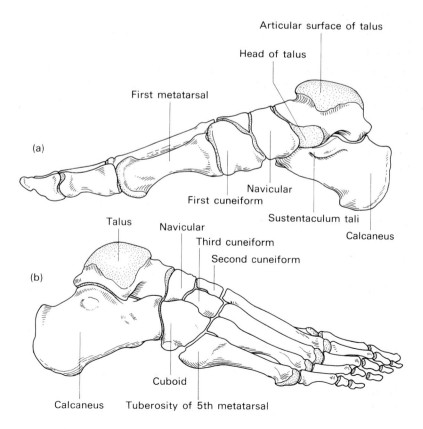

Fig. 172 The longitudinal arches of the right foot.
(a) Medial view.
(b) Lateral view.

arches. The *medial arch* comprises calcaneus, talus, navicular, the three cuneiforms and the three medial metatarsals; the apex of this arch is the talus. The *lateral arch*, which is lower, comprises the calcaneum, cuboid and the lateral two metatarsals.

The foot plays a double role; it functions as a rigid support for the weight of the body in the standing position, and as a mobile springboard during walking and running.

When one stands, the arches sink somewhat under the body's weight, the individual bones lock together, the ligaments linking them are at maximum tension and the foot becomes an immobile pedestal. When one walks, the weight is released from the arches, which unlock and become a mobile lever system in the spring-like actions of locomotion.

The arches are maintained by:
1 the shape of the interlocking bones;
2 the ligaments of the foot;
3 muscle action.

The ligaments concerned are (Fig. 173):
1 the dorsal, plantar and interosseous ligaments between the small bones of the forefoot;
2 the *spring ligament*, which passes from the *sustentaculum tali* of the calcaneus forward to the *tuberosity of the navicular* and which supports the inferior aspect of the head of the talus;
3 the *short plantar ligament* which stretches from the plantar surface of the calcaneus to the cuboid;

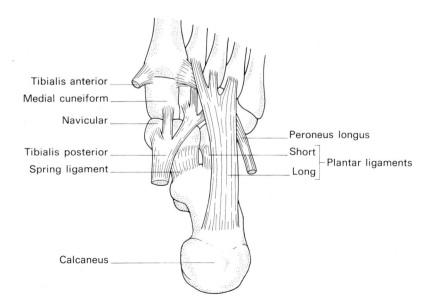

Fig. 173 Plantar aspect of the left foot to show the attachments of the important ligaments and long tendons.

4   the *long plantar ligament* which arises from the plantar surface of the calcaneus, covers the short plantar ligament, forms a tunnel for peroneus longus tendon with the cuboid, and is inserted into the bases of the 2nd, 3rd and 4th metatarsals.

These ligaments are reinforced in their action by the *plantar aponeurosis* which is the condensed deep fascia of the sole of the foot. This arises from the plantar aspect of the calcaneum and is attached to the deep transverse ligaments linking the heads of the metatarsals; it also continues forward into each toe to form the fibrous flexor sheaths, in a similar arrangement to that of the palmar fascia of the hand.

The principal muscles concerned in the mechanism of the arches of the foot are peroneus longus, tibialis anterior and posterior, flexor hallucis longus and the intrinsic muscles of the foot.

Peroneus longus tendon passes obliquely across the sole in a groove on the cuboid bone and is inserted into the lateral side of the base of the 1st metatarsal and the medial cuneiform. Into the medial aspect of these two bones is inserted the tendon of *tibialis anterior* so that these muscles form, in effect, a stirrup between them which supports the arches of the foot.

The medial arch is further reinforced by flexor hallucis longus, whose tendon passes under the sustentaculum tali of the calcaneus, and by *tibialis posterior*, two-thirds of whose fibres are inserted into the tuberosity of the navicular and support the spring ligament.

The longitudinally running intrinsic muscles of the foot also act as ties to the longitudinal arches.

The anatomy of walking

In the process of walking, the heel is raised from the ground, the metatarsophalangeal joints flex to give a 'push off' movement; the foot then leaves the ground completely and is dorsiflexed to clear the toes.

Just before the toes of one foot leave the ground, the heel of the other makes contact.

Forward progression is produced partly by the 'push off' of the toes, partly by powerful plantarflexion of the ankle and partly by the forward swing of the hips accentuated by swinging movements of the pelvis. Paraplegics can be taught to walk purely by this pelvic swing action, even though paralysed from the waist downwards.

When one foot is off the ground, dropping of the pelvis to the unsupported side is prevented by the hip abductors (gluteus medius and minimus and tensor fasciae latae). Their paralysis is one cause of a 'dipping gait' and of a positive Trendelenburg sign (see page 246).

Three important zones of the lower limb—the femoral triangle, adductor canal and popliteal fossa

The femoral triangle (Fig. 174)

This triangle is bounded:
- superiorly—by the inguinal ligament;
- medially—by the medial border of adductor longus;
- laterally—by the medial border of sartorius.

Its *floor* consists of iliacus, the tendon of psoas, pectineus and adductor longus.

The *roof* is formed by the superficial fascia, containing the superficial inguinal lymph nodes and the great saphenous vein with its tributaries, and the deep fascia (fascia lata), which is pierced by the saphenous vein at the saphenous opening.

The *contents* of the triangle are the femoral vein, artery and nerve together with the deep inguinal nodes.

Some of these structures must now be considered in greater detail.

The fascia lata

The deep fascia of the thigh, or fascia lata, extends downwards to

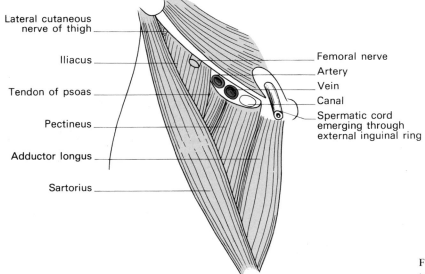

Lateral cutaneous nerve of thigh

Iliacus

Tendon of psoas

Pectineus

Adductor longus

Sartorius

Femoral nerve
Artery
Vein
Canal
Spermatic cord emerging through external inguinal ring

Fig. 174 The femoral triangle and its contents.

ensheath the whole lower limb except over the subcutaneous surface of the tibia (to whose margins it adheres), and at the saphenous opening. Above, it is attached all around to the root of the lower limb — that is to say, to the inguinal ligament, pubis, ischium, sacrotuberous ligament, sacrum and coccyx and the iliac crest. The fascia of the thigh is particularly dense laterally (the *iliotibial tract*), where it receives tensor fasciae latae, and posteriorly, where the greater part of gluteus maximus is inserted into it. The iliotibial tract, when tensed by its attached muscles, assists in the stabilization of the hip and the extended knee.

The tough lateral fascia of the thigh is an excellent source of this material for hernia and dural repairs.

The femoral sheath and femoral canal (Fig. 175)

The femoral artery and vein enter the femoral triangle from beneath the inguinal ligament within a fascial tube termed the *femoral sheath*. This is derived from the extraperitoneal intra-abdominal fascia, its anterior wall arising from the transversalis fascia and its posterior wall from the fascia covering the iliacus.

The medial part of the femoral sheath contains a small, almost vertically placed gap, the *femoral canal*, which is about 0.5 in (12 mm) in length and which just admits the tip of the little finger. The

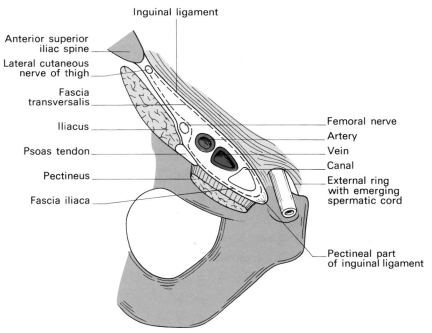

Inguinal ligament

Anterior superior iliac spine

Lateral cutaneous nerve of thigh

Fascia transversalis

Iliacus

Psoas tendon

Pectineus

Fascia iliaca

Femoral nerve

Artery

Vein

Canal

External ring with emerging spermatic cord

Pectineal part of inguinal ligament

Fig. 175 The femoral canal and its surrounds.

greater width of the female pelvis means the canal is somewhat larger in the female and femoral herniae are, consequently, commoner in this sex.

The boundaries of the femoral canal are:
- anteriorly—the inguinal ligament;
- medially—the sharp free edge of the pectineal part of the inguinal ligament, termed the *lacunar ligament* (*Gimbèrnat's ligament*);
- laterally—the femoral vein;
- posteriorly—the pectineal ligament (of Astley Cooper), which is the thickened periosteum along the pectineal border of the superior pubic ramus and which continues medially with the pectineal part of the inguinal ligament.

The canal contains a plug of fat and a constant lymph node—the *node of the femoral canal* or *Cloquet's gland*.

The canal has two functions: first, as a dead space for expansion of the distended femoral vein and, second, as a lymphatic pathway from the lower limb to the external iliac nodes.

Femoral hernia

The great importance of the femoral canal is, of course, that it is a potential point of weakness in the abdominal wall through which may develop a femoral hernia. Unlike the indirect inguinal hernia, this is never due to a congenital sac and, although cases do occur rarely in children, it is never found in the newborn.

As the hernia sac enlarges, it emerges through the saphenous opening then turns upwards along the pathway presented by the superficial epigastric and superficial circumflex iliac vessels so that it may come to project above the inguinal ligament. There should not, however, be any difficulty in differentiating between an irreducible femoral and inguinal hernia; the neck of the former must always lie below and lateral to the pubic tubercle whereas the sac of the latter extends above and medial to this landmark (Fig. 176).

The neck of the femoral canal is narrow and bears a particular sharp medial border; for this reason, irreducibility and strangulation occur more commonly at this site than at any other. In order to enlarge the opening of the canal at operation on a strangulated case, this sharp edge of Gimbernat's lacunar ligament may require incision; there is a slight risk of damage to the abnormal obturator artery in this manoeuvre and it is safer to enlarge the opening by making several small nicks into the ligament. The safe alternative is to divide the inguinal ligament, which can then be repaired.

Note. Normally there is an anastomosis between the pubic branch of the inferior epigastric artery and the pubic branch of the obturator

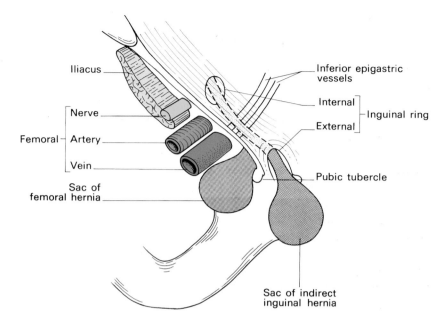

Fig. 176 The relationship of an indirect inguinal and a femoral hernia to the pubic tubercle; the inguinal hernia emerges above and medial to the tubercle, the femoral hernia lies below and lateral to it.

artery. Occasionally the obturator artery is entirely replaced by this branch from the inferior epigastric — the *abnormal obturator artery*. This aberrant vessel usually passes laterally to the femoral canal and is out of harm's way; more rarely, it passes behind Gimbernat's ligament and it is then in surgical danger.

The lymph nodes of the groin and the lymphatic drainage of the lower limb

The lymph nodes of the groin are arranged in a superficial and a deep group. The *superficial nodes* lie in two chains, a *longitudinal chain* along the great saphenous vein, receiving the bulk of the superficial lymph drainage of the lower limb, and a *horizontal chain*, just distal to the inguinal ligament. These horizontal nodes receive lymphatics from the skin and superficial tissues of:

1 the lower trunk and back, below the level of the umbilicus;
2 the buttock;
3 the perineum, scrotum and penis (or lower vagina and vulva) and the anus below its mucocutaneous junction.

In addition, some lymphatics drain via the round ligament to these nodes from the fundus of the uterus.

(All these sites, as well as the whole leg, must be examined carefully when a patient presents with an inguinal lymphadeno-pathy.)

The two groups of superficial nodes drain through the saphenous opening in the fascia lata into the *deep nodes* lying medial to the femoral vein, which also receive the lymph drainage from the tissues of the lower limb beneath the deep fascia. In addition, a small area of skin over the heel and lateral side of the foot drains by lymphatics along the small saphenous vein to nodes in the popliteal fossa and then, along the femoral vessels, directly to the deep nodes at the groin.

The deep groin nodes drain to the external iliac nodes by lymphatics which travel partly in front of the femoral artery and vein and partly through the femoral canal.

Clinical features

1   Minor sepsis and abrasions of the leg are so common that it is usual to find that the inguinal nodes are palpable in perfectly healthy people.
2   Secondary involvement of the inguinal nodes by malignant deposits may be dealt with by *block dissection of the groin*. This involves removal of the superficial and deep fascial roof of the femoral triangle, the saphenous vein and its tributaries and the fatty and lymphatic contents of the triangle, leaving only the femoral artery, vein and nerve. The inguinal ligament is detached so that, in addition, an extraperitoneal removal of the external iliac nodes can be carried out.
3   In making a differential diagnosis of a lump in the femoral triangle, think of each anatomical structure and of the pathological conditions to which it may give rise, thus:
• skin and soft tissues—lipoma, sebaceous cyst, sarcoma;
• artery—aneurysm of the femoral artery;
• vein—varix of the great saphenous vein;
• nerve—neuroma of the femoral nerve or its branches;
• femoral canal—femoral hernia;
• psoas sheath—psoas abscess;
• lymph nodes—any of the causes of lymphadenopathy.

The adductor canal (of Hunter) or subsartorial canal (Fig. 177)

This canal leads on from the apex of the femoral triangle. Its boundaries are:
• posteriorly—adductor longus and magnus;

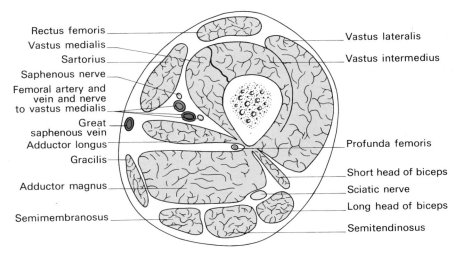

Fig. 177 Cross-section through the thigh in the region of the adductor, or subsartorial, canal of Hunter.

- anterolaterally—vastus medialis;
- anteromedially — the sartorius, which lies on a fascial sheet forming the roof of the canal.

The contents of the canal are the femoral artery, the femoral vein (which lies behind the artery) and the saphenous nerve.

John Hunter described the exposure and ligation of the femoral artery in this canal for aneurysm of the popliteal artery; this method has the advantage that the artery at this site is healthy and will not tear when tied, as may happen if ligation is attempted immediately above the aneurysm.

The popliteal fossa (Fig. 178)

The popliteal fossa is the distal continuation of the adductor canal. This 'fossa' is, in fact, a closely packed compartment which only becomes the rhomboid-shaped space of anatomical diagrams when opened up at operation or by dissection.

Its boundaries are:
- superolaterally—biceps tendon;
- superomedially — semimembranosus reinforced by semitendinosus;
- inferomedially and inferolaterally—the medial and lateral heads of gastrocnemius.

The *roof* of the fossa is deep fascia which is pierced by the small saphenous vein as this enters the popliteal vein.

Its *floor*, from above down, is formed by:

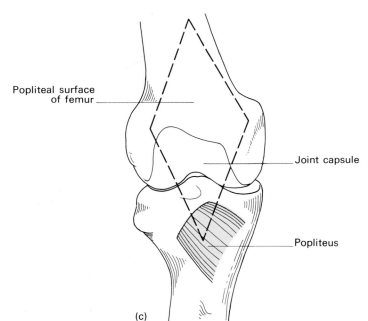

Fig. 178 The popliteal fossa. (a) Superficial dissection. (b) Deep dissection. (c) Floor.

- the popliteal surface of the femur;
- the posterior aspect of the knee joint;
- the popliteus muscle covering the upper posterior surface of the tibia.

From without in, the popliteal fossa contains nerves, vein and artery.

The *common peroneal nerve* passes out of the fossa along the medial border of the biceps tendon; the *tibial nerve* is first lateral to the popliteal vessels and then crosses superficially to these vessels to lie on their medial side.

The *popliteal vein* lies immediately superficial to the artery; the *popliteal artery* itself lies deepest of all in the fossa.

As well as these important structures, the fossa contains fat and the popliteal lymph nodes.

Clinical features

The popliteal fossa is another good example of the value of thinking anatomically when considering the differential diagnosis of a mass situated in a particular anatomical area.

When examining a lump in the popliteal region, let these possibilities pass through your mind:
- skin and soft tissues—sebaceous cyst, lipoma, sarcoma;
- vein—varicosities of the short saphenous vein in the roof of the fossa;
- artery—popliteal aneurysm;
- lymph nodes—infection secondary to suppuration in the foot;
- knee joint—joint effusion;
- tendons—enlarged bursae, especially those beneath semimembranosus and the heads of gastrocnemius;
- bones—a tumour of the lower end of femur or upper end of tibia.

The arteries of the lower limb

Femoral artery

The femoral artery is the distal continuation of the external iliac artery beyond the inguinal ligament. It traverses the femoral triangle and the adductor canal of Hunter, then terminates a hand's breadth above the adductor tubercle by passing through the hiatus in adductor magnus to become the popliteal artery (Fig. 153).

Throughout its course, the femoral artery is accompanied by its vein, which lies first on the medial side of the artery and then passes posteriorly to it at the apex of the femoral triangle.

Branches

In the groin, the femoral artery gives off:
1   the superficial circumflex iliac artery;
2   the superficial epigastric artery;
3   the superficial external pudendal artery.

These three vessels are encountered in the groin incision for repair for an inguinal hernia. Their corresponding veins drain into the great saphenous vein (see Fig. 180).

The *profunda femoris* arises posterolaterally from the femoral artery 2 in (5 cm) distal to the inguinal ligament. It is conventional to call the femoral artery above this branch the *common femoral*, and below it, the *superficial femoral artery*.

The profunda passes deep to adductor longus and gives off *medial* and *lateral circumflex branches* and four *perforating branches*. These are important both as the source of blood supply to the great muscles of the thigh and as collateral channels which link the rich arterial anastomoses around the hip and the knee.

Clinical features

1   Recapitulate the surface markings of the femoral artery — the upper two-thirds of a line connecting the mid-inguinal point with the adductor tubercle, the hip being held somewhat flexed and externally rotated (Fig. 153).

The femoral artery in the upper 4 in (10 cm) of its course lies in the femoral triangle where it is quite superficial and, in consequence, easily injured. A laceration of the femoral artery at this site is an occupational hazard of butchers and bullfighters.

2   The femoral artery at the groin is readily punctured by a hypodermic needle and is the most convenient site from which to obtain arterial blood samples. Arteriography of the peripheral leg vessels is also easily performed at this point. A Seldinger catheter can be passed proximally through a femoral artery puncture in order to carry out aortography or selective renal, coeliac and mesenteric angiography.

3   Arteriosclerotic changes, with consequent thrombotic arterial occlusion, frequently commence at the lower end of the femoral artery, perhaps as a result of compression of the diseased vessel by the margins of the hiatus in adductor magnus. Collateral circulation is maintained via anastomoses between the branches of profunda femoris and the popliteal artery. If arteriography demonstrates a patent arterial tree distal to the block, it is possible to bypass the

occluded segment by means of a graft between the common femoral and popliteal arteries.

Popliteal artery

The popliteal artery continues on from the femoral artery at the adductor hiatus and terminates at the lower border of the popliteus muscle. It lies deep within the popliteal fossa (see above), being covered superficially by the popliteal vein and, more superficially still, crossed by the tibial nerve.

The popliteal artery gives off muscular branches, geniculate branches (to the knee joint) and terminal branches, the anterior and posterior tibial arteries.

Clinical features

1   Aneurysm of the popliteal artery, once common, is now rare. Its frequency in former days was associated with the repeated traumata of horse-riding and the wearing of high riding-boots.

Pressure of the aneurysm on the adjacent vein may cause venous thrombosis and peripheral oedema; pressure on the tibial nerve may cause severe pain in the leg.

2   The popliteal artery is exposed by deep dissection in the midline within the popliteal fossa, care being taken not to injure the more superficial vein and nerve. It can also be exposed by a medial approach, which divides the insertion of adductor longus and detaches the origin of the medial head of gastrocnemius from the tibia.

Posterior tibial artery

The posterior tibial artery is the larger of the terminal branches of the popliteal artery. It descends deep to soleus, where it can be exposed by splitting gastrocnemius and soleus in the midline, then becomes superficial in the lower third of the leg and passes behind the medial malleolus between the tendons of flexor digitorum longus and flexor hallucis longus. It is accompanied by its corresponding vein and by the tibial nerve (Fig. 179).

Below the ankle, the posterior tibial artery divides into the *medial* and *lateral plantar arteries* which constitute the principal blood supply to the foot.

As well as branches to muscles and skin and a large nutrient branch to the tibia, the posterior tibial artery gives off the *peroneal*

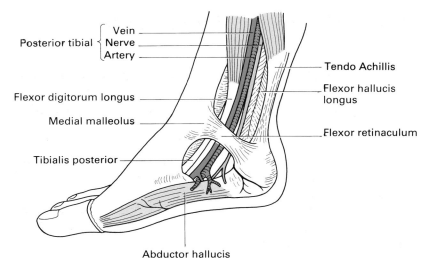

Posterior tibial { Vein
Nerve
Artery

Tendo Achillis

Flexor hallucis longus

Flexor digitorum longus

Medial malleolus

Flexor retinaculum

Tibialis posterior

Abductor hallucis

Fig. 179 The relations of the posterior tibial artery as it passes behind the medial malleolus.

artery about 1.5 in (4 cm) from it origin. The peroneal artery runs down the posterior aspect of the fibula, close to the medial margin of the bone, supplying adjacent muscles and giving a nutrient branch to the fibula. Above the ankle it gives off its *perforating branch* which pierces the interosseous membrane, descends over the lateral malleolus and anastomoses with the arteries of the dorsum of the foot.

Anterior tibial artery

The anterior tibial artery arises at the bifurcation of the popliteal artery. It passes forwards between the tibia and fibula over the upper margin of the interosseous membrane and descends on this structure in the anterior compartment of the leg.

At first deeply buried, it becomes superficial just above the ankle between the tendons of extensor hallucis longus and tibialis anterior, being crossed superficially by the former immediately proximal to the line of the ankle joint.

The artery continues over the dorsum of the foot as the *dorsalis pedis* (where its pulse may be palpated); this gives off the *arcuate artery* which, in turn, supplies cutaneous branches to the backs of the toes. Dorsalis pedis itself plunges between the 1st and 2nd metatarsals to join the lateral plantar artery in the formation of the *plantar arch*, from which branches run forwards to supply the plantar aspects of the toes.

The veins of the lower limb

The veins of the lower limb are divided into the deep and superficial groups according to their relationship to the investing deep fascia of the leg. The *deep veins* accompany the corresponding major arteries. The *superficial veins* are the great and *small* (or *long* and *short*) saphenous veins and their tributaries (Fig. 180).

The *small (short) saphenous vein* commences at the ankle behind the lateral malleolus where it drains the lateral side of the dorsal venous plexus of the foot. It courses over the back of the calf, perforates the deep fascia over the popliteal fossa and terminates in the popliteal vein. One or more branches run upwards and medially from it to join the long saphenous vein.

The *great (long) saphenous vein* drains the medial part of the venous plexus on the dorsum of the foot and passes upwards immediately in front of the medial malleolus (Fig. 154); here branches of the saphenous nerve lie in front of and behind the vein. The vein then ascends over the posterior parts of the medial condyles of the tibia

Superficial circumflex iliac vein
Superficial inferior epigastric vein
Saphenofemoral junction
Superficial external pudendal vein
Lateral accessory vein
Great saphenous vein
Great saphenous vein
Vein linking great and small saphenous veins
Small saphenous vein

Fig. 180 The superficial veins of the lower limb.

and femur to the groin where it pierces the deep fascia at the saphenous opening 1 in (2.5 cm) below the inguinal ligament, to enter the femoral vein.

The great saphenous vein is joined by one or more branches from the small saphenous, and by the *lateral accessory vein* which usually enters the main vein at the mid-thigh, although it may not do so until the saphenous opening is reached.

At the groin a number of tributaries from the lower abdominal wall, thigh and scrotum enter the great saphenous vein; these tributaries are variable in number and arrangement but usually comprise (Fig. 180):

1   the superficial epigastric vein;
2   the superficial circumflex iliac vein;
3   the superficial external pudendal vein.

The superficial epigastric vein communicates with the lateral thoracic tributary of the axillary vein via the *thoracoepigastric vein*. This dilates (and may become readily visible coursing over the trunk), following obstruction of the inferior vena cava. The great saphenous vein communicates with the deep venous system not only at the groin but also at a number of points along its course through *perforating veins*; one is usually present a hand's breadth above, another a hand's breadth below the knee.

The skin of the medial aspect of the leg is drained to the deep veins by two or three *direct perforating veins* which pierce the deep fascia behind the great saphenous vein.

Clinical features

1   We have already noted (under surface anatomy of the lower limb) the great importance of the constant position of the great saphenous vein lying immediately in front of the medial malleolus. Knowledge that a vein *must* be present at this site, even if not visible in an obese or collapsed patient, may be life-saving when urgent transfusion is requred. Occasionally, the immediately adjacent saphenous nerve is caught up by a ligature during this procedure — the patient, if conscious, will complain bitterly of pain if this is done.

2   The saphenous veins frequently become dilated, incompetent and varicose. Usually this is idiopathic but may result from the increased venous pressure caused by more proximal venous obstruction (a pelvic tumour or the pregnant uterus, for example) or may be secondary to obstruction of the deep venous pathway of the leg by thrombosis.

3   Stagnation of blood in the skin of the lower limb may result from venous thrombosis or valve incompetence; the skin, in consequence,

is poorly nourished and easily breaks down into a *varicose ulcer* if subjected to even minor trauma. This is especially liable to occur over the subcutaneous anteromedial surface of the tibia where the cutaneous blood supply is least generous.

4   In operating upon varicose veins it is important that all tributaries of the groin are ligated as well as the main saphenous trunk; if one tributary escapes, it in turn becomes dilated and produces recurrence of the varices.

The course and distribution of the principal nerves of the lower limb

The nerves of the lower limb are derived from the lumbar and sacral plexuses.

The lumbar plexus (Fig. 181)

The lumbar plexus originates from the anterior primary rami of L1–4.

The trunks of the plexus traverse psoas major and emerge from its lateral border. There are two exceptions: the obturator nerve appears

Subcostal nerve

Iliohypogastric nerve

Ilio-inguinal nerve

Genitofemoral nerve

Lateral cutaneous
nerve of thigh

Femoral nerve

T XII

L I

L II

L III

L IV

Contribution to
lumbo-sacral trunk

Obturator nerve

Fig. 181 Plan of the lumbar plexus (muscular branches have been omitted for clarity).

at the medial border of psoas tendon, and the genitofemoral nerve
emerges on the anterior aspect of the muscle.

The principal branches of the plexus are the femoral nerve and the
obturator nerve.

The *femoral nerve* (L2–4) passes through the substance of psoas
then under the inguinal ligament a finger's breadth lateral to the
femoral artery, to break up into its terminal branches after a course in
lower limb of only some 2 in (5 cm).

Its branches are:
- muscular—to the anterior compartment of the thigh (quadriceps,
sartorius and pectineus);
- cutaneous—the medial and intermediate cutaneous nerves of the
thigh and the saphenous nerve, which traverses the adductor canal to
supply the skin of the medial side of the leg, ankle and foot to the
great toe;
- articular—to the hip and knee joints.

The femoral nerve supplies the skin of the medial and anterior
aspects of the thigh via its medial and intermediate cutaneous
branches, but the lateral aspect is supplied by the *lateral cutaneous
nerve of the thigh*. This arises directly from the lumbar plexus and
enters the thigh usually by passing deep to the inguinal ligament.
Occasionally, the nerve pierces the ligament and may then be pressed
upon by it with resultant pain and anaesthesia over the upper
outer thigh (*meralgia paraesthetica*). This is relieved by dividing the
deeper fasciculus of the inguinal ligament where the nerve passes over
it.

The *obturator nerve* (L2–4) emerges from the medial aspect of the
psoas and runs downwards and forwards, deep to the internal iliac
vessels, to reach the superior part of the obturator foramen. This the
nerve traverses, in company with the obturator vessels, to enter the
thigh.

Its branches are:
- muscular — to obturator externus, the adductor muscles and
gracilis;
- cutaneous—to an area of skin over the medial aspect of the thigh;
- articular—to the hip and knee joints.

Clinical features

1   Spasm of the adductor muscles of the thigh in spastic paraplegia
can be relieved by division of the obturator nerve (*obturator neurec-
tomy*). This can be performed through a midline lower abdominal
incision exposing the nerve trunk extraperitoneally on each side as it
passes towards the obturator foramen.

2   Rarely, an *obturator hernia* develops through the canal where the obturator nerve and vessels traverse the membrane covering the obturator foramen. Pressure of a strangulated obturator hernia upon the nerve causes referred pain in its area of cutaneous distribution, so that intestinal obstruction associated with pain along the medial side of the thigh should suggest this diagnosis.

3   The femoral and obturator nerves, as well as the sciatic nerve and its branches, supply sensory fibres to both the hip and the knee; it is not uncommon for hip disease to present disguised as pain in the knee.

The sacral plexus (Fig. 182)

This plexus originates from the anterior primary rami of L4–5, S1–4.

Note that L4 is shared by both plexuses, a branch from it joining L5 to form the *lumbosacral trunk* which carries its contribution to the sacral plexus.

The sacral nerves emerge from the anterior sacral foramina and unite in front of piriformis where they are joined by the lumbosacral trunk.

Branches from the plexus supply:
• the pelvic muscles;
• the muscles of the hip;
• the skin of the buttock and the back of the thigh.

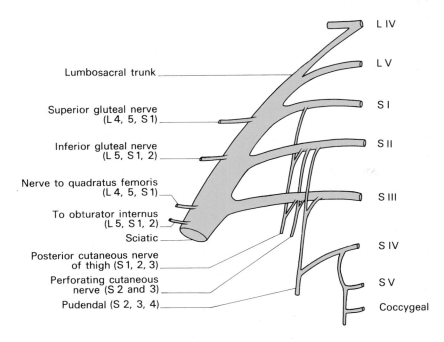

Fig. 182 Plan of the sacral plexus.

The plexus itself terminates as the pudendal nerve and the sciatic nerve.

The *pudendal nerve* (S2–4) provides the principal innervation of the perineum. It has a complex course, passing from the pelvis, briefly through the gluteal region, along the side-wall of the ischiorectal fossa and through the deep perineal pouch to end by supplying the skin of the external genitalia (Fig. 183).

It arises as the lower main division of the sacral plexus although it is dwarfed by the giant sciatic nerve. It leaves the pelvis through the greater foramen below the piriformis muscle. It crosses the dorsum of the ischial spine and immediately disappears through the lesser sciatic foramen into the perineum. The nerve now traverses the lateral wall of the ischiorectal fossa in company with the internal pudendal vessels, and lies within a distinct fascial compartment on the medial aspect of obturator internus termed the pudendal canal (Alcock's

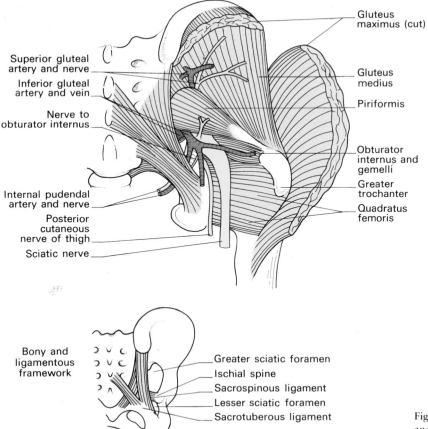

Fig. 183 The boundaries and contents of the sciatic foramina.

canal; see Fig. 99). Within the canal it first gives off the *inferior rectal nerve*, which crosses the fossa to innervate the external anal sphincter and the perianal skin, and then divides into the perineal nerve and the dorsal nerve of the penis (or clitoris).

The *perineal nerve* is the larger of the two. It bifurcates almost at once; its deeper branch supplies the sphincter urethrae and the other muscles of the anterior perineum (the ischiocavernosus, bulbospongiosus and the superficial and deep transverse perinei). Its more superficial branch innervates the skin of the posterior aspect of the scrotum.

The *dorsal nerve of the penis* (or *clitoris*) traverses the deep perineal pouch, pierces the perineal membrane and then penetrates the suspensory ligament of the penis to supply the dorsal aspect of this structure.

Clinical features

In obstetric practice the pudendal nerve can be blocked with local anaesthetic prior to forceps delivery by inserting a long needle through the vaginal wall and guided by a finger to the ischial spine, which can be palpated per vaginam. When the procedure is carried out bilaterally there is loss of the anal reflex (which is a useful test that a successful block has been achieved), relaxation of the pelvic floor muscles and loss of sensation to the vulva and lower one third of the vagina (see Fig. 99b).

The sciatic nerve

The sciatic nerve (L4, 5, S1–3) is the largest nerve in the body (Fig. 184). It is broad and flat at its origin, although peripherally it becomes rounded.

The nerve emerges from the greater sciatic foramen distal to piriformis and under cover of gluteus maximus, crosses the posterior surface of the ischium and descends on adductor magnus. Here it lies deep to the hamstrings and is crossed only by the long head of biceps.

The sciatic nerve terminates by dividing into the *tibial* and *common peroneal nerves* (see Fig. 178). The level of this division is variable—usually it is at the mid-thigh, but the two nerves may be separate even at their origins from the sacral plexus.

Branches

The trunk of the sciatic nerve supplies the hamstring muscles (biceps,

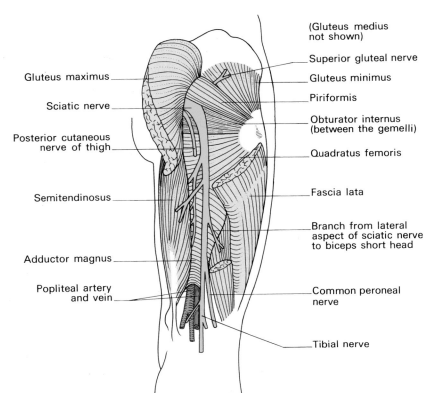

Gluteus maximus

Sciatic nerve

Posterior cutaneous
nerve of thigh

Semitendinosus

Adductor magnus

Popliteal artery
and vein

(Gluteus medius
not shown)

Superior gluteal nerve

Gluteus minimus

Piriformis

Obturator internus
(between the gemelli)

Quadratus femoris

Fascia lata

Branch from lateral
aspect of sciatic nerve
to biceps short head

Common peroneal
nerve

Tibial nerve

Fig. 184 Dissection of the sciatic nerve in the thigh and popliteal fossa. Note that gluteus medius has been removed to show the otherwise completely hidden gluteus minimus.

semimembranosus, semitendinosus) and also the adductor magnus, the latter being innervated also by the obturator nerve.

All the muscle branches apart from the one to the short head of biceps arise on the medial side of the nerve; its lateral border is therefore the side of relative safety in its operative exposure.

Clinical features

1   The sciatic nerve may be wounded in penetrating injuries or in posterior dislocation of the hip associated with fracture of the posterior lip of the acetabulum, to which the nerve is closely related (Fig. 168).

Damage to the sciatic nerve is followed by paralysis of the hamstrings and all the muscles of the leg and foot (supplied by its distributing branches); there is loss of all movements in the lower limb below the knee joint with foot drop deformity. Sensory loss is complete below the knee, except for an area along the medial side of the leg, over the medial malleolus and down to the hallux, which is innervated by the saphenous branch of the femoral nerve.

2   The sciatic nerve is accompanied by a companion artery (derived from the inferior gluteal artery) which bleeds quite sharply when the nerve is divided during an above-knee amputation. The artery must be neatly isolated and tied without any nerve fibres being incorporated in the ligature, since this would be followed by severe pain in the stump.

The tibial nerve

The tibial nerve (L4, 5, S1–3) is the larger of the two terminal branches of the sciatic nerve; it traverses the popliteal fossa superficial to the popliteal vein and artery, which it crosses from the lateral to the medial side.

Its branches in the popliteal fossa are:
- muscular—to gastrocnemius, soleus and popliteus;
- cutaneous—the *sural nerve*, which descends over the back of the calf, behind the lateral malleolus to the 5th toe; it receives a communicating branch from the common peroneal nerve and supplies the lateral side of the leg, foot and 5th toe;
- articular—to the knee joint.

It then descends deep to soleus, in company with the posterior tibial vessels, passes on their lateral side behind the medial malleolus to end by dividing into the *medial* and *lateral plantar* nerves.

Branches in the leg

The tibial nerve supplies flexor hallucis longus, flexor digitorum longus and tibialis posterior. Its terminal plantar branches supply the intrinsic muscles and skin of the sole of the foot, the medial plantar nerve having an equivalent distribution to that of the median nerve in the hand, the lateral plantar nerve being comparable to the ulnar nerve.

The common peroneal nerve

The common peroneal nerve (L4, 5, S1, 2) is the smaller of the terminal branches of the sciatic nerve. It enters the upper part of the popliteal fossa, passes along the medial border of the biceps tendon, then curves around the neck of the fibula where it lies in the substance of peroneus longus and divides into its terminal branches, the *deep peroneal* and *superficial peroneal nerves* (Fig. 155).

Branches

While still in the popliteal fossa, the common peroneal nerve gives off the *lateral cutaneous nerve of the calf*, a *peroneal (sural) communicating branch* and twigs to the knee joint, but has no muscular branches.

The deep peroneal nerve

The deep peroneal nerve pierces extensor digitorum longus, then

Fig. 185 The segmental cutaneous nerve supply of the skin.

descends, in company with the anterior tibial vessels, over the interosseous membrane and then over the ankle joint. Its branches are:
• muscular—to the muscles of the anterior compartment of the leg—extensor digitorum longus, extensor hallucis longus, tibialis anterior, peroneus tertius—and extensor digitorum brevis;
• cutaneous—to a small area of skin in the web between the 1st and 2nd toes.

The superficial peroneal nerve

The superficial peroneal nerve runs in the lateral compartment of the leg. Its branches are:
• muscular—to the lateral compartment muscles (peroneus longus and brevis);
• cutaneous — to the skin of the distal two-thirds of the lateral aspect of the leg and to the dorsum of the foot (apart from the small area between the 1st and 2nd toes supplied by the deep peroneal nerve).

Clinical features

The common peroneal nerve is in a particularly vulnerable position as it winds around the neck of the fibula. It may be damaged at this site by the pressure of a tight bandage or plaster cast or may be torn in severe adduction injuries to the knee. Damage to this nerve is followed by foot drop (due to paralysis of the ankle and foot extensors) and inversion of the foot due to paralysis of the peroneal muscles with unopposed action of the foot flexors and invertors). There is also anaesthesia over the anterior and lateral aspects of the leg and foot, although the medial side escapes, since this is innervated by the saphenous branch of the femoral nerve.

Segmental cutaneous supply of the lower limb
(Fig. 185)

The arrangement of root segments supplying the lower limb is as follows:
• L1, 2 and 3—supply the anterior aspect of the thigh from above down;
• L4—supplies the frontomedial aspect of the leg;
• L5—supplies the frontolateral aspect of the leg but also extends on to the medial side of the foot;
• S1—supplies the lateral side of the foot and the sole;

The surface anatomy of the neck

In the midline, from above down, can be felt (Fig. 186):
1   the *hyoid bone*—at the level of C3;
2   the notch of the *thyroid cartilage*—at the level of C4;
3   the *cricoid cartilage*—terminating in the trachea at C6;
4   the rings of the *trachea*, over the second and third of which can be rolled the *isthmus of the thyroid gland*;
5   the *suprasternal notch*.

Note that the lower border of the cricoid is an important level in the neck; it corresponds not only to the level of the 6th cervical vertebra but also to:
1   the junction of the larynx with the trachea;
2   the junction of the pharynx with the oesophagus;
3   the level at which the inferior thyroid artery and the middle thyroid vein enter the thyroid gland;
4   the level at which the vertebral artery enters the transverse foramen in the 6th cervical vertebra;
5   the level at which the superior belly of the omohyoid crosses the carotid sheath;
6   the level of the middle cervical sympathetic ganglion;
7   the site at which the carotid artery can be compressed against the transverse process of C6 (the carotid tubercle).

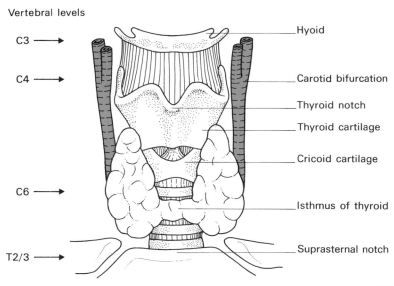

Fig. 186 Structures palpable on the anterior aspect of the neck, together with their corresponding vertebral levels.

Vertebral levels

C3 ⟶ Hyoid

C4 ⟶ Carotid bifurcation

Thyroid notch

Thyroid cartilage

Cricoid cartilage

C6 ⟶ Isthmus of thyroid

T2/3 ⟶ Suprasternal notch

By pressing the jaw laterally against the resistance of one's hand, the opposite *sternocleidomastoid* is tensed. This muscle helps define the posterior triangle of the neck, bounded by sternocleidomastoid, trapezius and the clavicle, and the anterior triangle, defined by sternocleidomastoid, the mandible and the midline (Fig. 187).

Violently clench the jaws; the platysma then comes into view as a sheet of muscle, passing from the mandible down over the clavicles, lying in the superficial fascia of the neck. The *external jugular vein* lies immediately deep to platysma, crosses the sternocleidomastoid into the posterior triangle, perforates the deep fascia just above the clavicle and enters the subclavian vein. It is readily visible in a thin subject on straining and is seen from the audience when a singer hits a sustained high note or when an orthopaedic surgeon reduces a fracture.

The *common carotid artery* pulse can be felt by pressing backwards against the long anterior tubercle of the transverse process of C6. The line of the carotid sheath can be marked out by a line joining a point midway between the tip of the mastoid process and the angle of the jaw to the sternoclavicular joint. Along this line, the carotid bifurcates into the *external* and *internal carotid arteries* at the level of the upper border of the thyroid cartilage; at this level the vessels lie just below the deep fascia where their pulsation is palpable and often visible.

The fascial compartments of the neck (Fig. 188)

The fascial planes of the neck are of considerable importance to the

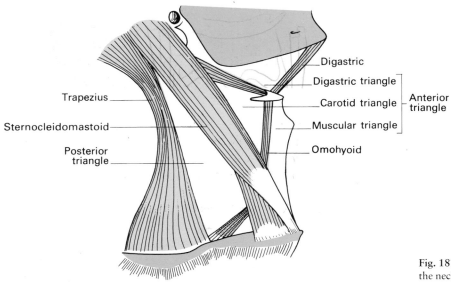

Fig. 187 The triangles of the neck.

Fig. 188 (a) Transverse section of the neck through C6—showing the fascial planes and also the contents of the pretracheal fascia (or 'visceral compartment of the neck'). (b) CT scan through the C6 level; compare this with the diagram.

surgeon; they form convenient lines of cleavage through which he may separate the tissues in operative dissections and they delimit the spread of pus in neck infections.

The *superficial fascia* is a thin fatty membrane enclosing the platysma.

The *deep fascia* can be divided into three layers.

1 The *enveloping fascia* invests the muscles of the neck. It is attached to all the bony landmarks at the upper and lower margin of the neck: above, to the mandible, zygomatic arch, mastoid process and superior nuchal line; below, to the manubrium, clavicle, acromion and scapular spine. Posteriorly, the ligamentum nuchae provides a longitudinal line of attachment for it.

This enveloping fascia splits to enclose the trapezius, the sterno-cleidomastoid, the strap muscles and the parotid and submandibular glands.

The external jugular vein pierces the deep fascia above the clavicle. If the vein is divided here, it is held open by the deep fascia which is attached to its margins, air is sucked into the vein lumen during inspiration and a fatal air embolism may ensue.

2 The *prevertebral fascia* passes across the vertebrae and prevertebral muscles behind the oesophagus, the pharynx and the great vessels. Above it is attached to the base of the skull. Laterally, the fascia covers the scalene muscles, and the emerging brachial plexus and subclavian artery. These structures carry with them a sheath formed from the prevertebral fascia, which becomes the axillary sheath.

Inferiorly, the fascia blends with the anterior longitudinal ligament of the upper thoracic vertebrae in the posterior mediastinum.

Pus from a tuberculous cervical vertebra bulges behind this dense fascial layer and may form a midline swelling in the posterior wall of the pharynx. The abscess may then track laterally, deep to the prevertebral fascia, to a point behind the sternocleidomastoid. Rarely, pus has tracked down along the axillary sheath into the arm.

3 The *pretracheal fascia* encloses the 'visceral compartment of the neck'. Extending from the hyoid above to the fibrous pericardium below, it encloses larynx and trachea, pharynx and oesophagus and the thyroid gland. A separate tube of fascia forms the *carotid sheath*, containing carotid, internal jugular and vagus nerve and bearing the cervical sympathetic chain in its posterior wall. (Some points of clinical significance concerning this fascia are to be found under 'The thyroid', page 286.)

The thyroid gland

The thyroid is made up of (Fig. 189):
1 the *isthmus*—overlying the 2nd and 3rd rings of the trachea;
2 the *lateral lobes*—each extending from the side of the thyroid cartilage downwards to the 6th tracheal ring;
3 an inconstant *pyramidal lobe* projecting upwards from the isthmus, usually on the left side, which represents a remnant of the embryological descent of the thyroid.

External carotid artery

Superior thyroid artery and vein

Internal jugular vein

Middle thyroid vein

Inferior thyroid artery

Thyrocervical trunk
Subclavian artery

Inferior thyroid vein

Left brachiocephalic (innominate) vein

Fig. 189 The thyroid and its blood vessels.

Relations (Fig. 188)

The gland is enclosed in the pretracheal fascia, covered by the strap muscles and overlapped by the sternocleidomastoids. The anterior jugular veins course over the isthmus. When the thyroid enlarges, the strap muscles stretch and adhere to the gland so that, at operation, they often appear to be thin layers of fascia.

On the deep aspect of the thyroid lie the larynx and trachea, with the pharynx and oesophagus behind and the carotid sheath on either side. Two nerves lie in close relationship to the gland; in the groove between the trachea and oesophagus lies the *recurrent laryngeal nerve* and deep to the upper pole lies the *external branch* of the *superior laryngeal nerve* passing to the cricothyroid muscle.

Blood supply

Three arteries supply and three veins drain the thyroid gland (Fig. 189):

* the *superior thyroid artery*—arises from the external carotid and passes to the upper pole;
* the *inferior thyroid artery*—arises from the thyrocervical trunk of the 1st part of the subclavian artery and passes behind the carotid sheath to the back of the gland;
* the *thyroidea ima artery*—is inconstant; when present, it arises from the aortic arch or the brachiocephalic artery;

• the *superior thyroid vein* — drains the upper pole to the internal jugular vein;
• the *middle thyroid vein* — drains from the lateral side of the gland to the internal jugular;
• the *inferior thyroid veins* — often several — drain the lower pole to the brachiocephalic veins.

As well as these named branches, numerous small vessels pass to the thyroid from the pharynx and trachea so that even when all the main vessels are tied, the gland still bleeds when cut across during a partial thyroidectomy.

Development

The thyroid develops from a bud which pushes out from the floor of the pharynx; this outgrowth then descends to its definitive position in the neck. It normally loses all connection with its origin which is commemorated, however, by the foramen caecum at the junction of the middle and posterior thirds of the tongue and by the inconstant pyramidal lobe on the isthmus (Fig. 190).

Clinical features

1   The development of the thyroid accounts for the rare occurrence of the whole or a part of the gland remaining as a swelling at the tongue base (*lingual thyroid*) and for the much commoner occurrence of a *thyroglossal cyst* or *sinus* along the pathway of descent. Such a sinus can be dissected from the midline of the neck along the front of the hyoid (in such intimate contact with it that the centre of the hyoid must be excised during the dissection) then backwards through the muscles of the tongue to the foramen caecum (Fig. 190).

Descent of the thyroid may go beyond the normal position in the neck down into the superior mediastinum (*retrosternal goitre*).

2   A benign enlargement of the thyroid may compress or displace any of its close relations; the trachea and oesophagus may be narrowed, with resulting difficulty in breathing and swallowing, and the carotid may be displaced posteriorly. A carcinoma of the thyroid invades its neighbours rather than displacing them — eroding into trachea or oesophagus, surrounding the carotid sheath and occasionally causing severe haemorrhage therefrom. The recurrent laryngeal nerve and the cervical sympathetic chain may be involved, producing changes in the voice and Horner's syndrome respectively.

3   We have already noted, in dealing with the fasciae of the neck, that the thyroid gland is enclosed in the pretracheal fascia. This thyroid capsule is much denser in front than behind and the enlarging

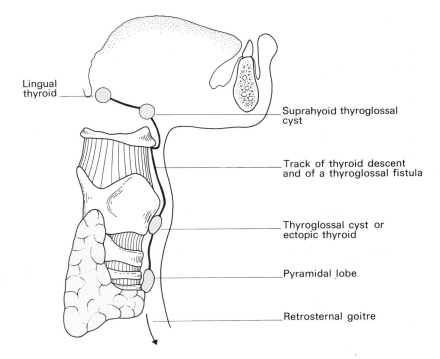

Lingual thyroid

Suprahyoid thyroglossal cyst

Track of thyroid descent and of a thyroglossal fistula

Thyroglossal cyst or ectopic thyroid

Pyramidal lobe

Retrosternal goitre

Fig. 190 The descent of the thyroid, showing possible sites of ectopic thyroid tissue or thyroglossal cysts, and also the course of a thyroglossal fistula. (The arrow shows the further descent of the thyroid which may take place retrosternally into the superior mediastinum.)

gland therefore tends to push backwards, burying itself round the sides and even the back of the trachea and oesophagus. Because of the attachments of its fascial compartment, a large goitre will also extend downwards into the superior mediastinum ('plunging goitre').

Above, the pretracheal fascia blends with the larynx, accounting for the upward movement of the thyroid gland with each act of swallowing.

4 *Thyroidectomy* is carried out through a transverse 'collar' incision, two fingers' breadth above the suprasternal notch. This lies in the line of the natural skin folds of the neck. Skin flaps are reflected, together with platysma, and the investing fascia opened longitudinally between the strap muscles and between the anterior jugular veins.

If more room is required in the case of a large goitre, the strap muscles are divided; this is carried out at their upper extremity because their nerve supply (the ansa hypoglossi) enters the lower part of the muscles and is hence preserved.

The pretracheal fascia is then divided, exposing the thyroid gland; unless this tissue plane deep to the fascia is found, dissection is a difficult and bloody procedure.

The thyroid is then mobilized and its vessels ligated *seriatim*. Both the recurrent and superior laryngeal nerves are at risk during this procedure and must be carefully avoided (Fig. 191).

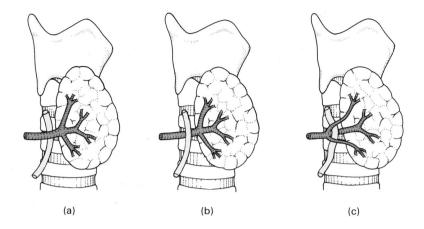

(a)   (b)   (c)

Fig. 191 The relationship of the recurrent laryngeal nerve to the thyroid gland and the inferior thyroid artery. (a) The nerve is usually deep to the artery but (b) may be superficial to it or (c) pass through its branches. In these diagrams the lateral lobe of the thyroid is pulled forwards, as it would be in a thyroidectomy.

The parathyroid glands (Fig. 192)

These are usually four in number, a superior and inferior on either side; however, the numbers vary from two to six. Ninety per cent are in close relationship to the thyroid, 10% are aberrant, the latter invariably being the inferior glands.

Each gland is about the size of a split pea and is of a yellowish-brown colour. The superior parathyroid is more constant in position than the inferior gland. It usually lies at the middle of the posterior border of the lobe of the thyroid above the level at which the inferior thyroid artery crosses the recurrent laryngeal nerve. The inferior parathyroid is most usually situated below the inferior artery near the lower pole of the thyroid gland. The next commonest site is within 1 cm of the lower pole of the thyroid gland. Aberrant inferior parathyroids may descend along the inferior thyroid veins in front of the trachea and may even track into the superior mediastinum in

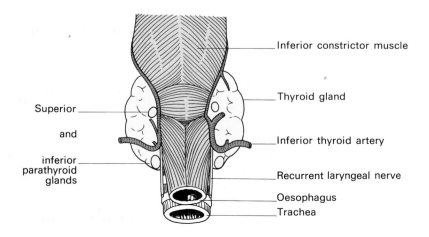

Inferior constrictor muscle

Thyroid gland

Superior

and

Inferior thyroid artery

inferior parathyroid glands

Recurrent laryngeal nerve

Oesophagus

Trachea

Fig. 192 The normal sites of the parathyroid glands (posterior aspect).

company with thymic tissues, for which there is an embryolog. explanation (see below). Less commonly, the inferior gland may lie behind and outside the fascial sheath of the thyroid and be found behind the oesophagus or even in the posterior mediastinum. Only on extremely rare occasions are the glands actually completely buried within thyroid tissue (Fig. 193).

Development

The superior parathyroids differentiate from the 4th branchial pouch. The inferior gland develops from the 3rd pouch in company with the thymus (Fig. 194). As the latter descends, the inferior parathyroid is dragged down with it.

It is thus easily understood that the inferior parathyroid may be dragged beyond the thyroid into the mediastinum and why, although very rarely, parathyroid tissue is found actually within the thymus.

Clinical features

1   These possible aberrant sites are, of course, of great importance in searching for a parathyroid adenoma in hyperparathyroidism.
2   The parathyroids are usually safe in subtotal thyroidectomy because the posterior rim of the thyroid is preserved. However, they may be inadvertently removed or damaged, with resultant tetany due to the lowered serum calcium.

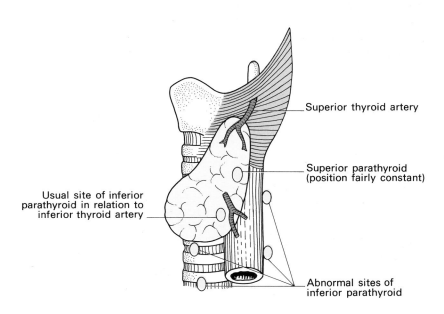

Fig. 193 Normal and abnormal sites of the parathyroid glands (lateral view).

Superior thyroid artery

Superior parathyroid (position fairly constant)

Usual site of inferior parathyroid in relation to inferior thyroid artery

Abnormal sites of inferior parathyroid

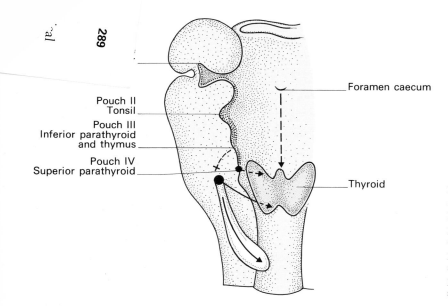

Pouch II
Tonsil

Pouch III
Inferior parathyroid
and thymus

Pouch IV
Superior parathyroid

Foramen caecum

Thyroid

289

Fig. 194 The derivatives of the branchial pouches. Note that the inferior parathyroid migrates downwards from the 3rd pouch whereas the superior parathyroid (4th pouch) remains stationary.

The palate

The palate separates the nasal and buccal cavities and comprises:
1   the *hard palate*—which is vault-shaped and made up of the palatine plate of the maxilla and the horizontal plate of the palatine bone; it is bounded by the alveolar margin anteriorly and laterally, and merges posteriorly with:
2   the *soft palate* — hanging as a curtain between the naso- and oropharynx; centrally it bears the *uvula* on its free posterior edge; laterally it blends into the anterior and posterior pillars of the fauces.

The hard palate is made up of bone, periosteum and a squamous mucosa in which are embedded tiny accessory salivary glands.

The framework of the soft palate is formed by the aponeurosis of the tensor palati muscle, which adheres to the posterior border of the hard palate. To this fibrous sheet are attached the palatine muscles covered by a mucous membrane, which is squamous on its buccal aspect and ciliated columnar on its nasopharyngeal surface.

The sensory supply of the palate is largely from the maxillary division of V but fibres of IX supply its most posterior part.

Motor innervation to the palatine muscles is from XI transmitted by the vagus fibres in the pharyngeal plexus. The tensor palati is the exception to this rule and is supplied by the mandibular division of V.

In speaking, swallowing and blowing, the soft palate closes off the nasopharynx from the buccal cavity. If the palate is paralysed, as may occur after diphtheria, the voice is impaired and fluids regurgitate through the nose on swallowing.

The development of the face, lips and palate with special reference to their congenital deformities (Fig. 195)

Around the primitive mouth, or stomodaeum, develop the following:

1   the *frontonasal process* which projects down from the cranium. Two olfactory pits develop in it and rupture into the pharynx to form the nostrils. Definitively, this process forms the nose, the nasal septum, nostril, the philtrum of the upper lip (the small midline depression) and the premaxilla—the V-shaped anterior portion of the upper jaw which usually bears the four incisor teeth;

2   the *maxillary processes* on each side, which fuse with the frontonasal process and become the cheeks, upper lip (exclusive of the philtrum), upper jaw and palate (apart from the premaxilla);

3   the *mandibular processes* which meet in the midline to form the lower jaw.

Abnormalities of this complex fusion process are numerous and constitute one of the commonest groups of congenital deformities. It is estimated that one child in 600 in England is born with some degree of either cleft lip or palate (Fig. 196).

Frequently, these anomalies are associated with other congenital conditions such as spina bifida, syndactyly (fusion of fingers or toes), etc. Indeed, it is good clinical practice to search a patient with any congenital defect for others.

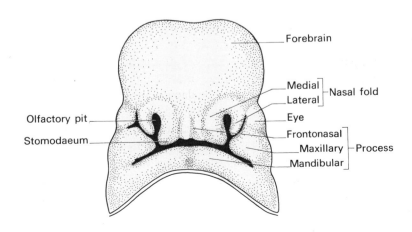

Fig. 195 The ventral aspect of a fetal head showing the three processes, frontonasal, maxillary and mandibular, from which the face, nose and jaws are derived.

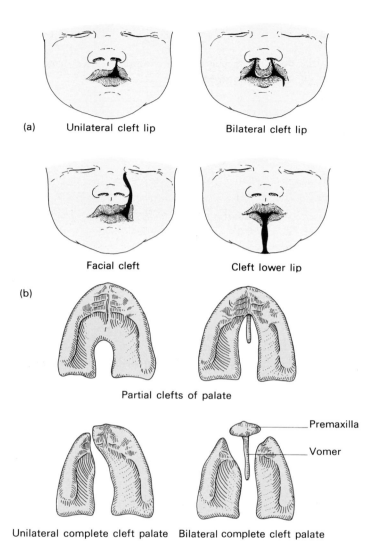

(a)   Unilateral cleft lip   Bilateral cleft lip

Facial cleft   Cleft lower lip

(b)

Partial clefts of palate

Premaxilla

Vomer

Unilateral complete cleft palate   Bilateral complete cleft palate

Fig. 196 Types of (a) cleft lip and (b) cleft palate.

The following anomalies are associated with defects of fusion of the face.

1 *Macrostoma* and *microstoma* are conditions where either too little or too great a closure of the stomodaeum occurs.

2 *Cleft upper lip* (or 'hare lip') — this is only very rarely like the upper lip of a hare, i.e. a median cleft, although this may occur as a failure of development of the philtrum from the frontonasal process. Much more commonly, the cleft is on one or both sides of the philtrum, occurring as failure of fusion of the maxillary and frontonasal processes. The cleft may be a small defect in the lip or may extend into the nostril, split the alveolus or even extend along the side of the nose as far as the orbit. There may be an associated cleft palate.

3 *Cleft lower lip* — occurs very rarely but may be associated with a cleft tongue and cleft mandible.

4 *Cleft palate* is a failure of fusion of the segments of the palate. The following stages may occur (Fig. 196):

(a) bifid uvula, of no clinical importance;

(b) partial cleft, which may involve the soft palate only or the posterior part of the hard palate also;

(c) complete cleft, which may be unilateral, running the full length of the maxilla and then alongside one face of the premaxilla, or bilateral in which the palate is cleft with an anterior V separating the premaxilla completely.

5 *Inclusion dermoids* may form along the lines of fusion of the face. The most common of these is the *external angular dermoid* at the lateral extremity of the upper eyebrow. Occasionally this dermoid extends through the skull to attach to the underlying dura.

The tongue and floor of the mouth

The tongue

The tongue consists of a buccal and a pharyngeal portion separated by a V-shaped groove on its dorsal surface, the *sulcus terminalis*. At the apex of this groove is a shallow depression, the *foramen caecum*, marking the embryological origin of the thyroid (see page 286). Immediately in front of the sulcus lie a row of large vallate papilliae.

The under aspect of the tongue bears the median *frenulum linguae*; the mucosa is thin on this surface and the lingual veins can thus be seen on either side of the frenulum. The lingual nerve and the lingual artery are medial to the vein but not visible. More laterally can be seen a fringed fold of mucous membrane termed the *plica fimbriata*. On either side of the base of the frenulum can be seen the orifice of the submandibular duct on its papilla. Inspect this in a mirror and note the discharge of saliva.

Structure

The thick stratified squamous mucosa of the dorsum of the tongue bears papillae over the anterior two-thirds back as far as the sulcus terminalis. These papillae (particularly the vallate) bear the taste buds. The posterior one-third has no papillae but carries numerous lymphoid nodules which, with the palatine tonsils and adenoids, make up the *lymphoid ring of Waldeyer*.

Small glands are scattered throughout the submucosa of the dorsum; these are predominantly serous anteriorly and mucous posteriorly.

The tongue is divided by a median vertical fibrous septum, as indicated on the dorsum by a shallow groove. On each side of this septum are the intrinsic and extrinsic muscles of the tongue (Fig. 197).

The *intrinsic muscles* are disposed in vertical, longitudinal and transverse bundles; they alter the shape of the tongue.

The *extrinsic muscles* move the tongue as a whole. They pass to the tongue from the symphysis of the mandible, the hyoid, styloid process and the soft palate, respectively the *genioglossus, hyoglossus, styloglossus* and *palatoglossus*. The functions of the individual extrinsic muscles can be deduced from their relative positions (Fig. 197). Genioglossus protrudes the tongue, styloglossus retracts it and hyoglossus depresses it. Palatoglossus is, in fact, a palatal muscle and helps to narrow the oropharynx in swallowing.

Blood supply

Blood is supplied from the lingual branch of the external carotid artery. There is little cross-circulation across the median raphe, which is therefore a relatively avascular plane.

Lymph drainage (Fig. 198)

The drainage zones of the mucosa of the tongue can be grouped into three:

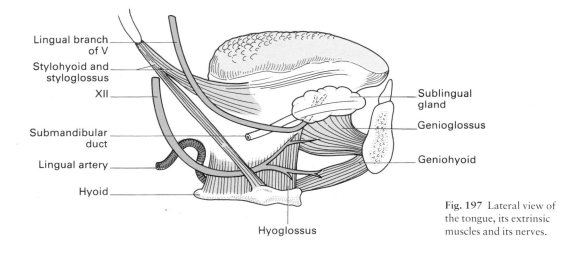

Lingual branch of V
Stylohyoid and styloglossus
XII
Submandibular duct
Lingual artery
Hyoid
Sublingual gland
Genioglossus
Geniohyoid
Hyoglossus

Fig. 197 Lateral view of the tongue, its extrinsic muscles and its nerves.

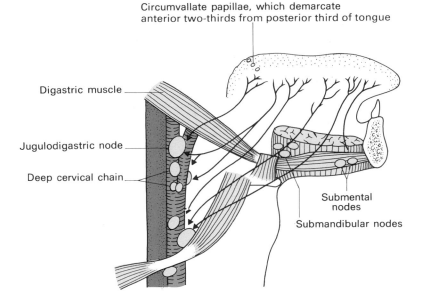

Circumvallate papillae, which demarcate
anterior two-thirds from posterior third of tongue

Digastric muscle

Jugulodigastric node

Deep cervical chain

Submental nodes

Submandibular nodes

Fig. 198 Diagram of the lymph drainage of the tongue. Note two points. (i) The anterior part of the tongue tends to drain to the nodes farthest down the deep cervical chain, whereas the posterior part drains to the upper chain. (ii) The anterior two-thirds of the tongue drain unilaterally, the posterior one-third bilaterally.

1   the tip drains to the submental nodes;
2   the anterior two-thirds drains to the submental and submandibular nodes and thence to the *lower* nodes of the deep cervical chain along the carotid sheath;
3   the posterior one-third drains to the *upper* nodes of the deep cervical chain.

There is a rich anastomosis across the midline between the lymphatics of the posterior one-third of the tongue so that a tumour on one side readily metastasizes to contralateral nodes. In contrast, there is little cross-communication in the anterior two-thirds, where growths more than 0.5 in (12 mm) from the midline do not metastasize to the opposite side of the neck till late in the disease.

Nerve supply

The anterior two-thirds of the tongue receives its sensory supply from the lingual branch of V which also transmits the gustatory fibres of the chorda tympani (VII).

Common sensation and taste to the posterior one-third, including the vallate papillae, are derived from IX. A few fibres of the superior laryngeal nerve (X) carry sensory fibres from the posterior part of the tongue.

All the muscles of the tongue except palatoglossus are supplied by XII; palatoglossus, a muscle of the soft palate, is innervated by XI via the pharyngeal branch of X.

Development (Fig. 199)

A small nodule, the *tuberculum impar*, is the first evidence of the developing tongue in the floor of the pharynx. This is soon covered over by the *lingual swellings*, one on each side, derived from the first branchial arch. These fuse in the midline to form the definitive anterior two-thirds of the tongue supplied by V and reinforced by chorda tympani.

Posteriorly, this mass meets the *copula* (or *hypobranchial eminence*), a central swelling in the pharyngeal floor which represents the 2nd, 3rd and 4th arches and which forms the posterior one-third of the tongue (nerve supply IX and X).

The tongue muscles derive from the occipital myotomes which migrate forward dragging with them their nerve supply (XII, the hypoglossal nerve).

Clinical features

1   Damage to the hypoglossal nerve is readily detected clinically by hemiatrophy of the tongue and deviation of the projected organ towards the paralysed side.

2   If the unconscious or deeply anaesthetized patient is laid on his back, the posterior aspect of the tongue drops back to produce a laryngeal obstruction. This can be prevented either by lying the patient on his side with the head down ('the tonsil position'), when the tongue flops forward with the weight of gravity, or by pushing the mandible forwards by pressure on the angle of the jaw on each side;

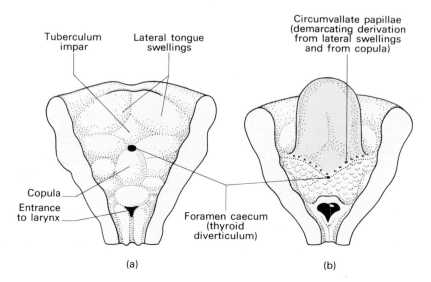

Tuberculum impar

Lateral tongue swellings

Circumvallate papillae (demarcating derivation from lateral swellings and from copula)

Copula

Entrance to larynx

Foramen caecum (thyroid diverticulum)

(a)

(b)

Fig. 199 Development of the tongue.

this is effective because genioglossus, attached to the symphysis menti, drags the tongue forward along with the lower jaw.

3   Although lymphatics pierce the floor of the mouth (i.e. the mylohyoid muscle) to reach the submental and submandibular lymph nodes, it is an interesting fact that these tissues are not affected by lymphatic spread of malignant cells (although they may be invaded by direct extension of growth). It seems that the nodes are involved by lymphatic emboli and not by a permeation of the lymphatic channels.

The bilateral lymphatic spread of growths of the posterior one-third of the tongue is one factor contributing to the poor prognosis of tumours at this site.

The floor of the mouth

The floor of the mouth is formed principally by the mylohyoid muscles. These stretch as a diaphragm from their origin along the mylohyoid line on the medial aspect of the body of the mandible on each side, to their insertion along a median raphe and into the hyoid bone. They support the tongue as a muscular sling (Fig. 200).

On the lower aspect of this diaphragm, on each side, are the anterior belly of the digastric muscle, the superficial part of the submandibular gland and the submandibular lymph nodes, all covered by deep fascia and platysma.

Lying above mylohyoid are the tongue muscles, as a central mass, with the sublingual salivary gland and the deep part of the sub-

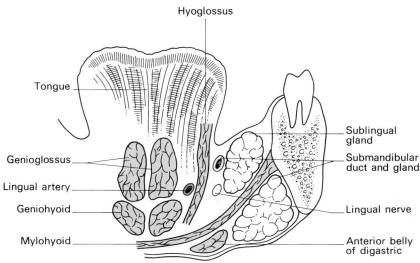

Fig. 200 Coronal section of the floor of the mouth.

mandibular gland and its duct lying beneath the mucosa of the mouth floor on either side.

Clinical features

Ludwig's angina is a cellulitis of the floor of the mouth, usually origi- nating from a carious molar tooth. The infection spreads above the mylohyoid; oedema forces the tongue upwards and the mylohyoid itself is pushed downwards so that there is swelling both below the chin and within the mouth. There is considerable danger of spread of infection backwards with oedema of the glottis and asphyxia.

Drainage is carried out by a deep incision below the mandible which must divide the mylohyoid muscle.

The pharynx

The pharynx is a musculofascial tube, incomplete anteriorly, which extends from the base of the skull to the oesophagus and which acts as a common entrance to the respiratory and alimentary tracts.

From above downwards, it is made up of three portions (Fig. 201):
1   the *nasopharynx* — lying behind the nasal fossae and above the soft palate;
2   the *oropharynx*—lying behind the anterior pillars of the fauces;
3   the *laryngopharynx*—lying behind the larynx.

The nasopharynx

The nasopharynx lies above the soft palate, which cuts it off from the rest of the pharynx during deglutition and therefore prevents regurgi- tation of food through the nose.

Two important structures lie in this compartment.

The *nasopharyngeal tonsil* ('the *adenoids*') consists of a collection of lymphoid tissue beneath the epithelium of the roof and posterior wall of this region. It helps to form a continuous lymphoid ring with the palatine tonsils and the lymphoid nodules on the dorsum of the tongue (*Waldeyer's ring*).

The *orifice of the pharyngotympanic or auditory tube* (*Eustachian canal*) lies on the side-wall of the nasopharynx level with the floor of the nose. The posterior lip of this opening is prominent, due to the underlying cartilage of the Eustachian tube, and is termed the

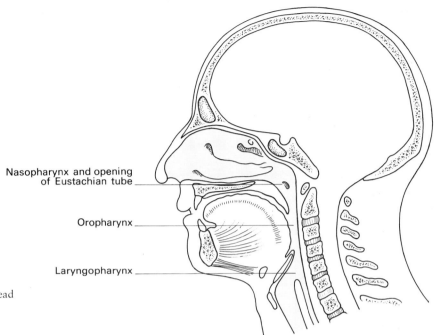

Nasopharynx and opening
of Eustachian tube

Oropharynx

Laryngopharynx

Fig. 201 A sagittal
section through the head
and neck to show the
subdivisions of the
pharynx.

Eustachian or *pharyngeal cushion*, behind which lies the slit-like *pharyngeal recess*.

Clinical features

1   The nasopharynx may be inspected indirectly by a mirror passed through the mouth (posterior rhinoscopy) or studied through a rhinoscope passed along the floor of the nose. Under anaesthesia, it can be palpated by a finger passed behind the soft palate.

2   The nasopharyngeal tonsils (adenoids) are prominent in children but usually undergo atrophy after puberty. When chronically inflamed they may all but fill the nasopharynx, causing mouth-breathing and also, by blocking the auditory tube, deafness and middle ear infection.

3   The Eustachian tube provides a ready pathway of sepsis from the pharynx to the middle ear and accounts for the frequency with which otitis media complicates infections of the throat.

4   The middle ear can be intubated through a catheter passed into the Eustachian tube. The catheter is passed along the nasal floor to the posterior wall of the nasopharynx. Its curved tip is then rotated laterally so that it lies in the pharyngeal recess; it is then withdrawn over the Eustachian cushion to slip into the orifice of the auditory tube.

The oropharynx

This part of the pharynx lies behind the mouth and tongue. Its anterior boundaries are the anterior pillars of the fauces and it extends from the uvula of the soft palate above to the tip of the epiglottis below. Its most important contents are the palatine tonsils.

The palatine tonsils

The palatine tonsil lies in the *tonsillar fossa* between the anterior and posterior pillars of the fauces. The anterior pillar, or *palatoglossal arch*, forms the boundary between the buccal cavity and the oropharynx; it fuses with the lateral wall of the tongue and contains the palatoglossus muscle. The posterior pillar, or *palatopharyngeal arch*, blends with the wall of the pharynx and contains the palatopharyngeus (Fig. 202).

The floor of the tonsillar fossa is formed by the superior constrictor of the pharynx separated from the tonsil by the *tonsillar capsule*, which is a thick condensation of the pharyngeal submucosa (the pharyngobasilar fascia). This capsule is itself separated from the superior constrictor by a film of loose areolar tissue.

The palatine tonsil consists of a collection of lymphoid tissue covered by a squamous epithelium, a unique histological combination which makes it easy to 'spot' in examinations. This epithelium is

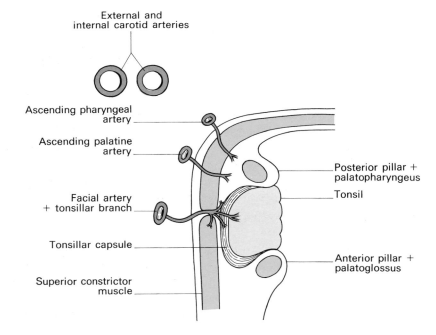

Fig. 202 Diagram of the palatine tonsil and its relations—in horizontal section.

pitted by *crypts*, up to twenty in number, and often bears a deep *intra-tonsillar cleft* in its upper part.

The lymphoid material may extend up to the soft palate, down to the tongue or into the anterior faucial pillar. From late puberty onwards this lymphoid tissue undergoes progressive atrophy.

Blood supply is principally from the tonsillar branch of the facial artery entering at the lower pole of the tonsil, although twigs are also derived from the lingual, ascending palatine and ascending pharyngeal arteries.

The venous drainage passes to the pharyngeal plexus. An important constant vein, the *paratonsillar vein*, descends from the soft palate across the lateral aspect of the tonsillar capsule. It is nearly always divided in tonsillectomy and may give rise to troublesome haemorrhage.

Lymph drainage is via lymphatics which pierce the superior constrictor muscle and pass to the nodes along the internal jugular vein, especially the *tonsillar or jugulodigastric node* at the angle of the jaw. Since this node is affected in tonsillitis it is the most common lymph node in the body to undergo pathological enlargement.

Embryologically, the tonsil derives from the second internal branchial pouch (Fig. 194).

Clinical features

1   *Tonsillectomy* may be carried out by dissection or by the guillotine; both depend on removing the lymphoid tissue and underlying fascial capsule from the loose areolar tissue clothing the superior constrictor in the floor of the tonsillar fossa. In dissection, an incision is made in the mucosa of the anterior pillar immediately in front of the tonsil; the gland is then freed by blunt dissection until it remains attached only by its pedicle of vessels near its lower pole. This pedicle is then crushed and divided by means of a wire snare.

In the second method, the guillotine is applied so that the tonsil bulges through the ring in the instrument. The tonsil is then removed by closing the blade of the guillotine.

Unless there have been repeated infections, the superior constrictor lies separated from the palatine tonsil and its capsule by loose areolar tissue which prevents the pharyngeal wall being dragged into danger during tonsillectomy.

Similarly, the internal carotid artery, although only 1 in (2.5 cm) behind the tonsil, is never injured in this operation since it lies safely freed from the pharynx by fatty tissue around the carotid sheath.

2   A *quinsy* is suppuration in the peritonsillar tissue secondary to

tonsillitis. It is drained by an incision in the most prominent part of the abscess where softening can be felt.

The laryngopharynx

The laryngopharynx extends from the level of the tip of the epiglottis to the termination of the pharynx in the oesophagus at the level of C6.

The inlet of the larynx, defined by the epiglottis, aryepiglottic folds and the arytenoids, lies anteriorly. The larynx itself bulges into this part of the pharynx leaving a deep recess anteriorly on either side, the *piriform fossa*, in which sharp ingested foreign bodies (for example, fish bones), may lodge.

The structure of the pharynx

The pharynx is made up of mucosa, submucosa, muscle and a loose areolar sheath. The mucosa is a ciliated columnar epithelium in the nasopharynx but elsewhere it is stratified and squamous. Beneath this, the submucosa is thick and fibrous (the *pharyngobasilar fascia*) and it is this layer which forms the capsule of the tonsil.

The three *pharyngeal constrictor muscles* (superior, middle and inferior) are arranged like flower pots placed one inside the other, but are open in front at the entries of the nasal, buccal and laryngeal cavities.

Each constrictor muscle is attached anteriorly to the side-wall of these cavities and fans out to insert into a median raphe along the posterior aspect of the pharynx, extending from the base of the skull to the oesophagus (Fig. 203).

Covering these muscles is an areolar sheath continuous with that covering the buccinator and hence termed the *buccopharyngeal fascia*.

Blood supply

The pharynx receives its arterial supply mainly from the superior thyroid and ascending pharyngeal branches of the external carotid.

A pharyngeal venous plexus lies in the areolar sheath of the pharynx and drains into the internal jugular vein.

Nerve supply

The pharyngeal branches of IX and X constitute the principal sensory

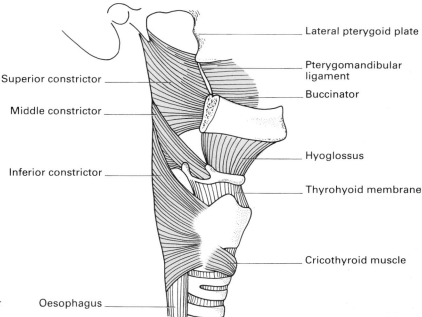

Lateral pterygoid plate

Pterygomandibular ligament

Buccinator

Superior constrictor

Middle constrictor

Hyoglossus

Inferior constrictor

Thyrohyoid membrane

Cricothyroid muscle

Oesophagus

Fig. 203 The constrictor muscles of the pharynx.

and motor supply of the pharynx respectively. The maxillary division of V supplies the sensory innervation of the nasopharynx.

The mechanism of deglutition

The act of swallowing not only conveys food down the oesophagus but also disposes of mucus loaded with dust and bacteria from the respiratory passages. Moreover, during deglutition, the Eustachian auditory tube is opened, thus equalizing the pressure on either side of the ear drum.

Deglutition is a complex, orderly series of reflexes. It is initiated voluntarily but is completed by involuntary reflex actions set up by stimulation of the pharynx. If the pharynx is anaesthetized then normal swallowing cannot take place. The reflexes are coordinated by the deglutition centre in the medulla, which lies near the vagal nucleus and the respiratory centres.

The food is first crushed by mastication and lubricated by saliva. It is a common experience that it is well nigh impossible to swallow a pill when the throat is dry. The bolus is then pushed back through the oropharyngeal isthmus by the pressure of the tongue against the palate, assisted by the muscles of the mouth floor.

During swallowing, the oral, nasal and laryngeal openings must be closed off to prevent regurgitation through them of food or fluid;

each of these openings is guarded by a highly effective sphincter mechanism.

The nasopharynx is closed by elevation of the soft palate, which shuts against a contracted ridge of superior pharyngeal constrictor, the *ridge of Passavant*. At the same time, the tensor palati opens the ostium of the Eustachian tube. The oropharyngeal isthmus is partially blocked by contraction of palatoglossus on each side, which narrows the space between the anterior faucial pillars. The residual gap is closed by the dorsum of the tongue wedging into it.

The protection of the larynx is a complex affair, brought about not only by closure of the sphincter mechanism of the larynx but also by tucking the larynx behind the overhanging mass of the tongue and by utilizing the epiglottis to guide the bolus away from the laryngeal entrance. The central nervous component of the swallowing reflex is depressed by narcotics, anaesthesia and cerebral trauma. In these circumstances aspiration of foreign material into the pulmonary tree becomes possible, particularly if the patient is lying on his back or in a head-up position.

The laryngeal sphincters are at three levels:
1 the aryepiglottic folds, defining the laryngeal inlet, which are apposed by the aryepiglottic and oblique arytenoid muscles;
2 the walls of the vestibule of the larynx, which are approximated by the thyroepiglottic muscles;
3 the vocal cords, which are closed by the lateral cricoarytenoid and interarytenoid mucles.

The larynx is elevated and pulled forward by the action of the thyrohyoid, stylohyoid, stylopharyngeus, digastric and mylohyoid muscles so that it comes into apposition with the base of the tongue, which is projecting backwards at this phase. While the larynx is raised and its entrance closed there is reflex inhibition of respiration.

The action of the epiglottis has been the subject of much speculation. As the head of the bolus reaches the epiglottis, the latter is first tipped backward against the pharyngeal wall and momentarily holds up the onward passage of the food. The larynx is then elevated and pulled forward, drawing with it the epiglottis so that it now stands erect, guiding the food bolus into streams along both piriform fossae and away from the laryngeal orifice, like a rock sticking up into a waterfall. Finally, the epiglottis is seen indeed to flap backwards as a cover over the laryngeal inlet, but this occurs only after the main bolus has passed beyond it. The epiglottis acts as a laryngeal lid at this stage to prevent deposition of fragments of food debris over the inlet of the larynx during re-establishment of the airway.

The cricopharyngeus then relaxes, allowing the bolus to cross the pharyngo-oesophageal junction. Fluids may shoot down the oesophagus passively under the initial impetus of the tongue action; semi-solid or solid material is carried down by peristalsis. The oesophageal transit time is about 15 seconds, relaxation of the cardia occurring just before the peristaltic wave reaches it. Gravity has little effect on the transit of the bolus, which occurs just as rapidly in the lying as in the erect position. It is, of course, quite easy to swallow fluid or solids while standing on one's head, a well-known party trick; here oesophageal transit is inevitably an active muscular process.

Clinical features

Pharyngeal pouch

The inferior constrictor muscle is made up of an upper oblique and a lower transverse part, the former arising from the side of the thyroid cartilage (the thyropharyngeus) and the latter from the cricoid (the cricopharyngeus).

Posteriorly, there is a potential gap between these two components termed the *pharyngeal dimple* or *Killian's dehiscence*. The mucosa and submucosa of the pharynx may bulge through this weak area to form a pharyngeal pouch (Fig. 204), possibly as a result of muscle incoordination or of spasm of the cricopharyngeus. This diverticulum first protrudes posteriorly; as it enlarges, backward extension is prevented by the prevertebral fascia and it therefore has to project to one side of the pharynx—usually to the more exposed left.

With further enlargement, the pouch pushes the oesophagus aside and lies directly in line with the pharynx; most food then passes into the pouch with resulting severe dysphagia and cachexia. Spill of the pouch contents into the larynx is very liable to cause inhalation of

Fig. 204 A pharyngeal pouch emerging between the two components of the inferior constrictor muscle.

Parts of the inferior constrictor {
Thyropharyngeal
Cricopharyngeal
}

Pharyngeal pouch

food material into the bronchi with respiratory infection and lung abscess as possible consequences.

The larynx

The larynx has a triple function, that of an open valve in respiration, that of a partially closed valve whose orifice can be modulated in phonation, and that of a closed valve, protecting the trachea and bronchial tree during deglutition. Coughing is only possible when the larynx can be closed effectively.

The structures which form its framework are the epiglottis, thyroid cartilage, cricoid and the arytenoids (Fig. 205).

The larynx is slung from the U-shaped hyoid bone by the *thyrohyoid membrane* and *thyrohyoid muscle*. The hyoid bone itself is attached to the mandible and tongue by the hyoglossus, the mylohyoid, geniohyoid and digastric muscles, to the styloid process by the stylohyoid ligament and muscle and to the pharynx by the middle constrictor. Three of the four strap muscles of the neck, the omohyoid, sternohyoid and thyrohyoid, find attachment to it, only the sternothyroid failing to gain it.

The *epiglottis* is a leaf-shaped elastic cartilage lying behind the root of the tongue. It is attached anteriorly to the body of the hyoid by the hyoepiglottic ligament and below to the back of the thyroid cartilage by the thyroepiglottic ligament immediately above the vocal

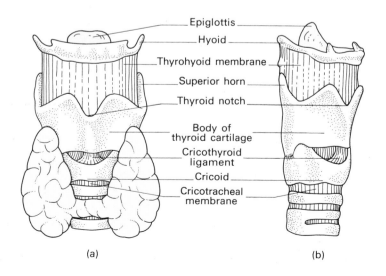

Epiglottis
Hyoid
Thyrohyoid membrane
Superior horn
Thyroid notch
Body of thyroid cartilage
Cricothyroid ligament
Cricoid
Cricotracheal membrane

(a)   (b)

Fig. 205 External view of the larynx: (a) anterior aspect; (b) anterolateral aspect.

cords. The sides of the epiglottis are connected to the arytenoids by the aryepiglottic folds which run backwards to form the margins of the entrance, or *aditus*, of the larynx.

The upper anterior surface of the epiglottis projects above the hyoid bone; the epiglottic mucosa is reflected forward to the base of the tongue and is raised up into a median glossoepiglottic fold and lateral pharyngoepiglottic folds. The depression on either side between these folds is termed the *vallecula*.

The *thyroid cartilage* is shield-like, being made up of two lateral plates meeting in the midline in the prominent 'V' of the 'Adam's apple', the laryngeal prominence, which is easily visible in the post-pubertal male.

The *cricoid* is signet-ring shaped, deepest behind. It is the only complete ring of cartilage throughout the respiratory tract. Inferiorly, it is attached to the trachea by the *cricotracheal membrane*.

The *arytenoids* sit one on each side of the posterior 'signet' of the cricoid cartilage.

In addition, there are two small nodules of cartilage at the inlet of the larynx; the *corniculate cartilage*, a nodule lying at the apex of the arytenoid, and the *cuneiform cartilage*, a flake of cartilage within the margin of the aryepiglottic fold. These are of no functional importance. They are, however, seen when the larynx is inspected through a laryngoscope (see Fig. 207) and, to the uninitiated, might mimic pathological nodules.

The *circothyroid membrane* (cricovocal membrane) connects the thyroid, cricoid and arytenoid cartilages. It is composed mainly of yellow elastic tissue. Its upper edge is attached anteriorly to the posterior surface of the thyroid cartilage and behind to the vocal process of the arytenoid. Between these two structures, the upper edge of the membrane is thickened slightly to form the *vocal ligament*. Anteriorly, the membrane thickens, as the *cricothyroid ligament*.

Passing forward from the arytenoid to the back of the thyroid cartilage, just below the epiglottic attachment, are two folds of mucosa. The upper is the *vestibular fold*, containing a small amount of fibrous tissue and forming on each side the *false vocal cord*. The lower fold (the *vocal fold* or *cord*) contains the vocal ligament (Fig. 206).

The mucosa is firmly adherent to the vocal ligament without there being any intervening submucosa. This accounts for the pearly white, avascular appearance of the vocal cords as seen on laryngoscopy. Oedema of the larynx cannot involve the true cords since there is no submucous tissue in which fluid can collect. The space between the vocal cords is the *rima glottidis*. These folds demarcate the larynx into three zones:

1   the supraglottic compartment (*vestibule*) above the false cords;

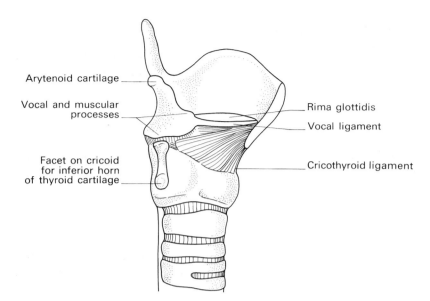

Arytenoid cartilage

Vocal and muscular
processes

Facet on cricoid
for inferior horn
of thyroid cartilage

Rima glottidis

Vocal ligament

Cricothyroid ligament

Fig. 206 The internal
structure of the larynx—
the lamina of the thyroid
cartilage has been cut
away.

2   the glottic compartment between the false and true cords;
3   the subglottic compartment between the true cords and the first
ring of the trachea.

On either side of the larynx the pharynx forms a recess, the *piriform fossa*, in which swallowed foreign bodies tend to lodge.

The muscles of the larynx function to open the glottis in inspiration, close the vestibule and glottis in deglutition and alter the tone of
the true vocal cords in phonation.

The cricothyroid is the only external muscle of the larynx and
tenses the vocal cord (the only muscle to do so), by a slight tilting
action on the cricoid. It is supplied by the superior laryngeal nerve.

The remaining muscles constitute a single encircling sheet whose
various attachments are denoted by the names of its separate parts:
the thyroarytenoid, posterior and lateral cricoarytenoid, the
aryepiglottic, thyroepiglottic and interarytenoid muscles. These are
all supplied by the recurrent laryngeal nerve.

All these muscles except one have a sphincter action; the exception is the posterior cricoarytenoid on each side which, by rotating
the arytenoids outwards, separates the vocal cords.

Blood supply

The larynx receives a superior and inferior laryngeal artery from the
superior and inferior thyroid artery respectively. These vessels accompany the superior and recurrent laryngeal nerves.

Lymph drainage

Above the vocal cords the larynx drains to the upper deep cervical and then to the mediastinal lymph nodes, some lymphatics passing via small nodes lying on the thyrohyoid membrane.

Below the cords, drainage is to the lower deep cervical nodes, partially via nodes on the front of the larynx and trachea.

The vocal cords themselves act as a complete barrier separating the two lymphatic areas, but posteriorly there is free communication between them; a laryngeal carcinoma may thus seed throughout the lymphatic drainage area of the larynx.

Nerve supply

The nerve supply of the larynx is of great practical importance and comprises the superior and recurrent laryngeal branches of the vagus nerve.

The *superior laryngeal nerve* passes deep to the internal and external carotid arteries where it divides; its internal branch pierces the thyrohyoid membrane together with the superior laryngeal vessels to supply the mucosa of the larynx down to the vocal cords. The external branch passes deep to the superior thyroid artery to supply the cricothyroid muscle.

The *recurrent laryngeal nerve* has a different course on each side. The right arises from the vagus as this crosses the front of the subclavian artery, passes deep to and behind this vessel, then ascends behind the common carotid to lie in the tracheo-oesophageal groove accompanied by the inferior laryngeal vessels (Fig. 188). The nerve then passes deep to the inferior constrictor muscle of the pharynx to enter the larynx behind the cricothyroid articulation.

The left nerve arises on the arch of the aorta, winds below it, deep to the ligamentum arteriosum, and ascends to the trachea. It then lies in the tracheo-oesophageal groove and is distributed as on the right side.

The recurrent nerves supply all the intrinsic laryngeal muscles, apart from the cricothyroid, and the mucosa below the vocal cords.

Clinical features

1   The laryngeal nerves bear relationships to the thyroid arteries which are of considerable practical importance in thyroidectomy. The external branch of the superior laryngeal nerve lies immediately deep to the superior thyroid artery and may be injured in ligating this vessel.

The recurrent laryngeal nerve, lying in the tracheo-oesophageal groove, is usually behind the terminal branches of the inferior thyroid artery. Occasionally, however, the nerve lies in front of these vessels or passes between them (Fig. 191). Moreover, when a large thyroid is pulled forward during thyroidectomy, the nerve becomes dragged forward with it and is therefore placed in further jeopardy. To avoid nerve damage during ligation of the inferior thyroid artery, this procedure should be carried out well laterally, just as the artery emerges from behind the carotid sheath and before it takes up its intimate and inconstant relationship to the nerve.

2   Damage to the superior nerve causes some weakness of phonation due to the loss of the tightening effect of the cricothyroid muscle on the cord.

3   Complete division of a recurrent laryngeal nerve causes the cord on the affected side to take up the neutral (or paramedian) position between abduction and adduction. Usually the other cord is able to compensate in a remarkable way and speech is not greatly affected; if both nerves are divided, however, the voice is completely lost and breathing becomes difficult through the only partially opened glottis.

4   If the recurrent nerve is only bruised or partially damaged, the abductors (posterior cricoarytenoids) are affected more than the adductors; this is known as *Semon's law*. The affected cord adopts the midline adducted position. In bilateral incomplete paralysis, therefore, the cords come together, stridor is intense and tracheotomy may become essential.

5   The left recurrent laryngeal nerve, in its thoracic course, may become involved in a bronchial or oesophageal carcinoma, or in a mass of enlarged mediastinal nodes, or may become stretched over an aneurysm of the aortic arch. The enlarged left atrium in advanced mitral stenosis may produce a recurrent laryngeal palsy by pushing up the left pulmonary artery which compresses the nerve against the aortic arch.

Either nerve, in the neck, may be damaged by an extending thyroid carcinoma or malignant lymph nodes. For these reasons, loss of voice must always be regarded as an ominous symptom requiring careful investigation.

6   The larynx can be inspected either directly, by means of the laryngoscope, or indirectly through a laryngeal mirror. The base of the tongue, valleculae, epiglottis, aryepiglottic folds and piriform fossae are viewed, then the false cords, which are red and widely apart, then, between these, the pearly white true cords (Fig. 207).

For the passage of the laryngoscope, endotracheal tube or bronchoscope it is essential to know the position which brings the axes of

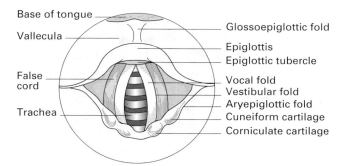

Base of tongue
Vallecula
False cord
Trachea

Glossoepiglottic fold
Epiglottis
Epiglottic tubercle
Vocal fold
Vestibular fold
Aryepiglottic fold
Cuneiform cartilage
Corniculate cartilage

Fig. 207 Diagram of the larynx as seen at laryngoscopy.

the mouth, oropharynx and laryngeal inlet into line; this is achieved by bringing the neck forward and at the same time extending the head fully at the atlanto-occipital joint — it is the position in which one sniffs at the fresh air after a long day in the operating theatre.

The salivary glands

The parotid gland

This is the largest of the salivary glands, lying wedged between the mandible and sternocleidomastoid and overflowing both these bounding structures (Fig. 208).

Relations

- Above—lie the external auditory meatus and temporomandibular joint.
- Below—it overflows the posterior belly of digastric.
- Anteriorly — it overflows the mandible with the overlying masseter.
- Medially—lies the styloid process and its muscles separating the parotid from the internal jugular vein, internal carotid artery, last four cranial nerves and the lateral wall of the pharynx.

The gland itself is enclosed in a split in the investing fascia, lying both on and below which are the parotid lymph nodes. Antero-inferiorly, this parotid fascia is thickened and is the only structure separating the parotid from the submandibular gland (the stylomandibular ligament).

Traversing the gland (from without in) are:

1   the facial nerve (see below);

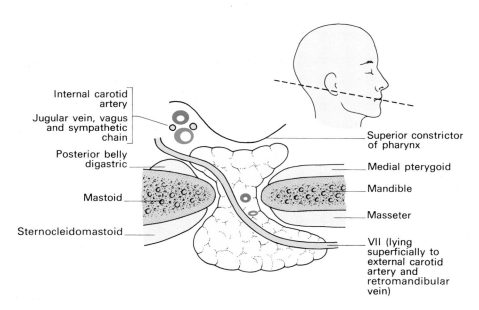

Fig. 208 The parotid and its surrounds in a schematic horizontal section—the facial nerve is the most superficial of the structures traversing the gland. (The line of section is shown in the inset head.)

2   the retromandibular (posterior facial) vein, formed by the junction of the superficial temporal and maxillary veins;
3   the external carotid artery, dividing at the neck of the mandible into its superficial temporal and maxillary terminal branches.

The *parotid duct (of Stensen)* is 2 in (5 cm) long. It arises from the anterior part of the gland, runs over the masseters a finger's breadth below the zygomatic arch to pierce the buccinator and open opposite the second upper molar tooth. The duct can easily be felt by a finger rolled over the masseter if this muscle is tensed by clenching the teeth.

The relations of the facial nerve to the parotid

The facial nerve is unique in traversing the substance of a gland, a fact of considerable importance to the surgeon. This coexistence is explained embryologically; the parotid gland develops in the crotch formed by the two major branches of the facial nerve. As the gland enlarges it overlaps these nerve trunks, the superficial and deep parts fuse and the nerve comes to lie buried within the gland. The fanciful comparison between the nerve and the two parotid lobes and sandwich–filling between two slices of bread is not valid because the two lobes of the parotid come to fuse intimately with each other both around and between the branches of the nerve.

The facial nerve emerges from the stylomastoid foramen, winds laterally to the styloid process and can then be exposed surgically in the inverted V between the bony part of the external auditory meatus and the mastoid process. This has a useful surface marking, the inter-tragic notch of the ear, which is situated directly over the facial nerve.

Just beyond this point the nerve dives into the posterior aspect of the parotid gland and bifurcates almost immediately into its two main divisions (occasionally it divides before entering the gland). The upper division divides into *temporal* and *zygomatic* branches; the lower division gives the *buccal, mandibular* and *cervical* branches (Fig. 209).

These two divisions may remain completely separate within the parotid, may form a plexus of intermingling connections, or, most usually, display a number of cross-communications which can be safely divided during dissection without jeopardy.

The branches of the nerve then emerge on the anterior aspect of the parotid to lie on the masseter, thence to pass to the muscles of the face. No branches emerge from the superficial aspect of the gland, which can therefore be completely exposed with impunity.

Clinical features

1   A malignant tumour of the parotid gland, unlike benign lesions, may involve VII and produce a facial palsy.

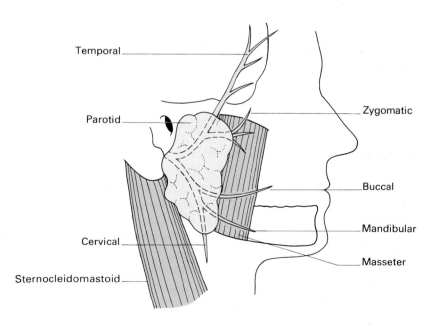

Fig. 209 The named branches of the facial nerve which traverse the parotid gland.

2   In removing a benign mixed salivary tumour of the parotid, the facial nerve is exposed posteriorly in the wedge-shaped space between the bony canal of the external auditory meatus and the mastoid process. It is then traced into the gland, its main divisions defined and the tumour excised with a wide margin of normal gland, carefully preserving the exposed nerves.

It is interesting that giant mixed tumours 'extrude' clear away from the facial nerve and can be excised with an adequate margin without even seeing the nerve.

3   The parotid duct and its ramifications can be demonstrated radiologically by injecting radio-opaque contrast through a cannula placed in the mouth of the duct (a parotid sialogram).

The submandibular gland

The submandibular gland is made up of a large superficial and a small deep lobe which connect with each other around the posterior border of the mylohyoid.

The superficial lobe of the gland lies at the angle of the jaw, wedged between the mandible and the mylohyoid and overlapping the digastric muscle (Fig. 200). Posteriorly it comes into contact with the parotid gland, separated only by a condensation of its fascial sheath (the stylomandibular ligament).

Superficially, the gland is covered by platysma and by its capsule of deep fascia, but it is crossed by the cervical branch of VII and by the facial vein. Its deep aspect lies against the mylohoid for the most part, but posteriorly the gland rests against the hyoglossus muscle and here comes into contact with the lingual and the hypoglossal nerve, both of which lie on hyoglossus as they pass forward to the tongue.

The facial artery also comes into close relationship with the gland, approaching it posteriorly, then arching over its superior aspect (which it grooves), to attain the inferior border of the mandible and thence to ascend on to the face in front of the masseter.

From the medial aspect of the superficial part of the gland projects its deep prolongation along the hyoglossus.

The *submandibular duct* (*Wharton's duct*) arises from this deep part of the gland and runs forward, beneath the mucosa of the floor of the mouth along the side of the tongue, to open immediately at the side of the frenulum linguae (Fig. 197). Here its orifice is readily visible and saliva can be seen trickling from it.

The sublingual gland (*vide infra*) lies immediately lateral to the submandibular duct.

The lingual nerve reaches the tongue by passing from the lateral side of the duct below and then medial to it—thus 'double-crossing' it.

The *submandibular lymph nodes* lie partly embedded within the gland and partly between it and the mandible.

Clinical features

1   The rather complex relations of this gland have been given at some length because excision of the gland for calculus or tumour is not uncommon. This operation is carried out through a skin crease incision below the angle of the jaw.

The mandibular branch of VII passes behind the angle of the jaw rather less than 1 inch from it before arching upwards over the body of the mandible to supply the depressor of the lip. The incision must therefore be placed rather more than 1 in (2.5 cm) below the angle of the jaw in order to preserve this nerve.

2   The presence of small lymph nodes actually within the substance of the gland makes removal of the gland an imperative part of block dissection of the neck.

3   In differentiating between an enlarged submandibular gland and a mass of submandibular lymph nodes, one remembers that the gland lies not only below the mandible but also extends into the floor of the mouth; it can therefore be palpated bimanually between a finger in the mouth and a finger below the angle of the jaw. Enlarged lymph nodes are felt only at the latter site.

4   A stone in Wharton's duct can be felt bimanually in the floor of the mouth and can be seen if sufficiently large.

The sublingual gland

This is an almond-shaped salivary gland lying immediately below the mucosa of the floor of the mouth and immediately in front of the deep part of the submandibular gland. Laterally, it rests against the sublingual groove of the mandible while medially it is separated from the base of the tongue by the submandibular duct and its close companion, the lingual nerve (Fig. 197).

The gland opens by a series of ducts into the floor of the mouth and also in the submandibular duct.

The sublingual gland produces a mucous secretion, the parotid a serous secretion and the submandibular gland a mixture of the two.

As well as these main salivary glands, small accessory glands are found scattered over the palate, lips, cheek, tonsil and tongue. These glands are occasional sites for development of a mixed salivary tumour.

The major arteries of the head and neck

The common carotid arteries

The *left common carotid artery* arises from the aortic arch in front and to the right of the origin of the left subclavian artery. It passes behind the left sternoclavicular joint, lying in its thoracic course at first in front and then to the left side of the trachea, with the left lung and pleura, the vagus and the phrenic nerve as its lateral relations.

The *right common carotid artery* begins behind the right sterno-clavicular joint at the bifurcation of the brachiocephalic artery.

In the neck, both common carotids have essentially similar courses and relationships; they ascend in the carotid fascial sheath which contains also the internal jugular vein laterally, and the vagus nerve between and rather behind the artery and vein. The cervical sympathetic chain ascends immediately posterior to the carotid sheath. These structures form a quartet which should always be considered in this inseparable manner; the relations of any one are those of the other three (Figs 188, 210).

In the neck, each common carotid artery lies on the cervical transverse processes, separated from them by the prevertebral muscles. Medially are larynx and trachea, pharynx and oesophagus, together with the thyroid gland, which overlaps on to the anterior aspect of the carotid. Superficially, the artery is covered by the sternocleidomastoid and, in its lower part, by the strap muscles.

The common carotid artery gives off no side branches but terminates at the level of the upper border of the thyroid cartilage (at the vertebral level C4) into the external and internal carotids, which are more or less equal in size.

The external carotid artery

This artery lies first deep to the anterior border of the sternocleido-mastoid and then quite superficially in the anterior triangle of the neck, where its pulsations are usually visible as well as palpable. At first it is slightly deep to the internal carotid, then passes anterior and lateral to it. The internal jugular vein is first lateral to the external carotid then posterior to it, coming into lateral relationship to the internal carotid. The pharynx lies medially.

The external carotid artery ascends beneath the XII nerve and the posterior belly of the digastric to enter the parotid gland, within which it lies deep to the facial nerve and the retromandibular vein (Fig. 208).

The artery ends within the parotid gland at the level of the neck of the mandible by dividing into the superficial temporal and internal maxillary arteries.

Branches (Fig. 210)

1 The *superior thyroid artery* (giving off the superior laryngeal artery).

2 The *lingual artery*, passing deep to the hyoglossus to supply the tongue.

3 The *facial artery*, which gives off its important branch to the palatine tonsil, loops over the submandibular gland, hooks round the mandible (against which it can be felt pulsating), and ascends on to the face.

4 The *occipital artery*, running along the inferior border of the digastric muscle's posterior belly to the back of the scalp, where its pulse is often palpable.

5 The *posterior auricular artery*.

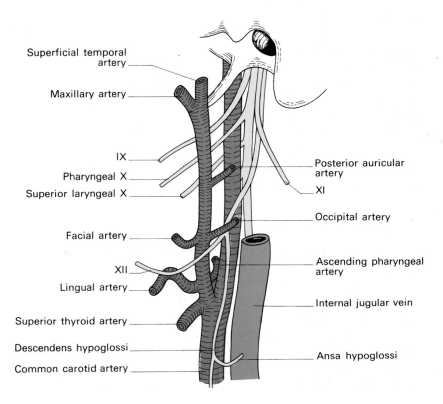

Fig. 210 The carotid arteries, their branches and their related nerves.

6   The *ascending pharyngeal artery*. Its terminal branches are:
   • the *superficial temporal artery*, which is palpable on the zygomatic process;
   • the *internal maxillary artery*, which supplies the upper and
   lower jaws, nasal cavity and the muscles of mastication and also
   gives off the middle meningeal artery. This small vessel ascends
   through the foramen spinosum and helps to supply the meninges.
   Its practical importance is that it may be torn in a skull fracture
   and result in the formation of an extradural haematoma.

The internal carotid artery

This artery commences at the bifurcation of the common carotid,
and, at its origin, is dilated into the carotid sinus. This area receives a
rich nerve supply from IX and acts as a pressor-receptor; through this
mechanism a rise of blood pressure brings about reflex slowing of the
heart and peripheral vasodilatation. Tucked deep to the bifurcation is
the small, yellowish *carotid body* which is also supplied by IX. This is
a chemoreceptor which produces a reflex increase in respiration in
response to any rise in CO_2 tension or fall in the oxygen tension of the
blood.

The internal carotid lies first lateral to the external carotid but
rapidly passes medial and posterior to it, to ascend along the side-wall
of the pharynx. It does so with the internal jugular vein, vagus and
cervical sympathetic chain in the same relationship to it that they bear
to the common carotid artery. At first the artery is covered superficially only by the sternocleidomastoid, XII, and the common facial
vein; it then passes under the posterior belly of the digastric muscle
and parotid gland to the base of the skull. It is separated from the
external carotid artery not only by the parotid but also by the styloid
process and the muscles arising from it, by IX and by the pharyngeal
branches of X.

At the base of the skull, the internal carotid artery enters the
carotid canal in the petrous temporal bone. Only at the skull base
does the internal jugular vein lose its close lateral relation to the internal carotid, passing posterior to the artery into the jugular foramen.
At this point the two vessels are separated by the emerging last four
cranial nerves.

The artery gives off no branches in the neck.

The internal carotid, on entering the skull, commences an extraordinary twisted course. It passes forwards through the temporal
bone, upwards into the cavernous sinus, forward in this, upwards
through the roof of the sinus to lie medial to the anterior clinoid
process, turns back on itself above the cavernous sinus, then passes up

once more, lateral to the optic chiasma, to end by dividing into the *anterior* and *middle cerebral arteries*. There are thus six bends in the intracranial course of this artery (readily appreciated by studying a lateral carotid arteriogram) which are believed to lessen the pulsating force of the arterial systolic blood pressure on the delicate cerebral tissues.

The *ophthalmic artery* originates from the internal carotid immediately after its emergence from the cavernous sinus, enters the orbit through the optic foramen below and lateral to the optic nerve and supplies the orbital contents and the skin above the eyebrow (via the supratrochlear and supra-orbital branches). Its most important branch, however, is the *central artery of the retina* which is the sole blood supply to this structure.

The two terminal branches of the internal carotid are distributed as follows (Fig. 211).

The *anterior cerebral artery* winds round the genu of the corpus callosum to supply the medial and superolateral aspect of the cerebral hemisphere.

The *middle cerebral artery* enters the lateral cerebral sulcus, gives off central branches to supply the internal capsule ('the *artery of cerebral haemorrhage*') and feeds most of the lateral aspect of the cerebral cortex.

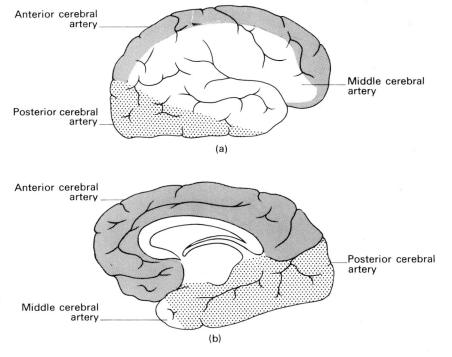

Fig. 211 The arterial supply of the cerebral cortex. (a) Lateral aspect. (b) Medial aspect.

The *arterial circle of Willis* (Fig. 212) is completed in front by the *anterior communicating artery*, which links the two anterior cerebral arteries, and behind by a *posterior communicating artery* on each side, passing backwards from the internal carotid to anastomose with the posterior cerebral, a branch of the *basilar artery*, the latter being formed by the junction of the two *vertebral arteries*.

Clinical features

The common carotid artery can be exposed through a transverse incision over the origin of the sternocleidomastoid immediately above the sternoclavicular joint. The carotid sheath lies immediately deep to the junction between the sternal and clavicular heads of the sternocleidomastoid and is revealed either by retracting this muscle laterally or by splitting between its heads. Opening the sheath then reveals the artery lying medial to the internal jugular vein.

Ligation of the common carotid artery may be performed for intracranial aneurysm arising on the internal carotid. This operation is effective because it lowers the blood flow through the aneurysm, allowing thrombosis to occur. Adequate blood supply to the brain on the affected side is provided by free communication between the branches of the external carotid arteries on each side. Within the cranium, cross-circulation occurs through the circle of Willis.

The internal and external carotids, as well as the terminal part of the common carotid artery, can be exposed through an incision along the anterior border of the sternocleidomastoid passing downwards from the angle of the jaw. The sternocleidomastoid is retracted, the

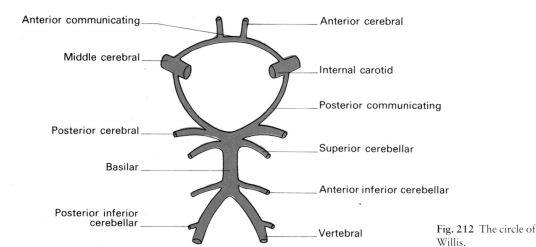

Fig. 212 The circle of Willis.

Recurrent laryngeal nerve
Brachial plexus
Dome of pleura
Vertebral artery
X
Scalenus anterior
Phrenic nerve
Thoracic duct
Subclavian artery and vein
Common carotid artery
Trachea on oesophagus
Sternohyoid on sternothyroid

Fig. 213 The root of the neck.

common facial vein divided, but the hypoglossal nerve, crossing the external and internal carotids just below the posterior belly of the digastric, is carefully preserved.

It may be surprisingly difficult to differentiate between the external and internal carotids at operation; the former is the anterior and rather deeper-placed vessel at origin and, moreover, is the only carotid in the neck which gives off branches.

The subclavian arteries (Fig. 213)

The *left subclavian artery* arises from the arch of the aorta, immediately behind the commencement of the left common carotid artery. It ascends against the mediastinal surface of the left lung and pleura laterally and the trachea and oesophagus medially to lie behind the sternoclavicular joint.

The *right subclavian artery* is formed behind the right sternoclavicular joint by the bifurcation of the brachiocephalic artery; beyond this point, the course of the two arteries is much the same.

The cervical course of the subclavian arteries is conveniently divided by the scalenus anterior muscle into three parts.

The first part arches over the dome of the pleura and lies deeply placed beneath the sternocleidomastoid and the strap muscles. It is crossed at its origin by the carotid sheath and, more laterally, by the phrenic and vagus nerves. At this site, on the right side, the vagus gives off its recurrent laryngeal branch which hooks behind the artery.

On the left side, the thoracic duct crosses the first part of the artery to open into the commencement of the left branchiocephalic vein.

The second part of the artery lies behind scalenus anterior which separates it from the subclavian vein. Behind lie scalenus medius and also the middle and upper trunks of the brachial plexus.

The *third* part extends to the lateral border of the first rib against which it can be compressed and its pulse easily felt, since here it is just below the deep fascia. Immediately behind the artery is the lower trunk of the brachial plexus which is, in fact, responsible for the 'subclavian groove' on the first rib.

Branches of the subclavian artery

- 1st part
  - 1 The vertebral artery
  - 2 The thyrocervical trunk:
    - (a) inferior thyroid artery
    - (b) transverse cervical artery
    - (c) suprascapular artery
  - 3 The internal thoracic artery
- 2nd part — the costocervical trunk (supplying deep structures of the neck via its deep cervical branch, and the superior intercostal artery, which gives off the 1st and 2nd posterior intercostal arteries).
- 3rd part—gives no constant branch.

The vertebral artery

This is the most important of the branches of the subclavian artery. It crosses the dome of the pleura, traverses the transverse foramina of the upper six cervical vertebrae, then turns posteriorly and medially over the posterior arch of the atlas to enter the cranial cavity at the foramen magnum by piercing the dura mater. It then runs on the anterolateral aspect of the medulla to join its fellow in front of the pons to form the *basilar artery* (Fig. 212).

The following are the important branches of the vertebral artery:
1 anterior and posterior spinal arteries;
2 posterior inferior cerebellar artery.
From the basilar:
3 anterior inferior cerebellar artery;
4 superior cerebellar artery;
5 posterior cerebral artery (supplying the occipital lobe and medial aspect of the temporal lobe; Fig. 211).

Clinical features

1   The right subclavian artery is grafted end-to-side into the right pulmonary artery to short-circuit the pulmonary stenosis of the tetralogy of Fallot (Blalock's operation) (see Fig. 33). It is important to note, therefore, that variations occur in the origins of the right subclavian artery, which may arise directly from the aortic arch either as its first or as its last branch. In the latter case, the right subclavian artery passes behind the trachea and oesophagus in the course to the neck; this vessel may then compress the oesophagus and produce difficulty in swallowing (dysphagia lusoria). Occasionally, the left subclavian artery has a common origin with the left carotid from the aortic arch.

2   An aneurysm of the subclavian artery is not rare; it never involves the thoracic part of the subclavian and its site of election is the third part of the artery. The close relation of the subclavian artery to the brachial plexus accounts for the pain, weakness and numbness in the arm which accompany this lesion. Oedema of the arm may result in compression of the subclavian vein.

3   A cervical rib may elevate the subclavian artery and render it unduly palpable; under these circumstances it may closely simulate an aneurysm and, in fact, there may be aneurysmal dilation of the artery distal to the edge of the cervical rib. Vascular changes in the arm associated with a cervical rib are probably due to peripheral emboli thrown off from thrombi forming on the walls of the compressed subclavian artery.

The veins of the head and neck

The cerebral venous system

The venous drainage of the brain follows two pathways:

1   the superficial structures, e.g. the cerebral and cerebellar cortices, drain to the nearest available dural sinus (see below) by thin-walled veins;

2   the deep structures drain through the *internal cerebral vein* on each side, which is formed at the interventricular foramen by the junction of the *choroid vein* (draining the choroid plexus of the lateral ventricle) with the *thalamostriate vein* (draining the basal ganglia).

The two internal cerebral veins unite to form the *great cerebral vein* (*the vein of Galen*) which emerges from under the splenium of the corpus callosum to join the inferior sagittal sinus in the formation of the straight sinus.

The venous sinuses of the dura (Fig. 214)

The venous sinuses lie between the layers of the dura. They receive the venous drainage of the brain and of the skull (the *diploic veins*) and disgorge ultimately into the internal jugular vein. They also communicate with the veins of the scalp, face and neck via *emissary veins* which pass through a number of the foramina in the skull.

The *superior sagittal sinus* lies along the attached edge of the falx cerebri and ends posteriorly (usually) in the right transverse sinus. Connecting with it are a number of venous lakes (*lacunae laterales*) into which project the *Pacchionian bodies* of arachnoid, filtering cerebrospinal fluid (C.S.F.) back into the blood.

The *inferior sagittal sinus* lies in the free margin of the falx cerebri and opens into the straight sinus.

The *straight sinus* lies in the tentorium cerebelli along the attach-

(a)

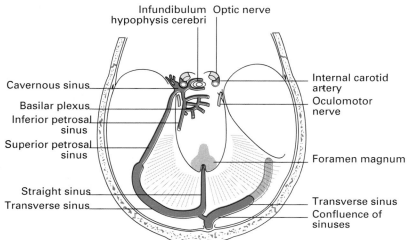

(b)

Fig. 214 The venous dural sinuses. (a) Lateral and (b) superior view.

ment of the flax cerebri. It is formed by the junction of the great cerebral vein with the inferior sagittal sinus and runs backwards to open (usually) into the left transverse sinus.

The *transverse sinuses* commence at the internal occipital protuberance and run in the tentorium cerebelli on either side along its attached margin. On reaching the mastoid part of the temporal bone each passes downwards, forwards and medially as the *sigmoid sinus* to emerge through the jugular foramen as the *internal jugular vein.*

The *cavernous sinuses* (Fig. 215) lie one on either side of the body of the sphenoid against the fibrous wall of the pituitary fossa and rest inferiorly on the greater sphenoid wing. They communicate freely with each other via the *intercavernous sinuses.*

Traversing the cavernous sinus are the carotid artery and the cranial nerves III, IV, V (ophthalmic and maxillary divisions) and VI. Lying above the cavernous sinus are three important structures—the optic tract, the uncus and the internal carotid artery, which first pierces the roof of the sinus then doubles back to lie against it.

The *ophthalmic veins* drain into the anterior aspect of the cavernous sinus which also links up, through these veins, with the pterygoid venous plexus and the anterior facial vein. The cavernous sinus also receives venous drainage from the brain (the *superficial middle cerebral vein*) and from the dura (the *sphenoparietal sinus*).

Posteriorly, the *petrosal sinuses* drain the cavernous sinus into the sigmoid sinus and into the commencement of the internal jugular vein.

Clinical features

1   The cavernous sinus is liable to sepsis and thrombosis as a result of spread of superficial infection from the lips and face via the ante-

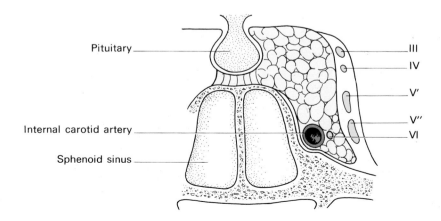

Fig. 215 The cavernous sinus—shown in coronal section.

rior facial and ophthalmic veins, or from deep infections of the face via the pterygoid venous plexus around the pterygoid muscles, or from suppuration in the orbit or accessory nasal sinuses along the ophthalmic vein and its tributaries. A characteristic picture results— blockage of the venous drainage of the orbit causes oedema of the conjunctiva and eyelids and marked exophthalmos, which demonstrates transmitted pulsations from the internal carotid artery. Pressure on the contained cranial nerves results in ophthalmoplegia. Examination of the fundus shows papilloedema, venous engorgement and retinal haemorrhages, all resulting from the acutely obstructed venous drainage.

2   Fractures of the skull or penetrating injuries of the skull base may rupture the internal carotid artery within the cavernous sinus. A caroticocavernous arteriovenous fistula results with pulsating exophthalmos, a loud bruit easily heard over the eye and, again, ophthalmoplegia and marked orbital and conjunctival oedema due to the venous pressure within the sinus being raised to arterial level.

3   The sigmoid and transverse sinuses are often together termed *the lateral sinus* by clinicians. Close relationship to the mastoid and middle ear renders these sinuses liable to infective thrombosis secondary to otitis media.

Spread of infection or thrombosis from the lateral sinus to the sagittal sinus may cause impaired C.S.F. drainage into the latter and therefore the development of a hydrocephalus — this syndrome of raised C.S.F. pressure associated with sinus thrombosis following ear infection is termed *otitic hydrocephalus.*

It is also possible for sagittal sinus thrombosis to follow infections of the skull, nose, face or scalp because of its diploic and emissary vein connections; if there were no emissary veins, infections of the face and scalp would never have achieved their sinister reputation.

The internal jugular vein

The internal jugular vein runs from its origin at the jugular foramen (where it continues the sigmoid sinus) to its termination behind the sternal extremity of the clavicle, where it joins the subclavian vein to form the brachiocephalic vein.

It lies lateral first to the internal and then to the common carotid artery within the carotid sheath and its relations are therefore identical with these vessels (Fig. 210). The deep cervical chain of lymph nodes lies close against the vein and, if involved by malignant or inflammatory disease, may become densely adherent to the vein. Tearing of the jugular vein for this reason is far from rare in dissections of tuberculous cervical lymph nodes.

Its tributaries are:

1 the pharyngeal venous plexus;
2 the common facial vein;
3 the lingual vein;
4 the superior and middle thyroid veins.

Superficial veins

The arrangement of the superficial veins of the head and neck are somewhat variable but the usual plan is as follows (Fig. 216).

The *superficial temporal* and *maxillary veins* join to form the *retromandibular vein*. This branches while traversing the parotid gland. Its posterior division, together with the *posterior auricular vein*, form the external jugular vein, whereas the anterior division joins the *facial vein* to form the *common facial vein* which opens into the internal jugular vein.

The *external jugular vein* crosses the sternocleidomastoid in the superficial fascia, traverses the roof of the posterior triangle then plunges through the deep fascia 1 in (2.5 cm) above the clavicle to enter the subclavian vein.

The *anterior jugular vein* runs down one on either side of the midline of the neck, crossing the thyroid isthmus. Just above the sternum it communicates with its fellow then passes outwards, deep to the sternocleidomastoid, to enter the external jugular vein.

The subclavian vein

This is the continuation of the axillary vein and extends from its com-

Fig. 216 The usual arrangement of the veins in the neck.

Superficial temporal

Maxillary (draining pterygoid plexus)

Posterior auricular

Retromandibular

Internal jugular

External jugular

Facial

Common facial

Anterior jugular

mencement at the outer border of the first rib to the medial border of scalenus anterior, where it joins the internal jugular vein to form the brachiocephalic vein. During its short course it crosses, and lightly grooves, the superior surface of the first rib. It arches upwards and then passes medially, downwards and slightly forwards to its termination behind the sternoclavicular joint. On the left side it receives the termination of the thoracic duct. Anteriorly the vein is related to the clavicle and subclavius muscle (Fig. 217).

Clinical features

Techniques of central venous catheterization are now of great clinical importance both to measure central venous pressure (c.v.p.), for practical purposes the pressure within the right atrium, and also to allow rapid blood replacement and long-term intravenous feeding by means of glucose, amino acids and fats. The internal jugular vein can be cannulated by direct puncture in the triangular gap between the sternal and clavicular heads of the sternocleidomastoid immediately above the clavicle. Feel this landmark on yourself. The needle is inserted near the apex of this triangle at an angle of 30–40° to the skin surface and is advanced caudally towards the inner border of the anterior end of the first rib behind the clavicle. A reflux of blood confirms venepuncture.

Subclavian venepuncture can be carried out most effectively by

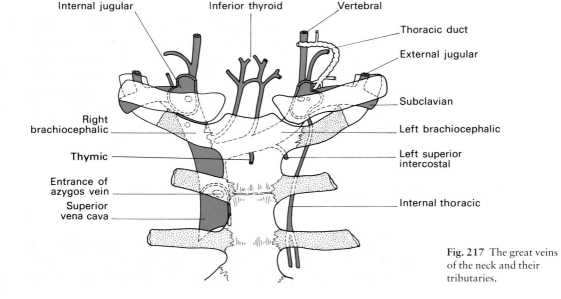

Fig. 217 The great veins of the neck and their tributaries.

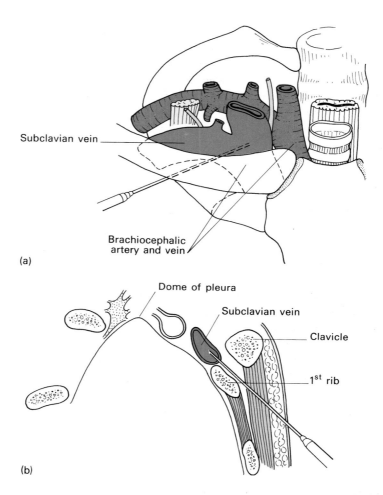

Fig. 218 The anatomy of the infraclavicular approach to the subclavian vein. (a) Anterior view. (b) In sagittal section.

the infraclavicular approach (Fig. 218). The needle is inserted below the clavicle of the junction of its medial and middle thirds. The needle is advanced medially and upwards behind the clavicle in the direction of the sternoclavicular joint to puncture the subclavian vein at its junction with the internal jugular. When a free flow of blood is obtained by syringe aspiration, a radio-opaque plastic catheter is threaded through the needle to pass into the brachiocephalic vein.

The lymph nodes of the neck

Although the lymph drainage of particular viscera is dealt with under appropriate headings (tongue, larynx, etc.), it is convenient to summarize here the arrangements of the lymph nodes of the head and neck as a whole (Fig. 219). These can be grouped into horizontal and vertically disposed aggregates.

The *horizontal nodes* form a number of groups which encircle the junction of the head with the neck and which are named, according to their position, the *submental, submandibular, superficial parotid* (or *preauricular*), *mastoid* and *suboccipital* nodes. These nodes drain the superficial tissues of the head and efferents then pass to the deep cervical nodes (although some lymph vessels pass direct to the cervical nodes, bypassing the horizontal nodes).

The *vertical nodes* drain the deep structures of the head and neck. The most important is the *deep cervical group*, which extends along the internal jugular vein from the base of the skull to the root of the neck (Fig. 198). The lymph then passes via the jugular trunk to the thoracic duct or the right lymphatic duct.

The *superficial cervical nodes* lie along the external jugular vein, serve the parotid and lower part of the ear and drain into the deep cervical group.

Along the front of the neck lies another group of vertically disposed nodes, the *infrahyoid* (on the thyrohyoid membrane), the *prelaryngeal* and the *pre-* and *paratracheal nodes*. These drain the thyroid, larynx, trachea and part of the pharynx and empty into the deep cervical group.

The *retropharyngeal nodes*, lying vertically behind the pharynx, drain the back of the nose, pharynx and Eustachian tube; their efferents pass to the upper deep cervical nodes.

Thus all structures in the head and neck drain through the deep cervical nodes either directly or ultimately.

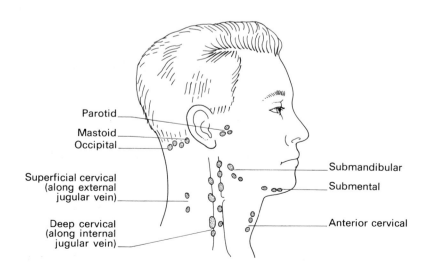

Fig. 219 Scheme of the lymph nodes of the head and neck.

Clinical features

1   *Block dissection of the neck* for malignant disease is the removal of the lymph nodes of the anterior and posterior triangles of the neck and their associated lymph channels, together with those structures which must be excised in order to make this lymphatic ablation possible. It is sometimes combined with en-bloc removal of the primary tumour.

The usual incision is Y-shaped, its centre being at the level of the upper border of the thyroid cartilage, its lower limb running downwards to the midpoint of the clavicle, its anterior limb extending to the symphysis menti and its posterior limb to the mastoid process. The block of tissue removed extends from the mandible above to the clavicle below and from the midline anteriorly to the anterior border of the trapezius behind. It consists of all the structures between the platysma and pretracheal fascia enclosed by these boundaries, preserving only the carotid arteries, the vagus trunk, the cervical sympathetic chain and the lingual and hypoglossal nerves. The sternocleidomastoid, omohyoid and digastric muscles are removed in the dissection. Excision also includes the external and internal jugular veins, around each of which lymph nodes are intimately related, and the submandibular gland and the lower pole of the parotid gland, since these both contain potentially involved lymph nodes.

The accessory nerve, passing across the posterior triangle, is usually sacrificed.

2   Tuberculous disease of the neck usually involves the upper part of the deep cervical chain (from tonsillar infection). These infected nodes may adhere very firmly to the internal jugular vein which may be wounded in the course of their excision.

The cervical sympathetic trunk

The sympathetic chain continues upwards from the thorax by crossing the neck of the first rib, then ascends embedded in the posterior wall of the carotid sheath to the base of the skull (Fig. 220). It bears three ganglia:

1   the *superior cervical ganglion* (the largest) lies opposite C2 and 3 vertebrae and sends grey rami communicantes to C1–4 spinal nerves;

2   the *middle ganglion* lies level with C6 vertebra and sends grey rami to C5 and 6 nerves;

3   the *inferior ganglion* lies level with C7 and is tucked behind the

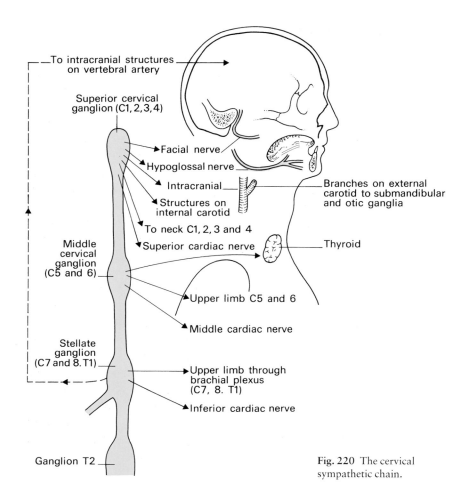

Fig. 220 The cervical
sympathetic chain.

vertebral artery. Frequently, it fuses with the first thoracic ganglion to
form the *stellate ganglion* at the neck of the first rib. Grey rami pass
from it to C7 and 8 nerves.

Note that these ganglia receive no white rami from the cervical
nerves; their preganglionic fibres originate from the upper thoracic
white rami and then ascend in the sympathetic chain.

As well as *somatic branches* transmitted with the cervical nerves,
the cervical chain gives off *cardiac branches* from each of its ganglia
and also *vascular plexuses* along the carotid, subclavian and vertebral
vessels. The sympathetic fibres to the dilator pupillae muscle travel in
this plexus along the internal carotid artery. Grey rami pass from the
superior ganglion to cranial nerves VII, IX and X.

Clinical features

1 'Cervical sympathectomy' is a misnomer; it is an upper thoracic sympathectomy carried out through a cervical incision. The sympathetic chain is divided below the 3rd thoracic ganglion and the grey and white rami to the 2nd and 3rd ganglia are also cut. In this way the sudomotor and vasoconstrictor pathways to the head and upper limb (from segments T2, 3 and 4) are divided, preserving the T1 connection and the stellate ganglion, which are the sympathetic connections to the eyelid and pupil. The upper thoracic chain can also be removed via a transthoracic transpleural approach through the second intercostal space, the incision being situated on the medial wall of the axilla. The lung is allowed to collapse and the chain identified as it lies on the heads of the upper ribs. Resection of the T2–4 segment results in a warm, dry hand.

2 *Horner's syndrome* results from interruption of the sympathetic fibres to the eyelids and pupil. The pupil is constricted (myosis, due to unopposed parasympathetic innervation via the oculomotor nerve), there is ptosis (partial paralysis of levator palpebrae) and the face on the affected side is dry and flushed (sudomotor and vasoconstrictor denervation). Enophthalmos is said to occur, but this is not confirmed by exophthalmometry. The syndrome may follow spinal cord lesions at the T1 segment (tumour or syringomyelia), closed, penetrating or operative injuries to the stellate ganglion or the cervical sympathetic chain, or pressure on the chain or stellate ganglion produced by enlarged cervical lymph nodes, an upper mediastinal tumour, a carotid aneurysm or a malignant mass in the neck.

The branchial system and its derivatives

Six visceral arches form on the lateral aspects of the fetal head separated, on the outside, by ectodermal branchial clefts and, on the inside, by five endodermal pharyngeal pouches (Fig. 194). In the human embryo the 5th and 6th arches do not appear externally and are represented only by a mesodermal core.

Each arch has its own nerve supply, cartilage, muscle an artery, although considerable absorption and migration of these derivatives occur in development. The 5th arch disappears entirely.

The embryological significance of many of the branchial derivatives has already been discussed under appropriate headings (the development of the face, tongue, thyroid, parathyroid and aortic arch) but Table 4 serves conveniently to bring these various facts together.

Branchial cyst and fistula

The second branchial arch grows downwards to cover the remaining arches, leaving temporarily a space lined with squamous epithelium. This usually disappears but may persist and distend with cholesterol-containing fluid to form a *branchial cyst*.

If fusion fails to occur distally, a sinus persists at the anterior border of the origin of the sternocleidomastoid; this *branchial fistula* can be traced upward between the internal and external carotids and may even open into the tonsillar fossa, demonstrating its association with the second branchial arch.

Table 4 Derivatives of the branchial system (Note the 5th arch disappears).

Arch	Nerve	Visceral	External cleft	Internal pouch	Floor	Cartilage	Muscle	Artery
I	V	Lower face	External auditory meatus	Eustachian tube, middle ear and mastoid antrum	Antr 2/3 tongue	Meckel's, incus and malleus	Muscles of mastication, antr belly digastric, tensor palati, tensor tympani	Disappears
II	VII	Grows down to cover remaining clefts to form skin of neck		Palatine tonsil	Contributes to antr tongue Thyroid forms as outgrowth between I and II	Stapes, styloid stylohyoid lig, upper body and lesser horn of hyoid	Muscles of facial expression, postr belly digastric, stylohyoid, stapedius	Disappears
III	IX			Thymus, inferior parathyroid	Postr 1/3 tongue	Greater horn and lower part of body of hyoid	Stylopharyngeus	Common and internal carotids
IV	X (superior laryngeal)			Superior parathyroid		Thyroid cartilage	Muscles of pharynx, larynx and palate	R— subclavian, L— aortic arch
VI	X (inferior laryngeal)				Outgrowth of lung buds	Cricoid cartilage		Pulmonary A. and ductus arteriosus

The surface anatomy and surface markings of the head

Many of the important landmarks of the skull are readily felt. Revise on your own skull the position of: the *external occipital protuberance* (the apex of this is termed the *inion*), the *nasion*, which is the depression between the two *supraorbital margins*, and the *glabella*, which is the ridge above the nasion. Feel the sharp edge of the lateral margin of the orbit which is formed by the frontal process of the zygomatic bone; behind the zygomatic bone is the *zygomatic arch* with the *superficial temporal artery* crossing its posterior extremity and forming a convenient pulse which the anaesthetist can reach. Rather less easily felt is the *jugal point*, the junction between the zygomatic bone and the zygomatic process of the frontal bone; it is the mass of bone encountered by the finger running forward along the upper border of the zygomatic arch, and it is a surface marking for the middle meningeal artery (*vide infra*). The anterior edge of the *mastoid* is easily palpable but its posterior aspect and its tip are rather obscured by the insertion of the sternocleidomastoid.

The whole of the superficial surface of the *mandible* is palpable apart from its coronoid process. The *condyloid process* can be felt by a finger placed immediately in front of, or within, the external auditory meatus while the mouth is opened and closed.

When the teeth are clenched, *masseter* and the *temporalis* can be felt contracting respectively over the ramus of the mandible and above the zygomatic arch. The *parotid duct* can be rolled over the tensed masseter and its orifice seen within the mouth at the level of the 2nd upper molar tooth.

The pulsation of the *facial artery* can be felt as it crosses the lower margin of the body of the mandible immediately in front of the masseter and again opposite the angle of the mouth. In the latter situation, if the cheek is gripped lightly with the finger placed within the mouth and the thumb placed on the skin surface, the pulse will be felt a little more than a centimetre from the angle of the mouth.

A line drawn vertically between the first and second premolar teeth passes through the mental foramen, the infraorbital foramen and the supraorbital notch. Through these three orifices, lying in plumb-line, pass branches from each of the divisions of the trigeminal nerve, respectively, the mental branch of the inferior alveolar nerve (V‴), the infra-orbital nerve (V″) and the supra-orbital nerve (V′).

The *middle meningeal artery* can be represented by a line drawn upwards and somewhat forwards from a point along the zygomatic arch, two fingers' breadths behind the jugal point. The posterior branch of this artery passes backwards a thumb's breadth above, and roughly parallel to, the zygomatic arch.

The *central sulcus* of the cerebrum corresponds to a line drawn downwards and forwards from a point 1 cm behind the midpoint between the nasion and the inion.

The scalp

The soft tissues of the scalp are arranged in five layers (Fig. 221), which may be remembered thus:

    S—skin;
    C—connective tissue;
    A—aponeurosis;
    L—loose connective tissue;
    P—periosteum.

Each of these layers has features of practical importance.

The *skin* of the scalp is richly supplied with sebaceous glands and is the commonest site in the body for sebaceous cysts.

The *subcutaneous connective tissue* consists of lobules of fat bound in tough fibrous septa, very much like the connective tissue of the palm and the sole. This dense encapsulation of fat makes it unsurprising that lipomata are extremely rare at these three sites, and also that excess fat does not collect in any of these places even in the grossly obese.

The blood vessels of the scalp lie in this layer. When the head is lacerated, the divided vessels retract between the fibrous septa and cannot be picked up individually by artery forceps in the usual way. Haemorrhage is arrested by pressing with the fingers firmly down on

Fig. 221 The layers of the scalp.

to the skull on either side of the wound (thus compressing the vessels), by placing series of artery forceps on the divided aponeurotic layer so that their weight again compresses these vessels and, finally, by suturing the laceration firmly in two layers (aponeurotic and cutaneous).

The haemorrhage from a scalp laceration or operation is profuse; this area has, in fact, the richest cutaneous blood supply of the body. For this reason, extensive avulsions of the scalp are usually viable providing even a narrow pedicle remains attached to the surrounding tissues.

The veins of the scalp connect with the intracranial venous sinuses via numerous *emissary veins* which pierce the skull and which also link these two venous systems with the *diploic veins* between the tables of the skull vault. A superficial infection of the scalp may spread via this system producing an osteitis of the skull, meningitis and venous sinus thrombosis.

The *aponeurotic layer* is the occipitofrontalis which is fibrous over the dome of the skull but muscular in the occipital and frontal regions. This muscle arises from the superior nuchal line of the occipital bone, gains a fascial insertion into the zygomatic arch, and inserts anteriorly into the subcutaneous tissues of the eyebrows and nose.

The layer of *loose connective tissue* beneath the aponeurosis accounts for the mobility of the scalp on the underlying bone; it is in this plane that the surgeon mobilizes scalp flaps, that machinery which has caught on to the hair avulses the scalp and that the Red Indians of bygone days scalped their victims.

Blood or pus collecting in this loose tissue tracks freely under the scalp but cannot pass into either the occipital or subtemporal regions because of the attachments of occipitofrontalis. Fluid can, however, track forward into the orbits and this accounts for the orbital haematoma that may form a few hours after a severe head injury or cranial operation.

The aponeurotic layer is under tension because of its muscular component and retracts on the underlying loose layer when divided; a gaping scalp wound must, therefore, have extended at least through the aponeurosis.

The *periosteum* adheres to the suture-lines of the skull; collections of pus or blood beneath this layer, therefore, outline the affected bone. This is particularly well seen in birth injuries involving the skull (*cephalohaematoma*).

The skull (Figs 222, 223, 224)

The important regional anatomy of the skull is dealt with under the appropriate headings (ear, nose, accessory sinuses, etc.). Collected together in this section are some general facts of clinical relevance.

   The bony vault of the skull is relatively elastic in consistency; thus a blow may injure the underlying brain without fracturing bone. Where the cranium is protected by thick muscle (the lower part of the occipital bone and the squamous temporal), the skull is correspondingly thin; if held up to the light it can be seen to be translucent at these sites.

   The palpable landmarks of the skull are enumerated in the section on the surface anatomy of the head. Radiologically, the *sutures* between the vault bones are important because they, as well as the vascular markings of the meningeal and diploic vessels, may be confused with fracture lines. The *coronal* suture divides the frontal from the parietal bones, the *sagittal* suture separates the parietal bones in the midline, the *lambdoid* suture marks off the occipital from the parietal and temporal bones and the *squamosal suture* separates the squamous temporal bone from the parietal bone and greater wing of sphenoid.

   In about 8% of cases the *metopic suture* persists in the midline

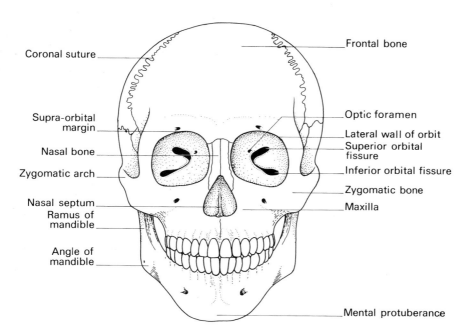

Coronal suture

Frontal bone

Supra-orbital margin

Optic foramen

Nasal bone

Lateral wall of orbit

Zygomatic arch

Superior orbital fissure

Inferior orbital fissure

Nasal septum

Zygomatic bone

Ramus of mandible

Maxilla

Angle of mandible

Mental protuberance

Fig. 222 The skull: anterior aspect.

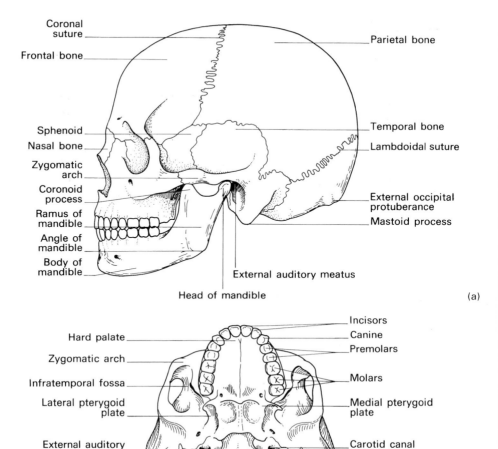

Fig. 223 The skull: (a) lateral aspect and (b) inferior aspect.

between the two frontal bones; in the rest, this suture fuses at about the 5th year.

Occasionally, small separate areas of ossification develop between the parietal and occipital bones termed *Wormian bones* which, again, may cause radiological confusion.

The *lambda* is the point of junction of the lambdoid and sagittal sutures (the *posterior fontanelle* of infancy).

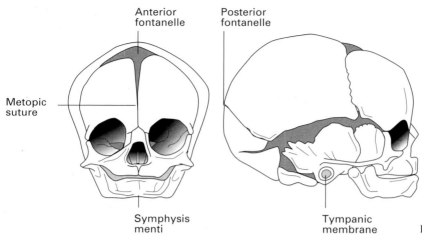

Anterior fontanelle

Posterior fontanelle

Metopic suture

Symphysis menti

Tympanic membrane

Fig. 224 The fetal skull.

The *bregma* is the junction of the sagittal and coronal sutures (the infant's *anterior fontanelle*).

The *diploë*, between the inner and outer tables of the skull vault, is one of the sites of persistent red marrow in the adult skeleton. This distinction it shares with the pelvis, vertebrae, ribs, sternum, upper end of humerus and upper end of femur—a doubtful honour since to these sites are almost confined secondary deposits of carcinoma in bone and multiple myelomata.

Development

The skull vault develops in membrane, the skull base in cartilage.

At birth (Fig. 224), the square anterior fontanelle and triangular posterior fontanelle are widely open. The posterior fuses at about 3 months, the anterior at about 18 months. Up till then, blood can be obtained by puncturing the sagittal sinus immediately below the anterior fontanelle in the midline, and C.S.F. aspirated by passing a needle obliquely into the lateral ventricle.

The face at birth is considerably smaller proportionally to the skull than in the adult; this is due to the teeth being non-erupted and rudimentary and the nasal accessory sinuses being undeveloped; the sinuses are evident at about 8 years but only fully developed in the late teens.

The mastoid and its air cells develop at the end of the 2nd year; until then the facial nerve is relatively superficial near its origin from the skull and may be damaged by quite trivial injuries.

With advancing age, the relative vertical measurement of the face again diminishes as a result of loss of teeth and subsequent absorption of the alveolar margins.

Development of the mandible and the teeth are considered on pages 347 and 348.

Clinical features

Fractures of the skull

Imagine the skull as a rather elastic sphere completely filled by semifluid material; a violent blow on such a structure will produce a splitting effect commencing at the site of the blow and tending to travel along lines of least resistance. The base of the skull is more fragile than the vault, and is thus commonly involved by such fractures. The petrous part of the temporal bone, however, forms a firm and rarely involved buttress of the skull base, the fracture line passing through less resistant areas, particularly the middle cranial fossa, the pituitary fossa and the various basal foramina.

A localized severe injury, in the adult, may produce a depressed comminuted fracture; the infant's skull is much more elastic and a similar injury here will result in a 'pond' depressed fracture, rather like the dimple produced by squeezing on a ping-pong ball.

Localizing signs in cranial fractures

Fractures of the anterior cranial fossa may involve the frontal, ethmoidal and sphenoidal sinuses and be accompanied by bleeding into the nose or mouth. In such cases C.S.F. leakage from the nose implies coexisting tearing of the meninges; the subarachnoid space is thus put in communication with the exterior via the nasal cavity with consequent risk of meningitis.

Fractures involving the roof of the orbit are frequently associated with blood tracking forward beneath the conjunctiva (subconjunctival haermorrhage); this must be differentiated from a small flame-shaped haemorrhage of the conjunctiva caused by direct injury to it.

A 'black eye' is not necessarily indicative of an anterior fossa fracture; it may be produced also by direct contusion of the soft tissues or by blood tracking down deep to the aponeurotic layer of the scalp (see 'The scalp', page 337).

Anterior basal fractures may involve the cribriform plate (with anosmia — loss of smell — due to rupture of fibres of the olfactory bulb) or the optic foramen (with primary optic atrophy and blindness).

Fractures of the middle fossa may produce bleeding into the mouth (sphenoid involvement), bleeding or C.S.F. leakage from the ear, and facial and auditory nerve injury. Aural bleeding may, of course, be produced by direct injury to the ear—for example, rupture of the drum—without necessarily implying a skull fracture. Because of its long course, the abducent (VI) nerve may be damaged with diplopia and paralysis of the lateral rectus muscle.

Posterior fossa fractures are occasionally accompanied by cranial nerve involvement. These fractures are suggested clinically by bruising over the mastoid region extending downwards over the sternocleidomastoid.

The accessory nasal sinuses

The nasal sinuses are air-containing sacs lined by ciliated epithelium and communicating with the nasal cavity through narrow, and therefore easily occluded, channels (Fig. 225). The maxillary antrum and sphenoid sinuses are present in a rudimentary state at birth, the rest become evident at about the 8th year, but all become fully formed only in adolescence.

The frontal sinuses

The frontal sinuses are contained in the frontal bone. They vary

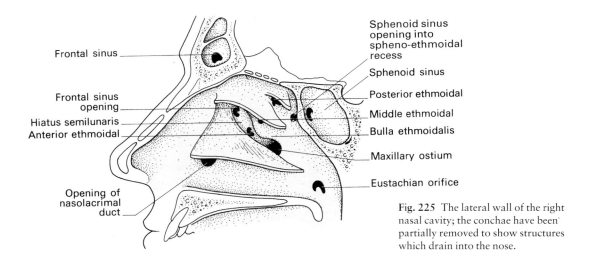

Frontal sinus

Frontal sinus opening

Hiatus semilunaris
Anterior ethmoidal

Opening of nasolacrimal duct

Sphenoid sinus opening into spheno-ethmoidal recess

Sphenoid sinus

Posterior ethmoidal

Middle ethmoidal

Bulla ethmoidalis

Maxillary ostium

Eustachian orifice

Fig. 225 The lateral wall of the right nasal cavity; the conchae have been partially removed to show structures which drain into the nose.

greatly in size and one or both are occasionally absent. In section each is roughly triangular, its anterior wall forming the prominence of the forehead, its posterosuperior wall lying adjacent to the frontal lobe of the brain, and its floor abutting against the ethmoid cells, the roof of the nasal fossa and the orbit.

The frontal sinuses are separated from each other by a median bony septum, and each in turn is further broken up by a number of incomplete septa. Each sinus drains into the anterior part of the middle nasal meatus via the infundibulum into the hiatus semilunaris.

Clinical features

1   The close relation of the frontal sinus to the frontal lobe of the brain explains how infection of this sinus may result in the development of a frontal lobe abscess.

2   A fracture involving the sinus, severe enough to tear the dura and pia-arachnoid, will place the subarachnoid space in communication with the nasal cavity and C.S.F. may then be detected trickling through the nostril, usually on the affected side (C.S.F. rhinorrhoea) although, as these sinuses may communicate, a contralateral leak sometimes occurs.

3   The neurosurgeon must take into account the considerable variations in size and extent of the frontal sinus when proposing to turn down a frontal skull flap; obviously, he will want to avoid opening the sinus because of the risk of infection. He therefore consults the radiographs of the patient's skull preoperatively, which will clearly show the configuration of the sinuses.

The maxillary sinus (antrum of Highmore)
(Fig. 226)

This is a pyramidal-shaped sinus occupying the cavity of the maxilla. Its medial wall forms part of the lateral face of the nasal cavity and bears on it the inferior concha. Above this concha is the opening, or *ostium*, of the maxillary sinus into the middle meatus in the hiatus semilunaris (Fig. 225). This opening, unfortunately, is inefficiently placed as an adequate drainage point.

The infra-orbital nerve lies in a groove which bulges down into the roof of the sinus, while its floor bears the impressions of the upper premolar and molar roots. These roots are separated only by a thin layer of bone which may, in fact, be deficient so that uncovered dental roots project into the sinus. Note that the floor of the sinus, therefore, corresponds to the level of the alveolus and not to the floor of the

(a)

(b)

Fig. 226 (a) The maxillary antrum in coronal section. (Note the inefficient drainage of this antrum and its close inferior relationship to the teeth.) (b) The corresponding CT scan.

nasal cavity—it actually extends about 0.5 in (12 mm) lower than the latter.

Clinical features

1   The maxillary sinus, or antrum, may become infected either from the nasal cavity or from caries of the upper molar teeth.

Antral puncture can be carried out using a trocar and cannula passed through the nasal cavity in an outward and backward direction below the inferior concha.

More adequate drainage may require removing a portion of the medial wall of the sinus below the inferior concha or fenestrating the antrum in the gingivolabial fold (Caldwell Luc operation). The old operation of draining the antrum via an extracted upper molar tooth is now seldom, if ever, performed.

2   The numerous symptoms and signs which may be produced by a carcinoma of the maxillary sinus are easily remembered anatomically.

(a)   Medial invasion encroaches on the nasal cavity, producing obstruction of the nares and epistaxis. Blockage of the naso-lacrimal duct in this wall may cause epiphorea (leakage of tears down the face).

(b)   Invasion of the orbit displaces the globe and causes diplopia. If the infra-orbital nerve becomes involved, there will be facial pain and then anaesthesia of the skin over the maxilla.

(c)   Invasion of the sinus floor may produce a visible bulge or even ulceration in the palatal roof.

(d)   Lateral spread may produce a swelling of the face or a palpable mass in the gingivolabial fold.

(e)   Posterior spread may involve the palatine nerves and produce severe pain referred to the teeth of the upper jaw.

The ethmoid sinuses

The ethmoid sinuses are made up of a group of 8–10 air cells within the lateral mass of the ethmoid and lie between the side-walls of the upper nasal cavity and the orbits (Fig. 226). Superiorly, they lie on each side of the cribiform plate and are related above to the frontal lobes of the brain. These cells drain into the superior and middle meatus (Fig. 225).

Clinical features

As with the frontal sinus, infection (ethmoiditis) may result in a frontal cerebral abscess and an ethmoidal fracture may cause a C.S.F. leakage into the nasal cavity.

The sphenoid sinuses

These lie one on either side of the midline, within the body of the sphenoid (Fig. 225). They vary a good deal in size and may extend laterally into the greater wing of the sphenoid or backwards into the basal part of the occipital bone.

Each sinus drains into the nasal cavity above the superior concha (the sphenoethmoidal recess).

Clinical features

The pituitary gland may be excised through a transsphenoidal approach in patients with pituitary tumour or disseminated breast cancer.

The mandible (Fig. 227)

The lower jaw comprises a horizontal *body* on each side which fuses at the *symphysis menti* (fusion occurring at the 2nd year). From the posterior part of the body projects the vertical *ramus* which bears an anterior *coronoid* and a posterior *condyloid* process or *head*. Between the two is the *mandibular notch*.

On the medial aspect of the ramus is the *mandibular foramen* for the inferior alveolar branch of the mandibular division of the trigeminal nerve, which traverses the body within the *mandibular canal*, then emerges as the mental nerve through the *mental foramen* on the

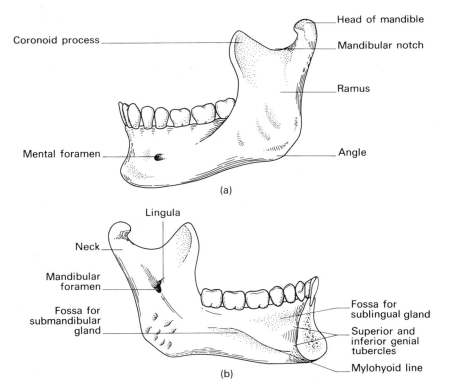

Fig. 227 The mandible: (a) lateral aspect, (b) medial aspect.

lateral surface of the body below and between the two premolars. The nerve supply to the incisors and canine runs forward within the mandible beyond this point in the *incisive canal*.

The upper border of the body bears the *alveolar border* with sixteen dental sockets or *alveoli*.

Development

The mandible develops as membrane bone in the fibrous sheath of Meckel's cartilage (the cartilage of the first branchial arch). The cartilage itself is completely absorbed. Bony union of the two halves of the mandible occurs in the 2nd year.

The temporomandibular joint

This joint lies between the condyloid process of the mandible and the articular fossa and articular eminence of the temporal bone. The articular surfaces are covered with fibrous (*not* hyaline) cartilage and there is also a fibrocartilaginous *articular disc* dividing the joint cavity into an upper and lower compartment.

The capsular ligament surrounding the joint is reinforced by a lateral *temporomandibular ligament* and by the *sphenomandibular ligament* which passes from the spine of the sphenoid to the *lingular process* immediately in front of the mandibular foramen; this ligament represents part of the primitive 1st arch, or Meckel's cartilage.

The lower jaw can be depressed, elevated, protruded, retracted and moved from side to side.

The muscles effecting these movements are:
* *elevation*—temporalis, masseter, medial pterygoid;
* *depression*—medial pterygoid, together with digastric, mylohyoid and geniohyoid (assisted by gravity—your jaw drops open when you fall asleep in a lecture);
* *retraction*—posterior fibres of temporalis;
* *protraction*—lateral and medial pterygoids together;
* *side to side*—lateral and medial pterygoids together, acting alternately on each side.

Clinical features

Dislocation of the jaw, when uncomplicated, occurs only in a forward direction. When the mouth is widely open, the condyloid process of the mandible slides forward on to the articular eminence; from thence, a blow, or even a yawn, may cause forward dislocation into the infratemporal fossa on one or both sides. Upward dislocation can occur only in association with extensive comminution of the skull base, and backward dislocation with smashing of the bony external auditory canal and tympanic cavity which lie immediately behind the joint.

Reduction is effected by pressing down on the molar teeth with the thumbs placed in the mouth, at the same time pulling up the chin; the former stretches the masseter and temporalis muscles which are in spasm, the latter levers the mandibular head back into place.

The teeth

There are twenty deciduous or 'milk' teeth replaced by thirty-two permanent teeth made up, in each half jaw, thus:
- *Deciduous*: 2 incisors, 1 canine, 2 molars;
- *Permanent*: 2 incisors, 1 canine, 2 premolars, 3 molars.

The times of eruption of the teeth are useful stepping-stones in a child's development as well as being of forensic interest.

As a rough guide, these times can be thought of in multiples of 6, thus:
- the 1st lower incisor deciduous tooth appears at 6 months
- all the deciduous teeth have appeared by   24 months
- the permanent 1st molar }
- the permanent 1st incisor } appear at   6 years
- the second permanent molar appears at 12 years (approx.)
- the third permanent molar appears at 18–24 years.

The lower teeth appear somewhat before their corresponding upper neighbours.

Each tooth is fixed in its socket by the *periodontal membrane* which is, in fact, periosteum. This layer is radiotranslucent and is the dark line seen around the root of each tooth on radiography.

Development

The enamel crown of the tooth develops from a downgrowth of the alveolar epithelium and represents the toughest tissue in the human body. The rest of the tooth (pulp, dentine and cement) differentiates from the underlying mesodermal connective tissue.

Clinical features

Osteomyelitis of the jaw following dental extractions is confined to the lower jaw and occurs only with the permanent dentition. The explanation of this is an anatomical one.

The lower jaw is supplied *only* by the inferior dental artery, which runs with the nerve in the mandibular canal; damage to this artery at extraction, or its thrombosis in subsequent infection, therefore, produces bone necrosis. The upper jaw, on the other hand, receives segmental vertical branches from the superior dental vessels and ischaemia does not follow injury to an individual artery. The deciduous teeth of the lower jaw are placed well clear of the mandibular canal which is, in any case, protected by the unerupted permanent teeth; damage to the artery cannot therefore occur during their removal.

Inferior alveolar nerve block

This is a useful procedure for the dental surgeon because it produces complete anaesthesia of all the lower teeth of one side of the mandible. The needle is passed deep to the last molar tooth on to the inner aspect of the ramus of the mandible. Anaesthesia is produced in the lower teeth, the skin and mucosa of the lower lip (via the mental branch of the inferior alveolar nerve) and often, because of spread of the anaesthetic solution, there is loss of sensation of the side of the tongue due to involvement of the lingual nerve, which lies immediately in front of the inferior alveolar nerve (see Fig. 259).

The vertebral column

The spinal, or vertebral, column is made up of thirty-three vertebrae, of which twenty-four are discrete vertebrae and nine are fused in the sacrum and coccyx.

In the embryo the spine is curved into a gentle C shape but, with the extension of the head and lower limbs that occurs when the child first holds up its head, then sits and then stands, secondary forward curvatures appear in the cervical and lumbar region, which produce the sinusoidal curves of the fully developed spinal column.

The basic vertebral pattern (Fig. 228) is that of a *body* and of a *neural arch* surrounding the *vertebral canal*.

The neural arch is made up of a *pedicle* on either side, each supporting a *lamina* which meets its opposite posteriorly in the midline. The pedicle bears a notch above and below which, with its neighbour, forms the *intervertebral foramen*. The arch bears a posterior *spine*, lateral *transverse processes* and upper and lower *articular facets*.

The intervertebral foramina transmit the segmental spinal nerves as follows: C1–7 pass over the superior aspect of their corresponding

Fig. 228 A 'typical' thoracic vertebra.

cervical vertebrae, C8 passes through the foramen between C7 and T1, and all subsequent nerves pass between the vertebra of their own number and the one below.

Now to consider the individual vertebrae in turn.

The cervical vertebrae (7)

These are readily identified by the *foramen transversarium* perforating the transverse processes. This foramen transmits the vertebral artery, the vein, and sympathetic nerve fibres. The spines are small and bifid (except C1 and C7 which are single) and the articular facets are relatively horizontal (Fig. 229).

The *atlas* (C1) (Fig. 230) has no body. Its upper surface bears a superior articular facet on a thick lateral mass on each side which articulates with the occipital condyles of the skull.

Just posteriorly to this facet, the upper aspect of the posterior arch of the atlas is grooved by the vertebral artery as it passes medially and upwards to enter the foramen magnum.

The *axis* (C2) (Fig. 231) bears the *dens* (*odontoid process*) on the superior aspect of its body, representing the detached centrum of C1.

Nodding and lateral flexion movements occur at the atlanto-occipital joint, whereas rotation of the skull occurs at the atlanto-axial joint around the dens, which acts as a pivot.

C7 is the *vertebra prominens*, so called because of its relatively long and easily felt non-bifid spine; it is the first clearly palpable spine on running one's fingers downwards along the vertebral crests, although the spine of T1 immediately below it is, in fact, the most prominent one.

Bifid spine

Lamina

Foramen transversarium

Posterior and anterior tubercles

Fig. 229 A 'typical' cervical vertebra.

Anterior tubercle

Anterior arch

Attachment of transverse ligament

Facet for odontoid

Foramen transversarium

Superior articular facet

Groove for vertebral artery

Fig. 230 The atlas in superior view.

Odontoid process

Facet for anterior arch of atlas

Foramen transversarium

Transverse process

Body

Fig. 231 The axis in oblique lateral view.

The vertebral artery enters its vertebral course nearly always at the foramen transversarium of C6; it is not surprising, therefore, that the foramen of C7, which transmits only the vein, is small or even sometimes absent.

The thoracic vertebrae (12)

These vertebrae are characterized by demifacets on the sides of their bodies for articulation with the heads of the ribs and by facets on their transverse processes (apart from those of the lower two or three vertebrae) for the rib tubercles. The spines are long and downward sloping and the articular facets are also relatively vertical.

The bodies of T5 and T8 are worth noting; they come into relationship with the descending aorta and are a little flattened by it on their left flank. If the descending aorta becomes aneurysmally dilated, these four vertebral bodies become eroded by its pressure, although their avascular intervertebral discs remain intact. You can make this diagnosis confidently when shown a specimen of four partly worn-away vertebrae with normal intervening discs.

The lumbar vertebrae (5)

These are of great size with strong, square, horizontal spines and with articular facets which lie in the sagittal plane (Fig. 232).

L5 is distinguished by its massive transverse process which connects with the whole lateral aspect of its pedicle and encroaches on its body; the transverse processes of the other lumbar vertebrae attach solely to the junction of pedicle with lamina.

The sacrum (5 fused)

The coccyx (3, 4, or 5 fused)

These are considered with the bony pelvis (see page 136).

Development

Each vertebra ossifies from three primary centres, one for each side of the arch and one for the body. The body occasionally develops from two centres and failure of one of these to form results in formation of a hemivertebra with a consequent congenital scoliosis. Failure of the two arch centres to fuse posteriorly results in the condition of spina bifida, which occurs particularly in the lumbar region. Usually this is not associated with any neurological abnormality (spina bifida

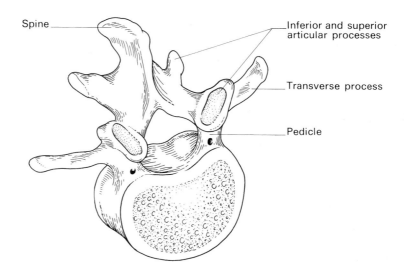

Spine

Inferior and superior
articular processes

Transverse process

Pedicle

Fig. 232 A lumbar
vertebra in anterosuperior
view.

occulta), although in such cases there is often an overlying dimple,
lipoma or tuft of hair to warn the observant of a bony abnormality
beneath. More rarely, there is a gross defect of one or several arches
with protrusion of the spinal cord or its coverings; this anomaly may
be associated with hydrocephalus.

L5 may occasionally fuse wholly or in part with the sacrum
(sacralization of the 5th lumbar vertebra) or, more rarely, the 1st
segment of the sacrum may differentiate as a separate vertebra.

The intervertebral joints

The spinal column is made up of individual vertebrae which articulate
body to body and their articular facets. Although movement between
adjacent vertebrae is slight, the additive effect is considerable. Movement particularly occurs at the cervicodorsal and dorsolumbar junc-
tions; these are the two common sites of vertebral injury.

The vertebral laminae are linked by the *ligamentum flavum* of
elastic tissue, the spines by the tough *supraspinous* and relatively
weak *interspinous ligaments*, and the articular facets by *articular ligaments* around their small synovial joints. All these ligaments serve to
support the spinal column when it is in the fully flexed position.

Running the whole length of the vertebral bodies, along their
anterior and posterior aspects respectively, are the tough *anterior* and
posterior longitudinal ligaments.

The vertebral bodies are also joined by the extremely strong *intervertebral discs* (Fig. 233). These each consist of a peripheral *annulus
fibrosus*, which adheres to the thin cartilage plate on the vertebral

(a)

(b)

Nucleus pulposus

Annulus fibrosus

A 'prolapsed disc'

Posterior longitudinal ligament

Ligamentum flavum

Body of L3

Intervertebral disc

Sacral promontory

Spine of L3

Fig. 233 (a) Longitudinal section through the lumbar vertebrae showing a normal and a prolapsed intervertebral disc. (b) MRI through a normal lumbar spine and sacrum. Note the excellent anatomical details.

body above and below, and which surrounds are gelatinous semifluid *nucleus pulposus*. The intervertebral discs constitute approximately a quarter of the length of the spine as well as accounting for its secondary curvatures.

In old age, the discs atrophy, with resulting shrinkage in height and return of the curvature of the spine to the C shape of the newborn.

Clinical features

1 Fractures of the spine most commonly involve T12, L1 and L2. The cause is usually a flexion–compression type of injury (for example, a fall from a height landing on the feet or buttocks, or a heavy weight falling on the shoulders), with resultant wedging of the involved vertebrae. If, in addition to compression, there is forceful forward movement, one vertebra may displace forward on its neighbour below with either dislocation or fracture of the articular facets between the two (fracture dislocation) and with rupture of the interspineous ligaments.

The cervical vertebrae may be fractured or, more commonly, dislocated by a fall on the head with acute flexion of the neck, as might happen on diving into shallow water. Dislocation may even result from the sudden forward jerk which may occur when a motorcar or aeroplane crashes. Note that the relatively horizontal intervertebral facets of the cervical vertebrae allow dislocation to take place without their being fractured, whereas the relatively vertical thoracic and lumbar interverbral facets nearly always fracture in forward dislocation of the dorsolumbar region.

2 The comparatively thin posterior part of the annulus fibrosus may rupture, either due to trauma or to degenerative changes, allowing the nucleus pulposus to protrude posteriorly into the vertebral canal—the so-called 'prolapsed intervertebral disc' (Fig. 233). This may sometimes occur at the lower cervical intervertebral discs (C5/6 and C6/7), very occasionally in the thoracic and upper lumbar region or, by far the most commonly, at the L4/5 or L5/S1 disc. The diagnosis of this and other spinal conditions has been greatly facilitated by the introduction of MRI scans which give excellent anatomical details of this region (Fig. 233b).

A prolapsed L4/5 disc produces pressure effects on the root of the 5th lumbar nerve, that of the L5/S1 disc on the 1st sacral nerve. Pain is referred to the back of the leg and foot along the distribution of the sciatic nerve. Hip flexion with the leg extended ('straight leg raising') is painful and limited due to the traction which this movement puts upon the already irritated and stretched nerve root. There may be a weakness of ankle dorsiflexion and numbness over the lower and lateral part of the leg and medial side of the foot (L5) or the lateral side of the foot (S1). If S1 is affected, the ankle jerk may be diminished or absent.

Occasionally the disc prolapses directly backwards, and, if this is extensive, may compress the whole cauda equina, producing paraplegia.

3 Lumbar puncture—see page 364.

Part 6
The Central Nervous System

The spinal cord

The spinal cord is 18 in (45 cm) long. It is continuous above with the medulla oblongata at the level of the foramen magnum and ends below at the lower level of the 1st, or the upper level of the 2nd lumbar vertebra. Inferiorly, it tapers into the *conus medullaris* from which a prolongation of pia mater, the *filum terminale*, descends to be attached to the back of the coccyx.

The cord bears a deep longitudinal *anterior fissure*, a narrower *posterior sulcus*, on either side, a *posterolateral sulcus* along which the *posterior (sensory) nerve roots* are serially arranged (Fig. 234).

These posterior roots each bear a ganglion which constitutes the first cell-station of the sensory nerves.

The *anterior (motor) nerve roots* emerge serially along the antero-lateral aspect of the cord on either side. Both the anterior and posterior nerves arise by a series of rootlets from the cord.

At each intervertebral foramen the anterior and posterior nerve roots unite to form a *spinal nerve* which immediately divides into its *anterior* and *posterior primary rami*, each transmitting both motor and sensory fibres.

The length of the roots increases progressively from above downwards due to the disparity between the length of the cord and the vertebral column; the lumbar and sacral roots below the termination of the cord at vertebral level L2 continue as a leash of nerve roots termed the *cauda equina*.

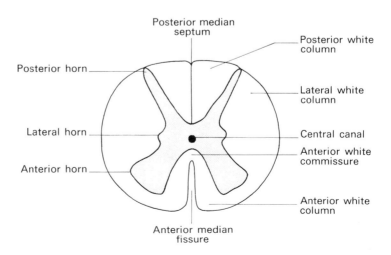

Fig. 234 The spinal cord —transverse section through a thoracic segment.

Posterior median septum

Posterior white column

Posterior horn

Lateral white column

Lateral horn

Central canal

Anterior white commissure

Anterior horn

Anterior white column

Anterior median fissure

Age differences

Up to the 3rd month of fetal life the spinal cord occupies the full extent of the vertebral canal. The vertebrae then outpace the cord in the rapidity of their growth so that, at birth, the cord reaches only the level of the 3rd lumbar vertebra (Fig. 235).

Further differential growth up to the time of adolescence brings the cord to its definitive position at the approximate level of the disc between the 1st and 2nd lumbar vertebrae (Fig. 236).

Structure (Fig. 234)

In transverse section of the cord is seen the *central canal* around which is the H-shaped *grey matter*, surrounded in turn by the *white matter* which contains the long ascending and descending tracts.

Within the *posterior horns* of the grey matter, capped by the *substantia gelatinosa*, terminate many of the sensory fibres entering from the posterior nerve roots. In the large *anterior horns* lie the motor cells which give rise to the fibres of the anterior roots.

Fig. 235 The relationship between the spinal cord and the vertebrae in the 3-month fetus and in the newborn child.

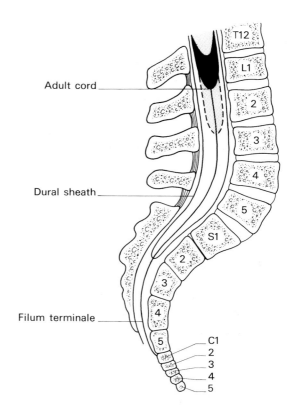

Adult cord

Dural sheath

Filum terminale

Fig. 236 The range of variation in the termination of the spinal cord in the adult.

In the thoracic and upper lumbar cord are found the *lateral horns* on each side, containing the cells of origin of the sympathetic system.

The more important long tracts in the white matter will now be dealt with.

Descending tracts (Fig. 237)

1 *The pyramidal (lateral cerebrospinal or crossed motor) tract.* The motor pathway commences at the pyramidal cells of the motor cortex, decussates in the medulla, then descends in the pyramidal tract on the contralateral side of the cord. At each spinal segment, fibres enter the anterior horn and connect up with the motor cells there — the tract therefore becomes progressively smaller as it descends.

2 *The direct pyramidal (anterior cerebrospinal or uncrossed motor) tract* is a small tract descending without medullary decussation. At each segment, however, fibres pass from it to the ventral horn (anterior) motor cells of the opposite side.

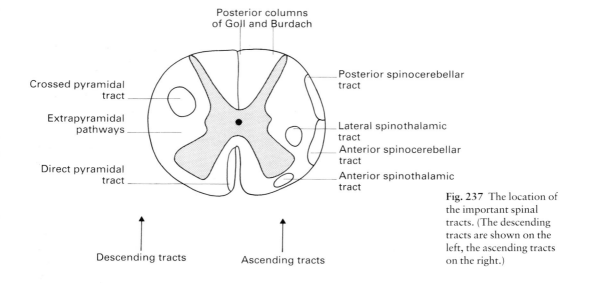

Fig. 237 The location of the important spinal tracts. (The descending tracts are shown on the left, the ascending tracts on the right.)

Ascending tracts (Fig. 237)

1   The *posterior* and *anterior spinocerebellar tracts* ascend on the same side of the cord and enter the cerebellum through the inferior and superior cerebellar peduncles respectively.

2   The *lateral* and *anterior spinothalamic tracts*. Pain and temperature fibres enter the posterior roots, ascend a few segments, relay in the substantia gelatinosa, then cross to the opposite side to ascend in these tracts to the thalamus, where they are relayed to the sensory cortex.

3   The *posterior columns* comprise a medial and lateral tract, termed respectively the *fasciculus gracilis (of Goll)* and *fasciculus cuneatus (of Burdach)*. They convey 1st order sensory fibres subserving fine touch and proprioception (position sense), mostly uncrossed, to the gracile and cuneate nuclei in the medulla where, after synapse, the 2nd order fibres decussate, pass to the thalamus and, after further synapse, 3rd order fibres are relayed to the sensory cortex. Some fibres pass from the medulla to the cerebellum along the inferior cerebellar peduncle.

Blood supply

The *anterior* and *posterior spinal arteries* descend in the pia from the intracranial part of the vertebral artery. They are reinforced serially by branches from the ascending cervical, the cervical part of the vertebral, the intercostal and the lumbar arteries.

Clinical features

1   Complete transection of the cord is followed by total loss of sensation in the regions supplied by the cord segments below the level of injury together with flaccid muscle paralysis. Voluntary sphincter control is lost but reflex emptying of bladder and rectum subsequently return, provided that the cord centres situated in the sacral zone of the cord are not destroyed.

2   Destruction of the centre of the cord, as occurs in syringomyelia and in some intramedullary tumours, first involves the decussating spinothalamic fibres so that initially there is bilateral loss of pain and temperature sense below the lesion; proprioception and fine touch are preserved till late in the uncrossed posterior columns.

3   Hemisection of the cord is followed by the *Brown-Séquard syndrome*; there is paralysis on the affected side below the lesion (pyramidal tract) and also loss of proprioception and fine discrimination (dorsal columns). Pain and temperature senses are lost on the *opposite* side below the lesion, because the affected spinothalamic tract carries fibres which have decussated below the level of cord hemisection.

4   *Tabes dorsalis*, which is a syphilitic degenerative lesion of the posterior columns and posterior nerve roots, is characterized by loss of proprioception; the patient becomes ataxic, particularly if he closes his eyes, because he has lost his position sense for which he can partially compensate by visual knowledge of his spatial relationship (Romberg's sign).

5   Intractable pain can be treated in selected cases by cutting the appropriate posterior nerve roots (posterior rhizotomy) or by division of the spinothalamic tract on the side opposite the pain (cordotomy). A knife passed 3 mm into the cord anterior to the denticulate ligament and then swept anteriorly from this point will sever the spinothalamic tract but preserve the pyramidal tract lying immediately posterior to it.

The membranes of the cord (Fig. 238)

The spinal cord, like the brain, is closely ensheathed by the *pia mater*. This is thickened on either side between the nerve roots to form the *denticulate ligament*, which passes laterally to adhere to the dura. Inferiorly, the pia continues as the *filum terminale*, which pierces the distal extremity of the dural sac and becomes attached to the coccyx.

The *arachnoid mater* lines the *dura matter*, leaving an extensive *subarachnoid space*, containing cerebrospinal fluid (C.S.F.), between

Fig. 238 The membranes
of the spinal cord.

it and the pia. Both pia and arachnoid are continued along the spinal nerve roots.

The dura itself forms a tough sheath to the cord. It ends distally at the level of the 2nd sacral vertebra. It also continues along each nerve root and blends with the sheaths of the peripheral nerves.

The *extradural* (or *epidural*) *space* is the compartment between the dural sheath and the spinal canal. It extends downwards from the foramen magnum (above which the dura becomes two-layered) to the sacral hiatus. It is filled with semiliquid fat and contains lymphatics (although there are no lymphatics within the nervous system deep to the dura), together with arteries and large, thin-walled veins. These can be considered equivalent to the cerebral venous sinuses which lie between the two layers of cerebral dura.

Whereas the arteries of this space are relatively insignificant, the extradural veins form a plexus which communicate freely and also receive the *basivertebral veins*, which emerge from each vertebral body on its posterior aspect. In addition, the veins link up with both the pelvic veins below and the cerebral veins above—a pathway for the spread of both bacteria and tumour cells. This accounts, for example, for the ready spread of prostatic cancer to the sacrum and vertebrae.

Clinical features

Lumbar puncture to withdraw C.S.F. from the spinal subarachnoid space must be performed well clear of the termination of the cord. A line joining the iliac crests passes through the 4th lumbar vertebra and therefore the intervertebral spaces immediately above or below this landmark can be used with safety. The spine must be fully flexed (with the patient either on his side or seated) so that the vertebral inter-spinous spaces are opened to their maximum extent (Fig. 239). The

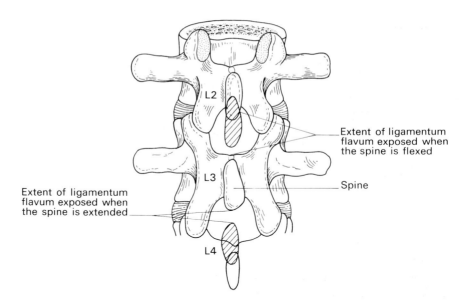

Fig. 239 The lumbar interlaminar gap when the spine is flexed; this anatomical fact makes lumbar puncture possible. The locations of the spines of L2 and L4 in the extended position are shown cross-hatched.

needle is passed inwards and somewhat cranially exactly in the midline and at right angles to the spine; the supraspinous and interspinous ligaments are traversed and then the dura is penetrated, the latter with a distinct 'give'. Occasionally root pain is experienced if a root of the cauda equina is impinged upon, but usually these float clear of the needle.

At spinal puncture C.S.F. can be obtained for examination; antibiotics, radio-opaque contrast medium or anaesthetics may be injected into the subarachnoid space, and the C.S.F. pressure can be estimated (normal, 80–180 mm C.S.F.). A block in the spinal canal above the point of puncture, produced, for example, by a spinal tumour, can be revealed by *Queckenstedt's test* as follows.

Pressure is applied to the neck in order to compress the internal jugular veins; this reduces venous outflow from the cranium and raises the intracranial pressure. Consequently, C.S.F. is displaced into the spinal sac and the C.S.F. pressure, as determined by lumbar puncture and manometry, rises briskly by at least 40 mm. This rise in pressure is not seen if a spinal block is present.

Extradural block. The extradural space can be entered by a needle passed either between the spinal laminae or via the sacral hiatus (caudal or sacral anaesthesia, see page 142).

The brain

The medulla

The medulla is 1 in (25 mm) in length and about 0.75 in (18 mm) in diameter. It is continuous below, through the foramen magnum, with the spinal cord and above with the pons; posteriorly, it is connected with the cerebellum by way of the inferior cerebellar peduncles.

External features (Fig. 240)

The anterior surface of the medulla is grooved by an anteromedian fissure, on either side of which are the swellings due to the pyramidal tracts. These *pyramids*, in turn, are separated from the *olivary eminences* by the anterolateral sulcus along which the rootlets of the XIIth cranial nerve emerge. Between the olive and the inferior cerebellar peduncle there is yet another groove corresponding to the posterolateral sulcus of the spinal cord; emerging from this groove are the rootlets of cranial nerves IX, X and XI (see Fig. 242). The posteromedian sulcus of the cord is continued half-way up the medulla, where it widens out to form the posterior part of the IVth ventricle. On either side of the fissure the posterior columns of the spinal cord expand to form two distinct tubercles, corresponding to the gracile and cuneate nuclei.

Deep structure

The deep structure of the medulla is best shown by reference to diagrams representing the cross-sectional appearance of the medulla at the level of the sensory decussation and the lower part of the IVth ventricle (Figs 241, 242).

The *blood supply* of the medulla is derived from the vertebral arteries directly and from their posterior inferior cerebellar branches.

Clinical features

The medulla contains the respiratory, cardiac and vasomotor centres — the 'vital centres'. The respiratory centre is particularly vulnerable to compression, injury or poliomyelitis with consequent respiratory failure.

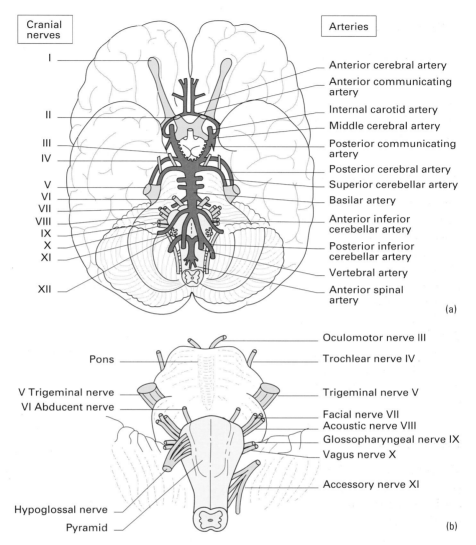

Cranial nerves		Arteries

(a)

I
II
III
IV
V
VI
VII
VIII
IX
X
XI
XII

Anterior cerebral artery
Anterior communicating artery
Internal carotid artery
Middle cerebral artery
Posterior communicating artery
Posterior cerebral artery
Superior cerebellar artery
Basilar artery
Anterior inferior cerebellar artery
Posterior inferior cerebellar artery
Vertebral artery
Anterior spinal artery

(b)

Oculomotor nerve III
Pons
Trochlear nerve IV
V Trigeminal nerve
Trigeminal nerve V
VI Abducent nerve
Facial nerve VII
Acoustic nerve VIII
Glossopharyngeal nerve IX
Vagus nerve X
Accessory nerve XI
Hypoglossal nerve
Pyramid

Fig. 240 The base of the brain showing the cranial nerve roots and their relationships to the circle of Willis.

The pons

External features (Fig. 240)

The pons lies between the medulla and the midbrain and is connected to the cerebellum by the middle cerebellar peduncles. It is 1 in (25 mm) in length and 1.5 in (38 mm) in width. Its ventral surface presents a shallow median groove and numerous transverse ridges, which are continuous laterally with the middle cerebellar peduncle.

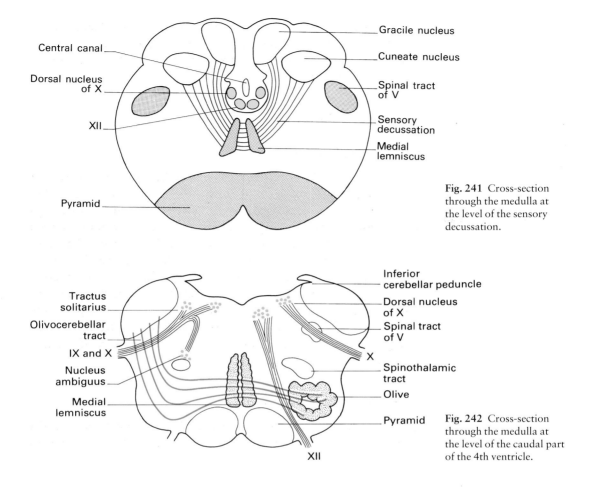

Fig. 241 Cross-section through the medulla at the level of the sensory decussation.

Fig. 242 Cross-section through the medulla at the level of the caudal part of the 4th ventricle.

The dorsal surface of the pons forms the upper part of the floor of the IVth ventricle. Its junction with the medulla is marked close to the ventral midline by the emergence of the VIth cranial nerves and, in the angle between the pons and the cerebellum, by the VIIth and VIIIth nerves. Both the motor and sensory roots of V leave the lateral part of the pons near its upper border.

Internal structure

The pons consists for the most part of a number of cell masses (the *pontine nuclei*), scattered amongst the long ascending and descending pathways and the decussating pontocerebellar fibres, the pontine *tegmentum* (the pontine component of the reticular formation) and the central connections of the Vth, VIth and VIIth cranial nerves.

A typical cross-section through the pons is shown in Fig. 243.

The *blood supply* of the pons is derived from the basilar artery (Fig. 240), formed by the junction of the two vertebral arteries, by way of a number of small pontine branches.

The cerebellum

External features

The cerebellum is the largest part of the hind-brain and occupies most of the posterior cranial fossa. It is made up of two lateral *cerebellar hemispheres* and a median *vermis*. Inferiorly, the vermis is clearly separated from the two hemispheres and lies at the bottom of a deep cleft, the *vallecula*; superiorly, it is only marked off from the hemispheres as a low median elevation. A small ventral portion of the hemisphere lying on the middle cerebellar peduncle is almost completely separated from the rest of the cerebellum as the *flocculus*. The surface of the cerebellum is divided into numerous narrow *folia* and, by a few deep fissures, into a number of lobules. The effect of this fissuring is to give the cerebellum in section the appearance of a many-branched tree (the *arbor vitae*).

Internal structure

The structure of the cerebellum is remarkably uniform. It consists of a *cortex* of grey matter (in which all the afferent fibres terminate) cover-

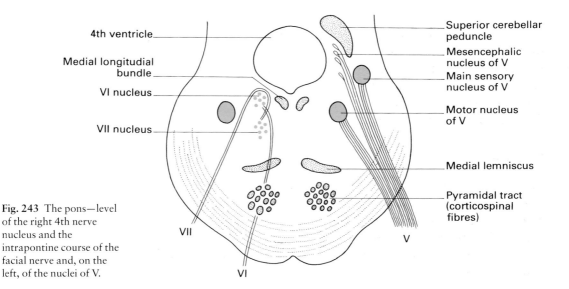

Fig. 243 The pons—level of the right 4th nerve nucleus and the intrapontine course of the facial nerve and, on the left, of the nuclei of V.

ing a mass of white matter, in which deep nuclei of grey matter are buried. Of these, the *dentate nucleus* is by far the largest and occupies the central area of each hemisphere. The other nuclei are *emboliformis*, *globosus* and *fastigii*.

The cerebellum is connected to the brainstem by way of three pairs of *cerebellar peduncles*. The *inferior peduncles* connect it to the dorsolateral aspect of the medulla; the *middle cerebellar peduncles* to the pons, and the *superior peduncles* to the caudal of the midbrain. Ventrally, the cerebellum is related to the 4th ventricle and to the medulla and pons; laterally, to the sigmoid sinus and the mastoid antrum and air cells; while dorsally, it is separated from the cerebral hemispheres by the *tentorium cerebelli.*

The *blood supply* of the cerebellum is derived from three pairs of arteries (Fig. 212); the *posterior inferior cerebellar* branches of the vertebral arteries supply the posterior aspect of the vermis and hemispheres, and the *anterior inferior* and *superior cerebellar branches of the basilar artery* supply the anterolateral part of the under surface and the superior aspect of the cerebellum respectively.

Connections of the cerebellum

The principal afferent and efferent pathways of the cerebellum are set out in Table 5.

Clinical features

1   The cerebellum is principally concerned with balance and the regulation of posture, muscle tone and muscular co-ordination; consequently, cerebellar lesions result in some disturbance of one or more

Table 5 The cerebellar connections.

Peduncle	Afferent pathway	Efferent pathway
Superior	Anterior spinocerebellar (uncrossed)	From dentate nucleus (crossed) to: 1 thalamus 2 cerebral cortex 3 red nucleus
Middle	Pontocerebellar (crossed)—relays from cerebral cortex via pontine nuclei	
Inferior	1 Vestibulocerebellar (uncrossed) 2 Posterior spinocerebellar (uncrossed) 3 Olivocerebellar (crossed)— function unknown	From cerebellar cortex and fastigial nucleus to, vestibular nuclei

of these motor functions in the form of an unsteady gait, hypotonia, tremor, nystagmus and dysarthria. Lesions of the cerebellum give rise to symptoms and signs on the *same side of the body*. Destruction of the dentate nucleus or the superior cerebellar peduncle results in almost as severe a disability as ablation of the entire cerebellar hemisphere.

2   Thrombosis of the posterior inferior cerebellar artery gives rise to a characteristic syndrome marked by ataxia and hypotonia of the homolateral limbs due to involvement of the inferior cerebellar peduncle and cortex, signs of cranial nerve involvement (V to X) and contralateral loss of pain and thermal sensibility (spinothalamic involvement).

The midbrain

The midbrain is the shortest part of the brainstem; it is just under 1 in (25 mm) long and connects the pons and cerebellum to the diencephalon. It lies in the gap in the tentorium cerebelli and is largely hidden by the surrounding structures.

External features (Fig. 240)

The only parts of the midbrain visible from the ventral aspect of the brain are the two *cerebral peduncles*, which emerge from the substance of the cerebral hemisphere and pass downwards and medially, connecting the internal capsule to the pons. The fibres of the 3rd nerves emerge between the two cerebral peduncles in the *interpeduncular fossa*. Viewed from the lateral aspect, the midbrain can be seen to consist of three distinct portions: the basis pedunculi ventrally, the midbrain *tegmentum* centrally and the *tectum* dorsally. The trochlear nerve (IV), the optic tract and the posterior cerebral artery wind around this aspect of the midbrain. The dorsal surface of the midbrain presents the four *colliculi* (or *corpora quadrigemini*) and the superior *medullary velum* between the two superior cerebellar peduncles. The *pineal gland* rests between the two superior colliculi and is attached by a stalk to the posterior dorsal thalamus.

Internal structure

The internal structure of the midbrain is again best described by reference to cross-sectional diagrams at representative levels: viz. at the level of the inferior and the superior colliculi (Figs 244, 245).

Observe that these sections pass through the midbrain at the level of the decussation of the superior cerebellar peduncle and the nucleus

Fig. 244 The midbrain—level of the inferior colliculus and decussation of the superior cerebellar peduncle.

Fig. 245 The midbrain—level of the superior colliculus and the red nucleus.

of the 4th nerve, on the one hand, and through the red nucleus and the nucleus of III on the other. The three subdivisions of the midbrain are also clearly seen in these figures. Above the level of the cerebral aqueduct lies the *tectum* and between the aqueduct and the basis pedunculi is the grey matter of the *tegmentum* separated from basis pedunculi by the deeply pigmented lamina of the *substantia nigra*. This pigment is neuromelanin, contained within the neurons of the substantia nigra.

Clinical features

When calcified, the pineal gland is easily identified on skull radiographs. It may then give the important radiological sign of lateral displacement by a space-occupying lesion of the cerebral hemisphere.

The diencephalon

The diencephalon comprises the hypothalamus and thalamus. It is that part of the brain surrounding the 3rd ventricle (Fig. 246).

The hypothalamus

The hypothalamus forms the floor of the 3rd ventricle. It includes, from before backwards, the *optic chiasma*, the *tuber cinereum*, the *infundibular stalk* (leading down to the posterior lobe of the pituitary), the *mamillary bodies* and the posterior *perforated substance*. In each of these there is a number of cell masses or nuclei surrounded by a fibre pathway — the medial forebrain bundle — which runs throughout the length of the hypothalamus and serves to link it with the midbrain posteriorly and the basal forebrain areas anteriorly.

From a functional point of view, the hypothalamus is largely concerned with autonomic activity and can be divided into a posteromedial sympathetic area and an anterolateral area concerned with parasympathetic activity.

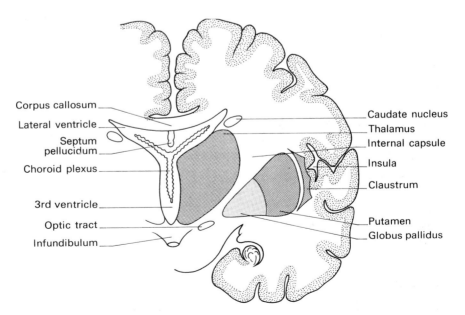

Corpus callosum
Lateral ventricle
Septum pellucidum
Choroid plexus

3rd ventricle
Optic tract
Infundibulum

Caudate nucleus
Thalamus
Internal capsule
Insula
Claustrum
Putamen
Globus pallidus

Fig. 246 The thalamus and 3rd ventricle in coronal section.

Sherrington described the hypothalamus as the head ganglion of the autonomic system. In addition, the hypothalamus plays an important part in endocrine control by the formation of releasing factors or release-inhibiting factors. These substances, following their secretion into the hypophyseal portal vessels, influence the production by the cells of the anterior pituitary of corticotrophin, follicle-stimulating hormone, luteinizing hormone, prolactin, somatotrophin, thyrotrophin and melanocyte-stimulating hormone.

The hormones oxytocin and vasopressin (anti-diuretic hormone, ADH) are produced by neurones in the hypothalamus and released at their axon terminals in the posterior pituitary.

Clinical features

1   Lesions of the hypothalamus may result in a variety of autonomic disturbances, e.g. somnolence, disturbances of temperature regulation and obesity, as well as a variety of endocrine abnormalities, e.g. hypogonadism and hypothyroidism.
2   Damage to the supraoptic nuclei or the infundibular stalk leads to diabetes insipidus.

The pituitary gland (hypophysis cerebri)

This is an example of a 'two in one' organ of which nature is so keen; compare the two glandular components of the adrenal cortex and medulla, and the exocrine and endocrine parts of the pancreas and testis. The pituitary comprises a larger anterior and smaller posterior lobe, the latter connected by the hollow infundibulum (pituitary stalk) to the tuber cinereum in the floor of the 3rd ventricle. The two lobes are connected by a narrow zone termed the pars intermedia.

The pituitary lies in the cavity of the pituitary fossa covered over by the diaphragma sellae, which is a fold of dura mater. This fold has a central aperture through which passes the infundibulum. Below is the body of the sphenoid, laterally lies the cavernous sinus and its contents separated by dura mater (Fig. 215), with intercavernous sinuses communicating in front, behind and below. The optic chiasma lies above, immediately in front of the infundibulum.

Structure

The anterior lobe is extremely cellular and consists of chromophobe, eosinophilic and basophilic cells. The pars intermedia contains large colloid vesicles reminiscent of the thyroid. The posterior lobe is made up of nerve fibres whose cell stations lie in the hypothalamus.

Development

The posterior lobe is a cerebral diverticulum. The anterior lobe and the pars intermedia develop from Rathke's pouch in the roof of the primitive buccal cavity. Occasionally a tumour grows from remnants of the epithelium of this pouch (craniopharyngioma). These tumours are often cystic and calcified.

Clinical features

Tumours of the pituitary, as well as forming intracranial space-occupying lesions, may have two special features; their endocrine disturbances and their relationship to the optic chiasma.

Chromophobe adenoma is the commonest pituitary tumour. As it enlarges it expands the pituitary fossa (sella turcica) and this may be demonstrated radiologically. Compression of the optic chiasma produces the very rapid typical bitemporal hemianopia (see 'The optic nerve', page 394). The tumour itself is non-secretory and gradually destroys the normally functioning gland. The patient develops hypopituitarism with loss of sex characteristics, hypothyroidism and hypoadrenalism. In childhood there is an arrest of growth. As the tumour extends there may be involvement of the hypothalamus with diabetes insipidus and obesity.

The *eosinophil adenoma* secretes the pituitary growth hormones. If it occurs before puberty, which is unusual, it produces gigantism; after puberty it results in acromegaly.

The *basophil adenoma* is small, produces no pressure effects and may be associated with Cushing's syndrome, although this more often results from hyperplasia or tumour of the adrenal cortex.

Pituitary tumours may be approached through a frontal bone flap. Occasionally when access is difficult the surgeon may have to divide the optic nerve and he therefore chooses to attack the tumour on the side with the more serious visual defect.

The close relationship of the pituitary to the sphenoid sinus makes it possible to insert a trocar into the pituitary gland by a transnasal, transsphenoidal approach. Radioactive material can then be introduced into the pituitary fossa to destroy the gland in the treatment of advanced malignant disease of the breast. At present, most centres perform open transsphenoidal hypophysectomy using the operating microscope.

The thalamus (Figs 246, 248)

The thalamus is an oval mass of grey matter which forms the lateral wall of the 3rd ventricle; it extends from the interventricular foramen rostrally to the midbrain caudally. Laterally, it is related to the internal capsule (and through it to the basal ganglia), and dorsally to the body of lateral ventricle. Medially, it is frequently connected with its fellow of the opposite side through the *massa intermedia* (interthalamic connexus). Posteriorly, it presents three distinct eminences, the *pulvinar*, and the *medial* and *lateral geniculate bodies*.

The thalamus is largely composed of relay nuclei, which forward impulses either from the main sensory pathways or from suprasegmental levels on to the cerebral cortex. This it does by way of a number of *thalamic radiations* in the internal capsule.

The *blood supply* of the thalamus is derived principally from the posterior cerebral artery through its thalamostriate branches, which pierce the posterior perforated substance to supply also the posterior part of the internal capsule.

The cerebral hemispheres

The cerebral hemispheres, which in man have developed out of all proportion to the rest of the brain, comprise the cerebral cortex, the basal ganglia, their afferent and efferent connections and the lateral ventricles.

The cerebral cortex

The cortex of the cerebral hemispheres is divided on topographical and functional grounds into four lobes (Fig. 247).

Frontal lobe

This includes all the cortical areas anterior to the *central sulcus of Rolando*. The important cortical areas of the frontal lobe are as follows.

1 *Motor cortex*. The motor area occupies a large part of the precentral gyrus; it receives afferents from the thalamus and cerebellum and is concerned with voluntary movement.

2 *Premotor cortex*. This occupies the anterior part of the precentral gyrus and the adjoining lower part of the frontal gyri. It too is concerned with voluntary movement.

3 *Eye motor field*. This lies in the cortex anterior to the premotor area; lesions in this area result in impaired eye movement.

4 *Broca's speech area*. Lesions of the area around the posterior part of the inferior frontal gyrus of the dominant (usually the left) hemisphere were shown by Broca to affect the motor element in speech.

Fig. 247 Localization of function in the cerebral cortex. (a) Lateral aspect. (b) Medial aspect.

5   *Frontal association cortex* (clinically called the *prefrontal cortex*). This comprises a considerable part of the frontal lobe and is one of the remarkable developments of the human brain. Its afferents are derived from the thalamus, limbic area and also from other cortical areas; it probably sends efferents to the thalamus and hypothalamus. From a functional point of view the lateral aspect of the frontal lobe appears to be related to 'intellectual activity', the medial and orbital surfaces to affective (or emotional) behaviour and to the control of autonomic activity.

Parietal lobe

The parietal lobe is bounded anteriorly by the central sulcus and behind by a line drawn from the parieto-occipital sulcus to the posterior end of the lateral sulcus. The important cortical areas of the parietal lobe are as follows.

1   The *somatic sensory cortex*. The postcentral gyrus receives afferent fibres from the thalamus and is concerned with all forms of somatic sensation.

2   The *parietal association cortex*, comprising the remainder of the parietal lobe, is concerned largely with the recognition of somatic sensory stimuli and their integration with other forms of sensory

information. It receives its afferents from the thalamus and, when damaged, gives rise to a peculiar inability to recognize somatic stimuli called *astereognosis*; put a key or a penny into the patient's hand—he is aware of the object but cannot identify it.

The temporal lobe

This is arbitrarily separated from the occipital lobe by a line drawn vertically downwards from the upper end of the lateral sulcus.

The important cortical areas of the temporal lobe are the following.

1   The *auditory cortex*. This lies in the superior temporal gyrus on the lateral and superior surfaces of the hemisphere. Its afferent fibres are from the medial geniculate body and it is concerned with the perception of auditory stimuli.

2   The *temporal association cortex*. The area surrounding the auditory cortex is responsible for the recognition of auditory stimuli and for their integration with other sensory modalities. Lesions of this area result in auditory agnosia, i.e. the inability to recognize or to understand the significance of meaningful sounds. The cortical region just behind and above this area on the dominant hemisphere is of considerable importance in the sensory aspect of the speech mechanism.

The rest of the temporal cortex is unique in that it does not appear to receive afferent fibres from the thalamus. Its significance is unknown but there is some evidence that it may be concerned with memory.

The parahippocampal gyrus

The cortex of the most medial part of the undersurface of the temporal lobe is known as the parahippocampal gyrus, much of which is referred to as the *entorhinal cortex*. It receives widespread association cortical afferents and is a significant source of inputs to the hippocampus. Anteriorly, it is related to the olfactory cortex of the uncus but it is not known whether the two are connected. Medially, it is in direct continuity with the layer of in-rolled cortex which is the *hippocampus* and which is one of the most important sources of afferents to this obscure cortical structure. The hippocampus occupies the whole length of the floor of the inferior horn of the lateral ventricle and extends into the amygdala. It sends its efferents into the overlying layer of white matter known as the *alveus*. The fibres of the alveus collect on the medial margin of the hippocampus to form a compact bundle, the *fimbria*, which, as it arches under the corpus callosum, becomes known as the *fornix*. The fornix passes forwards and then downwards in front of the interventricular foramen and finally backwards into the hypothalamus to terminate in the *mamillary body*. It also gives fibres to the thalamus and the hypothalamus.

This complex system was formerly referred to as the 'rhinencephalon' but there is little evidence that it is concerned with olfaction. The term *limbic system* is now given to this area. Its exact function in man is still controversial but it appears to be the anatomical substrate for emotions, behaviour and memory patterns. An intact limbic system with the hypothalamus and brainstem appears to be necessary for fully integrated brief and prolonged homeostasis under a wide variety of environmental conditions.

The amygdaloid nuclear complex

The amygdaloid nuclear complex is also a prominent temporal lobe structure, situated immediately rostral to the hippocampus. It is conveniently divided into three groups of nuclei: corticomedial, central and basolateral, which receive largely olfactory, gustatory, and association neocortical afferents respectively. These divisions also have more or less separable projections to the hypothalamus and septum, brainstem autonomic centres and ventral striatum. The amygdala has long been at the heart of the 'limbic system' concept and has been implicated in the control of emotional behaviour. Its neuroanatomical connections are clearly appropriate for such a role, since it is in a position to affect emotional responses in endocrine, autonomic and motor domains. Destruction of the amygdala is particularly associated with reduced aggressive behaviour, whilst the very high density of benzodiazepine receptors here has suggested amygdaloid involvement in anxiety and stress and their treatments.

Occipital lobe

The occipital lobe lies behind the parietal and temporal lobes. On its medial aspect it presents the Y-shaped *calcarine* and *postcalcarine sulci* (Fig. 247). The following cortical areas are noteworthy.

1 The *visual cortex* surrounds the calcarine and postcalcarine sulci and receives its afferent fibres from the lateral geniculate body of the thalamus of the same side; it is concerned with vision of the opposite half field of sight.

2 The *occipital association cortex* lies anteriorly to the visual cortex. This area is particularly concerned with the recognition and integration of visual stimuli.

The insula (Fig. 246)

If the lips of the lateral sulcus are separated, it is seen that there is a considerable area of cortex buried in the floor of this sulcus. This area is known as the *insula of Reil*. It is divided into a number of small gyri and is crossed by the middle cerebral artery. Apart from its upper part, which abuts on the sensory cortex and probably represents the taste area of the cerebral cortex, the function of the insula is unknown. Its stimulation excites visceral effects such as belching, increased salivation, gastric movements and vomiting.

The connections of the cerebral cortex

As has been indicated, most areas of the cerebral cortex receive their main afferent input from the thalamus, but, in addition to this, there are well-established *commissural connections* with the corresponding area of the opposite hemisphere by way of the *corpus callosum*. *Associational connections* also link neighbouring cortical areas on the same side and, in some cases, connect distant cortical areas; thus, the frontal, occipital and temporal lobes are directly connected by long association pathways.

Clinical features

It is convenient to summarize here the clinical effects of lesions affecting the principal cortical areas.

1   *Frontal cortex* — impairment of higher mental functions and emotions.

2   *Precentral (motor) cortex* — weakness of the opposite side of the body; lesions low down the cortex affecting the face and arm, high lesions affecting the leg. Midline lesions (meningioma, sagittal sinus thrombosis or a gunshot wound) may produce paraplegia by involving both leg areas.

3   *Sensory cortex* — contralateral hemianaesthesia (distributed in the same pattern as the motor cortex) affecting especially the higher sensory modalities such as stereognosis and two-point position sense.

4   Occipital cortex — contralateral homonymous hemianopia.

5   Lesions adjacent to the lateral sulcus in either the frontal, parietal or temporal lobes of the dominant hemisphere result in dysphasia.

The basal ganglia (Figs 246, 248)

These compact masses of grey matter are situated deep in the substance of the cerebral hemisphere and comprise the *corpus striatum*

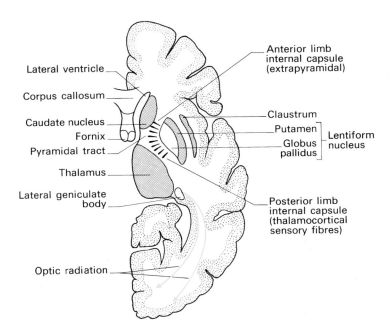

Fig. 248 The basal
ganglia and internal
capsule shown in
horizontal section
through the cerebrum.

(composed of the caudate nucleus, the putamen and the globus pallidus) and the *claustrum*. It should be noted that the details of the complex interconnections of the basal ganglia and of their functions are still at an elementary stage.

The corpus striatum

The *caudate nucleus* is a large homogeneous mass of grey matter consisting of a *head*, anterior to the interventricular foramen and forming the lateral wall of the anterior horn of the lateral ventricle; a *body*, forming the lateral wall of the body of the ventricle; and an elongated *tail*, which forms the roof of the inferior (temporal) horn of the ventricle. It is largely separated from the putamen by the *internal capsule*, but the two structures are connected anteriorly. The *putamen* is a roughly ovoid mass closely applied to the lateral aspect of the *globus pallidus*; together, they are called the *lentiform nucleus*. The basal ganglia receive their afferent connections from the cerebral cortex and the thalamus and send their efferents to the globus pallidus. Efferent fibres pass thence to the thalamus, hypothalamus, red nucleus, substantia nigra and inferior olivary nucleus (Figs 242 and 245).

The long ascending and descending pathways

The somatic afferent pathways (Fig. 249)

1 Proprioceptive and tactile impulses pass uninterruptedly through
the posterior root ganglia, through the ipsilateral *posterior columns*

Sensory cortex

Cerebrum

Thalamus

Midbrain

Medial lemniscus
and spinal lemniscus

Pons

Spinal lemniscus

Medulla

Sensory decussation (from
nucleus gracilis and
nucleus cuneatus)

Medial lemniscus

Lateral spinothalamic tract –
crossed (pain, temperature)

Dorsal column – direct
(deep sensation, tactile
discrimination)

Cord

Fig. 249 The long ascending pathways of the dorsal columns (yellow lines) and
spinothalamic tracts (red lines).

of the spinal cord to the *gracile* and *cuneate nuclei* in the lower part of the medulla. In the posterior columns there is a fairly precise organization of the afferent fibres; those from sacral and lumbar segments are situated medially in the tracts while fibres from thoracic and cervical levels are successively added to their lateral aspect. This arrangement according to body segments is maintained in the gracile and cuneate nuclei and in the efferents from these nuclei to the contralateral thalamus. The fibres arising from the gracile and cuneate nuclei immediately cross over to the opposite side in the *sensory decussation* of the medulla (Fig. 241) and continue up to the thalamus as a compact contralateral bundle—the *medial lemniscus*.

2   Dorsal root fibres subserving pain and temperature, together with some tactile afferents, end ipsilaterally in the *substantia gelatinosa* of the posterior horn. From the posterior horn afferent fibres are relayed up to the contralateral thalamus in the *anterior* and *lateral spinothalamic tracts* (or *anterolateral columns*), the fibres crossing in the anterior white commissure of the spinal cord. In the brainstem these fibres come to lie immediately lateral to the medial lemniscus and are sometimes known as the *spinal lemniscus* (Figs 249, 258). They terminate in the thalamus.

These somatic afferents are relayed from the thalamus, through the posterior limb of the internal capsule (Fig. 248) to the somatic sensory cortex of the *postcentral gyrus*. In the internal capsule the fibres are arranged in the sequence 'face, arm, trunk and leg' from before backwards, and this segregation persists in the sensory cortex, where the leg is represented on the dorsal and medial part of the cortex, the trunk and arm in its middle portion and the face most inferiorly. Since the size of the area cortical representation reflects the density of the peripheral innervation rather than the area of the receptive field, there is a good deal of distortion of the body image in the cortex, the cortical representation of the face and hand being much greater than that of the limbs and trunk.

Clinical features

1   Lesions of the sensory pathway most commonly occur in the internal capsule following some form of cerebrovascular accident. If complete, these result in a total hemianaesthesia of the opposite side of the body. In partial lesions the area of sensory loss will be determined by the site of the injury in the internal capsule and, from a knowledge of the sensory (and motor) loss, it is usually possible to determine with some degree of accuracy the site of a lesion in the capsule.

2   Since there is modality segregation below the decussation of the medial lemniscus, lesions of the sensory pathways at cord level result

in dissociation of sensation, with an area of analgesia contralaterally together with impairment of tactile sensibility ipsilaterally.

The auditory, visual and olfactory pathways are dealt with later under the appropriate cranial nerves.

The motor pathways (Fig. 250)

It is customary to divide the motor pathways of the brain and spinal cord into pyramidal and extrapyramidal systems. Although the latter is an imprecise concept, it provides a useful collective term for the many motor structures not confined to the pyramidal tracts in the medulla.

The pyramidal tract

The pyramidal system is the main 'voluntary' motor pathway and derives its name from the fact that projections to the motor neurons in the spinal cord are grouped together in the medullary pyramids. The fibres in this pathway arise from a wide area of the cerebral cortex. About two-thirds derive from the motor and premotor cortex of the frontal lobes; however, about one-third arises from the primary somatosensory cortex. In both the motor and premotor cortex there is an organization comparable to that seen in the sensory area. Again, the body is inverted so that the 'leg area' is situated in the dorsomedial part of the precentral gyrus encroaching on the medial surface of the hemisphere, the 'face area' is near the lateral sulcus, while the 'arm area' occupies a central position. Again, the body-form is greatly distorted so that the area representing the hand, lips, eyes and foot are exaggerated out of all proportion to the rest of the body.

From the cortex, the motor fibres pass through the posterior limb of the internal capsule (Fig. 248) where they are again organized in the sequence of 'face, arm, leg', anteroposteriorly. From the internal capsule the fibres form a compact bundle which occupies the central third of the cerebral peduncle. Hence they pass through the ventral pons, where they are broken up into a number of small bundles between the cells of the pontine nuclei and the transversely disposed pontocerebellar fibres. Near the lower end of the pons they again collect to form a single bundle which comes to lie on the ventral surface of the medulla and forms the elevation known as the 'pyramid'. As it passes through the brainstem, the pyramidal system gives off, at regular intervals, contributions to the somatic and branchial arch efferent nuclei of the cranial nerves. Most of these *corticobulbar fibres* cross over in the brainstem, but many of the cranial nerve nuclei are bilaterally innervated.

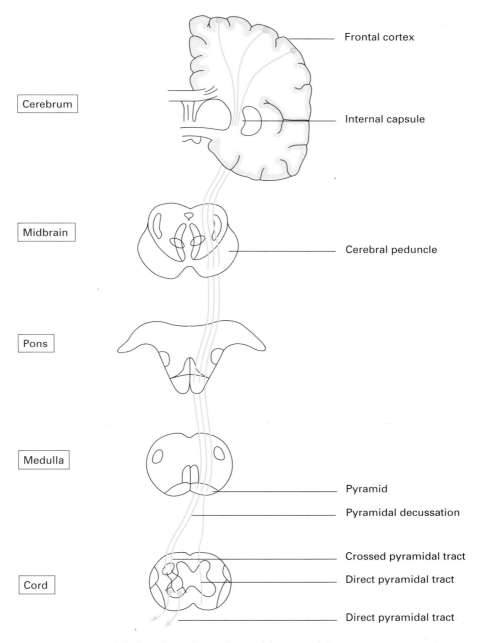

Fig. 250 The long descending pathway of the pyramidal tract.

Near the lower end of the medulla the great majority of the pyramidal tract fibres cross over to the opposite side and come to occupy a central position in the lateral white column of the spinal cord. This is the so-called 'crossed pyramidal tract' shown in Fig. 237. A small proportion of the fibres of the medullary pyramid, however, remain uncrossed until they reach the segmental level at which they finally terminate. This is the *direct* or *uncrossed pyramidal tract*, which runs downwards close to the anteromedian fissure of the cord, with fibres passing from it at each segment to the opposite side.

In view of the frequent involvement of the pyramidal tract in cerebrovascular accidents, its blood supply is listed here in some detail:

- *motor cortex* — leg area: anterior cerebral artery; face and arm areas: middle cerebral artery;
- *internal capsule*—branches of the middle cerebral artery;
- *cerebral peduncle*—posterior cerebral artery;
- *pons*—pontine branches of basilar artery;
- medulla—anterior spinal branches of vertebral artery;
- spinal cord—segmental branches of anterior and posterior spinal arteries.

Clinical features

1   It is important to remember that, in the motor cortex, movements are represented rather than individual muscles; lesions of this pathway result in paralysis of voluntary movement on the opposite side of the body although the muscles themselves are not paralysed and may cause involuntary movements. This is the essential difference between an 'upper motor neuron' lesion (i.e. a lesion of the central motor pathway) and a 'lower motor neuron' lesion (i.e. a lesion affecting the cranial nerve nuclei, or the anterior horn cells or their axons). In both types of lesion muscular paralysis results; in the latter, reflex activity is abolished, flaccidity and muscular atrophy follow, whereas, in pyramidal lesions, there is spasticity, increased tendon reflexes and an extensor plantar response.

2   Experimental lesions strictly confined to the pyramidal tract are not followed by increased muscular tone in the affected part (spasticity), but clinically this is a feature of upper motor neuron lesions; it is attributable to concomitant involvement of the extrapyramidal system.

3   The pyramidal tract is most frequently involved in cerebrovascular accidents where it passes through the internal capsule. Indeed, the artery supplying this area—the largest of the perforating branches of the middle cerebral artery — has been termed the *artery of cerebral haemorrhage.*

4   A list of the more important related signs is given here for involvement of the pyramidal tract at each level.

- *Cortex* — isolated lesions may occur here, resulting in loss of voluntary movement in, say, only one contralateral limb, but often the

sensory cortex is also involved. Dysphasia in dominant hemisphere lesions is common (Broca's area).
- *Internal capsule*—usually all parts of the tract are involved, giving a complete contralateral hemiplegia with associated sensory loss. The lesion may extend back to involve the visual radiation, giving a contralateral homonymous field defect (hemianopia).
- *Cerebral peduncle and midbrain* — the fibres from the 3rd nerve are often concomitantly involved so that there are the associated signs of a 3rd nerve palsy.
- *Pons*—here the 4th nerve is often involved, alone or together with VII. There may then be a hemiplegia affecting the arm and leg of the opposite side and an abducens and a facial palsy of the lower motor neuron type on the same side as the lesion.
- *Medulla* — because of the proximity of the pyramids to one another, medullary lesions often affect both sides of the body. Paralysis of the tongue on the side of the lesion is due to involvement of the 12th nerve or its nucleus. The respiratory, vasomotor and swallowing centres may also be affected.
- *Spinal cord*—the paralysis following lesions of the spinal cord is ipsilateral and accurately depends on the level at which the pyramidal tract is involved. The proximity of the pyramidal tracts to the ascending sensory pathways accounts for the concomitant sensory changes which are usually found.

The extrapyramidal system

The extrapyramidal motor system should, by definition, include all those motor projections which do not pass physically through the medullary pyramids. It was once thought to control movement in parallel with and, to a large extent, independently of the pyramidal motor system and the pyramidal/extrapyramidal division was used clinically to distinguish between two motor syndromes: one characterized by spasticity and paralysis whereas the other involved involuntary movements, or immobility without paralysis. It is now clear that many 'extrapyramidal' structures, particularly the basal ganglia, actually control movement by altering activity in the motor cortex and, thus, its pyramidal motor projections. This clearly emphasizes the blur between the two systems.

Components of the extrapyramidal system include the red nuclei, vestibular nuclei and reticular formation in the brain stem, all of which project via discrete pathways to influence spinal cord motor neurons. Cerebellar projections (see page 370) are also included since they influence not only these brainstem motor pathways, but also the motor cortex itself via the dentatothalamic projection.

Perhaps the most important structures to retain an extrapyramidal definition are the basal ganglia. The neostriatum (caudate and putamen) receives widespread cortical afferents, including those from high order sensory association and motor areas, and projects mainly to the globus pallidus. The latter nucleus is the major outflow for the basal ganglia and, via the ventral anterior thalamus, exerts its major influence on prefrontal and premotor cortices. This pattern of connections suggests that the basal ganglia are involved in complex aspects of motor control, including motor planning and the initiation of movement.

A variety of motor disorders are associated with basal ganglia pathology and, in some instances, neuroanatomically discrete deficits in specific neurotransmitters. For example, Parkinson's disease is now known to involve the degeneration of dopaminergic neurons in the substantia nigra in the midbrain. This pigmented nucleus provides the neostriatum with a dense dopaminergic innervation which may be completely lost in severe cases of Parkinsonism. Knowledge of this selective chemical neuropathology has resulted in the development of a treatment of the disease which involves the oral administration of the dopamine precursor L-dopa.

The membranes of the brain

The three membranes surrounding the spinal cord, the dura mater, arachnoid mater and pia mater, are continued upwards as coverings to the brain.

The *dura* is a dense membrane which, within the cranium, is made up of two layers. The outer layer is intimately adherent to the skull; the inner layer is united to the outer layer except where separated by the great dural venous sinuses and where it projects to form four sheets (Fig. 214):

- the falx cerebri;
- the falx cerebelli;
- the tentorium cerebelli;
- the diaphragma sellae.

The *arachnoid* is a delicate membrane separated from the dura by the potential *subdural space*. It projects only into the longitudinal fissure and the stem of the lateral fissure.

The *pia* is closely moulded to the outline of the brain; it dips down into the cerebral sulci leaving the *subarachnoid space* between it and the arachnoid. This space is broken up by trabeculae of fine fibrous strands and contains the cerebrospinal fluid.

The ventricular system and the cerebrospinal fluid circulation

The cerebrospinal fluid (C.S.F.) is formed by the secretory activity of the epithelium covering the choroid plexuses in the lateral, 3rd and 4th ventricles; it circulates through the ventricular system of the brain and drains into the subarachnoid space from the roof of the 4th ventricle before being reabsorbed into the dural venous system.

The general appearance of the ventricular system is indicated in Fig. 251. The two *lateral ventricles*, which are by far the largest components of the system, occupy a considerable part of the cerebral hemispheres. Each has an *anterior horn* (in front of the interventricular foramen), a *body*, above and medial to the body of the caudate nucleus, a *posterior horn* in the occipital lobe and an *inferior horn* reaching down into the temporal lobe. The choroid plexuses of the lateral ventricles, which are responsible for the production of most of the C.S.F., extend from the inferior horn, through the body, to the interventricular foramen where they become continuous with the plexus of the 3rd ventricle (Fig. 246).

The *3rd ventricle* is a narrow midline slit-like cavity between the two thalami in its upper portion and the hypothalamus in its lower part. Its floor is formed by the hypothalamus. From the 3rd ventricle the C.S.F. passes through the narrow *cerebral aqueduct* (*of Sylvius*) in the midbrain to reach the 4th ventricle.

The *4th ventricle* is diamond-shaped when viewed from above and tent-shaped as seen from the side. Its floor is formed below by the medulla and above by the pons. Its roof is formed by the cerebellum and the superior and inferior medullary vela. The C.S.F. escapes from the 4th ventricle into the subarachnoid space by way of the *median* and *lateral apertures* (of *Magendie* and *Luschka* respectively) and then flows over the surface of the brain and spinal cord.

Body of lateral ventricle
Anterior horn
3rd ventricle
4th ventricle
Interventricular foramen
Posterior horn
Aqueduct
Inferior horn

Fig. 251 The ventricular system.

In certain areas the subarachnoid space is considerably enlarged to form distinct cisterns. The most important of these are: the *cisterna magna* between the cerebellum and the dorsum of the medulla; the *cisterna pontis* over the ventral surface of the pons, the *interpeduncular cistern* between the two cerebral peduncles, the *cisterna ambiens* between the splenium of the corpus callosum and the superior surface of the cerebellum (containing the great cerebral vein and the pineal gland), and the *chiasmatic cistern* around the optic chiasma. Reabsorption of C.S.F. is principally by way of the superior longitudinal and the other dural sinuses, the modified arachnoid of the arachnoid granulations piercing the dura and bringing the C.S.F. into direct contact with the sinus mesothelium. Along the superior sagittal sinus these granulations (or arachnoid villi) clump together to form the Pacchionian bodies, which produce the pitted erosions readily seen along the median line of the inner aspect of the skull cap.

About one-fifth of the C.S.F. is absorbed along similar spinal villi or escapes along the nerve sheaths into the lymphatics. This absorption of C.S.F. is passive, depending on its hydrostatic pressure being higher than that of the venous blood.

Clinical features

1   Computerized axial tomography (CT scanning) has quite revolutionized the investigation of intracranial space-occupying lesions (post-traumatic haematoma, abscess and neoplasms), both by delineating the lesion itself and by demonstrating displacement of the ventricular system. Figures 252 and 253 are representative transverse cuts through the skull to illustrate normal anatomical features; note that the details of the anatomy of the ventricles are clearly visualized.
2   Magnetic resonance imaging (MRI) is particularly valuable in producing high-quality images of the central nervous system, although at present the technique is slower and much more expensive than CT.
3   The C.S.F. probably serves several purposes, including the provision of a protective water-jacket and a regulating mechanism of intracranial pressure with changing cerebral blood flow.
4   The total capacity of the C.S.F. in the adult is about 150 ml, of which some 25 ml is contained within the spinal theca; it is normally under a pressure of about 100 mm of water (with a range of 80–180) in the lateral horizontal position. The dural theca acts as a simple hydrostatic system, so that when the patient sits up, the C.S.F. pressure in the lumbar theca rises to between 350 and 550 mm, whereas the ventricular fluid pressure falls to below atmospheric.
5   Certain parts of the C.S.F. pathway are narrow and easily

Genu of corpus callosum

Septum pellucidum

Lateral ventricle

Posterior horn

Skull

(a)

Frontal lobe

Anterior horn of lateral ventricle

Septum pellicidum

Calcified pineal gland

Calcified choroid plexus in posterior horn

Occipital lobe

(b)

Fig. 252 (a) CT scan of skull through the level of the bodies of the lateral ventricles. (b) CT scan cut through the level of the anterior horns of the lateral ventricles.

obstructed. These sites are the interventricular foramina, the 3rd ventricle, the aqueduct, the exit foramina of the 4th ventricle and the subarachnoid space around the midbrain in the tentorial notch. Obstruction to the system causes increased intracranial pressure and ventricular dilatation (hydrocephalus).

Frontal lobe

Anterior horn of lateral ventricle

Septum pellucidum

3rd Ventricle

Fornix

Occipital lobe

Fig. 253 CT scan cut through the level of the third ventricle.

Corpus callosum

Aqueduct

Fornix

Thalamus

Cerebellum

Pituitary gland

4th Ventricle

Pons

Medulla

Fig. 254 MRI sagittal section of the head. Note the fine details of brain structure which can be visualized by this technique.

6   The meningeal coverings, together with the subarachnoid space, are prolonged along the optic nerve. Raised C.S.F. pressure is transmitted along this space and may compress the venous drainage of the eye, thus producing *papilloedema*. This swelling of the optic disc can be detected by ophthalmoscopic examination of the fundus.

7   Lumbar puncture—see page 364.

The cranial nerves

The olfactory nerve

The fibres of the olfactory nerve, unlike other afferent fibres, are unique in being the central processes of the olfactory cells and not the peripheral processes of a central group of ganglion cells.

The central processes of the olfactory receptors pass upwards from the olfactory mucosa in the upper part of the superior nasal concha and septum, through the cribriform plate of the ethmoid bone to end by synapsing with the dendrites of mitral cells in the olfactory bulb. The mitral cells in turn send their axons back in the olfactory tract to terminate in the cortex of the uncus, the adjacent inferomedial temporal cortex and the region of the anterior perforated space. The further course of the olfactory pathway is uncertain in man, but it is now clear that the hippocampus–fornix system is not directly concerned with olfaction.

Clinical features

1 The sense of smell is not highly developed in man and is easily disturbed by conditions affecting the nasal mucosa generally (e.g. the common cold). However, unilateral anosmia may be an important sign in the diagnosis of frontal lobe tumours. Tumours in the region of the uncus may give rise to the so-called 'uncinate' type of fit, characterized by olfactory hallucinations associated with impairment of consciousness and involuntary chewing movements.
2 Bilateral anosmia due to interruption of the 1st nerve is common after head injuries, particularly in association with anterior cranial fossa fractures, when leakage of C.S.F. through the cribriform plate may present as rhinorrhoea.

The optic nerve and the visual pathway

(For a description of the eye itself see the section on special senses, page 418.)

The optic nerve is the nerve of vision. It is not a true cranial nerve but should be thought of as a brain tract which has become drawn out from the cerebrum. Embryologically, it is developed, together with the retina, as a lateral diverticulum of the forebrain. Devoid of neurilemmal sheaths, its fibres, like other brain tissues, are incapable of regeneration after division.

From a functional point of view the retina can be regarded as consisting of three cellular layers: a layer of receptor cells—the *rods* and *cones*—an intermediate layer of *bipolar cells*, and a layer of *ganglion cells*, whose axons form the optic nerve (Fig. 255). From all parts of the retina these axons converge on the *optic disc* whence they pierce the sclera to form the optic nerve.

The optic nerve passes backwards and medially to the optic foramen through which it reaches the optic groove on the dorsum of the body of the sphenoid. Here, all the fibres from the medial half of the retina (i.e. those concerned with the temporal visual field) cross over in the *optic chiasma* to the optic tract of the opposite side, while the fibres from the lateral half of the retina (nasal visual field) pass back in the optic tract of the same side. The great majority of the fibres in the optic tract end in the six-layered *lateral geniculate body* of the thalamus, but a small proportion, subserving pupillary, ocular and head and neck reflexes, bypass the geniculate body to reach the superior colliculus and pretectal area. From the lateral geniculate body the fibres of the optic radiation sweep laterally, and some forwards into the temporal lobe before passing backwards to the occipital visual cortex (the striate area surrounding the calcarine fissure) where they terminate in such a way that the upper and lower halves of the retina are represented on the upper and lower lips of the fissure respectively (Figs 247, 256).

Clinical features

1   Lesions of the retina or optic nerve result in *ipsilateral blindness* in the affected segment, but lesions of the optic tract and central parts of visual pathway result in contralateral *homonymous defects*. Similarly, lesions of the optic chiasma (e.g. from an expanding pituitary

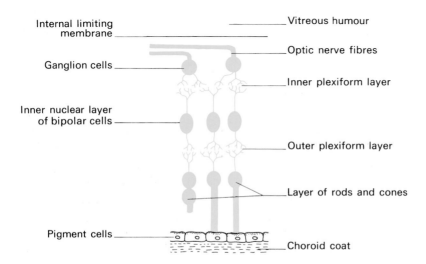

Internal limiting membrane

Ganglion cells

Inner nuclear layer of bipolar cells

Pigment cells

Vitreous humour

Optic nerve fibres

Inner plexiform layer

Outer plexiform layer

Layer of rods and cones

Choroid coat

Fig. 255 The layers of the retina.

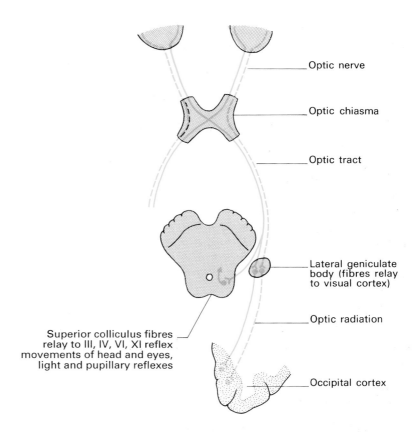

Optic nerve

Optic chiasma

Optic tract

Lateral geniculate body (fibres relay to visual cortex)

Optic radiation

Superior colliculus fibres relay to III, IV, VI, XI reflex movements of head and eyes, light and pupillary reflexes

Occipital cortex

Fig. 256 Diagram of the optic pathway.

tumour) will give rise to a *bitemporal hemianopia*, i.e. there will be a loss of vision in both temporal eye-fields.

2   The lesion responsible for the *Argyll Robertson pupil* is thought to be in the vicinity of the pretectal area, but there is no satisfactory explanation why the pupillary reaction to light should be abolished while the convergence–accommodation reflex is preserved.

The oculomotor nerve

In addition to supplying most of the extrinsic eye muscle, the oculomotor nerve conveys the preganglionic parasympathetic fibres for the sphincter of the pupil via the ciliary ganglion. Its nucleus of origin lies in the floor of the cerebral aqueduct at the level of the superior colliculus and consists essentially of two components: the *somatic efferent nucleus*, which supplies the ocular muscles, and the *Edinger–Westphal nucleus* from which the preganglionic parasympathetic fibres are derived.

From these nuclei, fibres pass vertically through the midbrain tegmentum to emerge just medial to the cerebral peduncle. Passing forwards between the superior cerebellar and posterior cerebral arteries, the nerve pierces the dura mater to run in the lateral wall of the cavernous sinus (Fig. 257) as far as the superior orbital fissure. Before entering the fissure it divides into a superior and inferior branch; both branches enter the orbit through the tendinous ring from which the recti arise (see Fig. 262). The superior branch passes lateral to the optic nerve to supply the superior rectus muscle and levator palpebrae superioris; the inferior branch supplies three muscles, the medial rectus, the inferior rectus and the inferior oblique, the nerve to the last conveying the parasympathetic fibres to the ciliary ganglion.

The ciliary ganglion

This small but important ganglion lies near the apex of the orbit just lateral to the optic nerve. It receives, in addition to the preganglionic parasympathetic fibres from the Edinger–Westphal nucleus, a sympathetic (postganglionic) root ultimately from the plexus on the internal carotid artery, and a sensory root from the nasociliary nerve. Of these fibres, only the *parasympathetic* synapse in the ganglion, the others pass directly through it. The postganglionic efferent fibres from the ganglion pass to the ciliary muscle and the muscles of the iris by way of about ten short *ciliary nerves*. Stimulation results in pupillary constriction and in accommodation of the lens. The sympathetic and sensory fibres are, respectively, vasoconstrictor and pupillodilator, and sensory to the globe of the eye.

(Note that the majority of sympathetic dilator pupillae nerve fibres are transmitted to the eye in the long ciliary branches of the nasociliary nerve.)

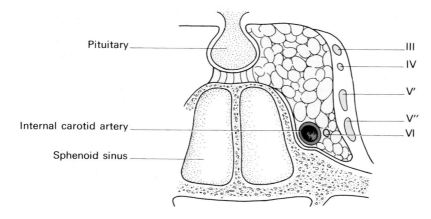

Fig. 257 The cavernous sinus—showing the relations of the 3rd, 4th, 5th and 6th cranial nerves.

Clinical features

Complete division of the 3rd nerve results in a characteristic group of signs:
- ptosis—due to paralysis of the levator palpebrae superioris;
- a *divergent squint*—due to the unopposed action of the superior oblique and lateral rectus muscles, rotating the eyeball laterally;
- *dilatation of the pupil*—the dilator action of the sympathetic fibres being unopposed;
- *loss of the accommodation–convergence* and *light reflexes*—due to constrictor pupillae paralysis;
- *double vision.*

The trochlear nerve

The trochlear nerve is the most slender of the cranial nerves and supplies only one eye muscle, the *superior oblique*. Its nucleus of origin lies in a similar position to that of the 3rd nerve at the level of the inferior colliculus, but from here its fibres pass dorsally around the cerebral aqueduct and decussate in the superior medullary velum (Fig. 258).

Emerging on the dorsum of the pons (being the only cranial nerve to arise from the dorsal aspect of the brainstem), the nerve winds round the cerebral peduncle and then passes forwards between the superior cerebellar and posterior cerebral arteries to pierce the dura. It then runs forwards in the lateral wall of the cavernous sinus (Fig. 257) between the oculomotor and ophthalmic nerves to enter the orbit through the superior orbital fissure, lateral to the tendinous ring

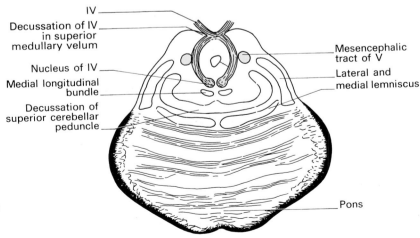

Fig. 258 Section through the upper pons to show the nucleus of nerve IV.

IV

Decussation of IV in superior medullary velum

Nucleus of IV

Medial longitudinal bundle

Decussation of superior cerebellar peduncle

Mesencephalic tract of V

Lateral and medial lemniscus

Pons

from which the recti take origin. It then passes medially over the optic nerve to enter the superior oblique muscle.

Clinical features

A lesion of the trochlear nerve results in paralysis of the superior oblique muscle with the result that diplopia occurs when the patient attempts to look downwards and laterally.

The trigeminal nerve (Fig. 259)

As the name suggests, this nerve consists of three divisions. Together they supply sensory fibres to the greater part of the skin of the head and face, the mucous membranes of the mouth, nose and paranasal air sinuses and, by way of a small motor root, the muscles of mastication. In addition it is associated with four autonomic ganglia, the ciliary, pterygopalatine, otic and submandibular.

The trigeminal ganglion

This ganglion, which is also termed the *semilunar ganglion*, is equivalent to the dorsal sensory ganglion of a spinal nerve. It is crescent-shaped and is situated within an invaginated pocket of dura in the

Mesencephalic nucleus

Motor nucleus

Sensory nucleus of the spinal tract

Ophthalmic

Maxillary and Mandibular divisions

Trigeminal ganglion

Main sensory nucleus

Fig. 259 Plan of the trigeminal nerve and its nuclei in dorsal view.

middle cranial fossa. It lies near the apex of the petrous temporal bone, which is somewhat hollowed for it. The motor root of the trigeminal nerve and the greater superficial petrosal nerve both pass deep to the ganglion. Above lies the hippocampal gyrus of the temporal lobe of the cerebrum; medially lies the internal carotid artery and the posterior part of the cavernous sinus. The trigeminal ganglion represents the 1st cell station for all sensory fibres of the trigeminal nerve except those subserving proprioception.

V_1: The ophthalmic division

This is the smallest division of the trigeminal nerve; it is wholly sensory and is responsible for the innervation of the skin of the forehead, the upper eyelid, cornea and most of the nose. Passing forwards from the trigeminal ganglion, it immediately enters the lateral wall of the cavernous sinus where it lies beneath the trochlear nerve (Fig. 257). Just before entering the orbit it divides into three branches, frontal, lacrimal and nasociliary.

The *frontal nerve* runs forward just beneath the roof of the orbit for a short distance before dividing into its two terminal branches, the *supratrochlear* and *supra-orbital nerves*, which supply the upper eyelid and the scalp as far back as the lambdoid suture.

The *lacrimal nerve* supplies the lacrimal gland (with postganglionic parasympathetic fibres from the pterygopalatine ganglion which reach it by way of the maxillary nerve) and the lateral part of the conjunctiva and upper lid.

The *nasociliary nerve* gives branches to the ciliary ganglion, the eyeball, cornea and conjunctiva the medial half of the upper eyelid, the dura of the anterior cranial fossa, and to the mucosa and skin of the nose.

V_2: The maxillary nerve

The maxillary nerve is again purely sensory. Passing forwards from the central part of the trigeminal ganglion, close to the cavernous sinus, it leaves the skull by way of the *foramen rotundum* and emerges into the upper part of the pterygopalatine fossa. Here it gives off a number of branches before continuing through the inferior orbital fissure and the infra-orbital canal as the *infra-orbital* nerve which supplies the skin of the cheek and lower eyelid.

The maxillary nerve has the following named branches:
1   the *zygomatic nerve*, whose zygomaticotemporal and zygomaticofacial branches supply the skin of the temple and cheek respectively;

2   *superior alveolar (dental) branches* to the teeth of the upper jaw; and

3   the *branches from the pterygopalatine ganglion*, which run a descending course and are distributed as follows: the *greater* and *lesser palatine nerves*, which pass through the corresponding palatine foramina to supply the mucous membrane of the hard and soft palates, the uvula and the tonsils, and the mucous membrane of the nose and a *pharyngeal branch* supplying the mucosa of the nasopharynx. The *nasopalatine nerve* (long sphenopalatine) supplies the nasal septum then emerges through the incisive canal of the hard palate to supply the gum behind the incisor teeth. The *posterior superior lateral nasal nerves* (short sphenopalatine) supply the posterosuperior lateral wall of the nose.

The pterygopalatine ganglion

Associated with the maxillary division of V as it lies in the pterygopalatine fossa is the relatively large pterygopalatine ganglion. This receives its parasympathetic or secretomotor root from the greater superficial petrosal branch of VII, its *sensory* component from two pterygopalatine branches of the maxillary nerve and its *sympathetic* root from the internal carotid plexus. Its parasympathetic efferents pass to the lacrimal gland through a communicating branch to the lacrimal nerve. Sensory and sympathetic (vasoconstrictor) fibres are distributed to nose, nasopharynx, palate and orbit.

V₃: The mandibular nerve

This is the largest of the three divisions of the trigeminal nerve. In addition to supplying the skin of the temporal region, part of the auricle and the lower face, the mucous membrane of the anterior two-thirds of the tongue and the floor of the mouth, it also conveys the motor root to the muscles of mastication and secretomotor fibres to the parotid gland.

Passing forwards from the trigeminal ganglion, it almost immediately enters the foramen ovale through which it reaches the infratemporal fossa. Here it divides into a small anterior and a larger posterior trunk, but before doing so it gives off the *nervus spinosus* to supply the dura mater and the *nerve to the medial pterygoid muscle* from which the *otic ganglion* is suspended and through which motor fibres are transmitted to tensor palati and tensor tympani.

The anterior trunk gives off:

1   a sensory branch, the *buccal nerve*, which supplies part of the skin of the cheek and the mucous membrane on its inner aspect; and

2   motor branches to the masseter, temporalis and lateral pterygoid muscles.

The posterior trunk, which is principally sensory, divides into three branches:

1   the *auriculotemporal nerve*, which conveys sensory fibres to the skin of the temple and auricle and secretomotor fibres from the otic ganglion to the parotid gland;

2   the *lingual nerve*, which passes downwards under cover of the ramus of the mandible to the side of the tongue (Fig. 197), where it supplies the mucous membrane of the floor of the mouth, the anterior two-thirds of the tongue (including the taste buds by way of fibres which join it from the chorda tympani), and the sublingual and submandibular salivary glands;

3   the *inferior alveolar (dental) nerve*, which passes down into the mandibular canal and supplies branches to the teeth of the lower jaw. It then emerges from the mental foramen to supply the skin of the chin and lower lip. This branch also conveys the only motor component of the posterior trunk: the *nerve to the mylohyoid*, supplying the muscle of that name and the anterior belly of the digastric.

The otic ganglion

The otic ganglion is unique among the four ganglia associated with the trigeminal nerve in having a motor as well as parasympathetic, sympathetic and sensory components. It lies immediately below the foramen ovale as a close medial relationship to the mandibular nerve.

Its *parasympathetic* fibres reach the ganglion by the lesser superficial petrosal branch of the glossopharyngeal nerve; these relay in the ganglion and pass via the auriculotemporal nerve to the parotid gland, and are its secretomotor supply. The *sympathetic* fibres are derived from the superior cervical ganglion along the plexus which surrounds the middle meningeal artery. The *sensory* fibres arrive from the auriculotemporal nerve; they are, respectively, vasoconstrictor and sensory to the parotid gland.

Motor fibres pass through the ganglion from the nerve to the medial pterygoid (a branch of the mandibular nerve) and supply the tensor tympani and tensor palati muscles.

The submandibular ganglion

This is suspended from the lower aspect of the lingual nerve. Its *parasympathetic* supply is derived from the chorda tympani branch of the facial nerve (see Fig. 263) by which it is conveyed to the lingual nerve; it carries the secretomotor supply to the submandibular and sublingual salivary glands.

Sympathetic fibres are transmitted from the superior cervical ganglion via the plexus on the facial artery and supply vasoconstrictor fibres to these same two salivary glands. The *sensory* component is contributed by the lingual nerve itself, which provides sensory fibres to these salivary glands and also to the mucous membrane of the floor of the mouth.

The central connections of the trigeminal nerve

The central processes of the trigeminal ganglion cells enter the lateral aspect of the pons and divide into ascending and descending branches which terminate in one or other component of the sensory nucleus of V (Figs 243, 260). This nucleus consists of three parts, each of which appears to subserve different sensory modalities: a chief sensory nucleus in the pontine tegmentum concerned with touch; a descending, or spinal, nucleus subserving pain and temperature; and a mesencephalic nucleus receiving proprioceptive afferents. The *motor root of the trigeminal nerve* lies just medial to the sensory nucleus in the

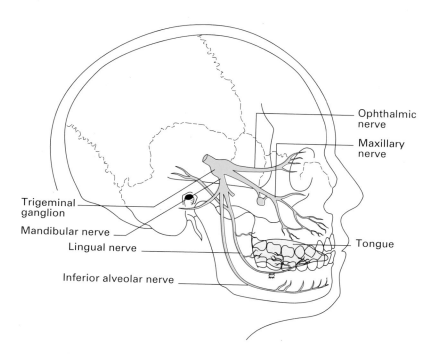

Ophthalmic nerve

Maxillary nerve

Trigeminal ganglion

Mandibular nerve

Lingual nerve

Tongue

Inferior alveolar nerve

Fig. 260 Distribution of the trigeminal nerve.

upper part of the pons; its efferents pass out with the sensory fibres and are distributed by way of the mandibular division of the nerve.

Clinical features

1   Section of the whole trigeminal nerve results in unilateral anaesthesia of the face and anterior part of the scalp, the auricle and the mucous membranes of the nose, mouth and anterior two-thirds of the tongue, together with paralysis and wasting of the muscles of mastication on the affected side. Lesions of separate divisions give rise to corresponding sensory and motor deficits in the area of distribution of the affected nerve.

2   Trigeminal neuralgia may affect any one or more of the three divisions, giving rise to the characteristic pain over the appropriate area (Fig. 261). Surgical treatment for this condition may involve:

- alcohol injection of the trigeminal ganglion;
- section of the central root of the nerve or of the appropriate division; or
- section of the descending tract of V as this lies superficially in the medulla—a procedure known as *medullary tractotomy*.

3   Pain is frequently referred from one segment to another. Thus, a patient with a carcinoma of the tongue (lingual nerve) frequently complains bitterly of earache (auriculotemporal nerve). The classical description of such a case is an old gentleman sitting in out-patients spitting blood and with a piece of cotton wool in his ear.

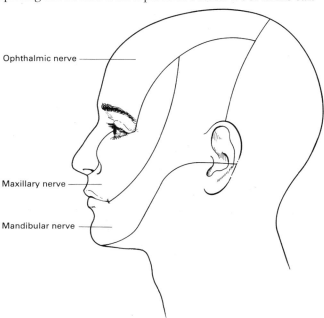

Ophthalmic nerve

Maxillary nerve

Mandibular nerve

Fig. 261 Areas of the face and scalp supplied by the three divisions of the trigeminal nerve.

The abducent nerve

Like the trochlear nerve, the abducent nerve supplies only one eye muscle, the *lateral rectus*. Its nucleus lies in the caudal part of the pons (Fig. 243) and from there its fibres pass through the pontine tegmentum to emerge on the base of the brain at the junction of the pons and medulla. The nerve then passes forwards to enter the cavernous sinus (Fig. 257). Here it lies lateral to the internal carotid artery and medial to the 3rd, 4th and 5th nerves. Passing through the tendinous ring just below the 3rd nerve, it enters the orbit to pierce the deep surface of the lateral rectus (Fig. 262).

Clinical features

On account of its long intracranial course, the 6th nerve is frequently involved in injuries to the base of the skull. When damaged, it gives rise to diplopia and a convergent squint.

The facial nerve

In addition to supplying the muscles of facial expression, the facial nerve conveys secretomotor fibres to the sublingual and sub-mandibular salivary glands and the lacrimal gland as well as the nasal mucosa; it also carries taste fibres from the anterior two-thirds of the tongue.

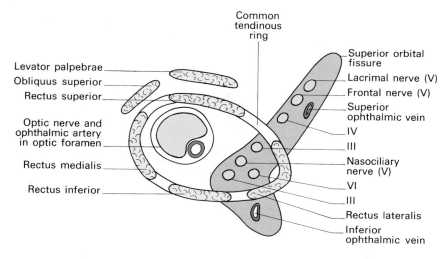

Common
tendinous
ring

Levator palpebrae
Obliquus superior
Rectus superior

Optic nerve and
ophthalmic artery
in optic foramen

Rectus medialis

Rectus inferior

Superior orbital
fissure
Lacrimal nerve (V)
Frontal nerve (V)
Superior
ophthalmic vein
IV
III
Nasociliary
nerve (V)
VI
III
Rectus lateralis
Inferior
ophthalmic vein

Fig. 262 The superior orbital fissure and tendinous ring of origin of the extrinsic orbital muscles, showing the relations of the cranial nerves as they enter the orbit.

The fibres innervating the facial muscles have their nucleus of origin in the ventral part of the caudal pons; the secretomotor fibres for the salivary glands are derived from the superior salivary nucleus. The sensory fibres associated with the nerve have their cells of origin in the facial (geniculate) ganglion.

From the motor nucleus, fibres of the facial nerve run a devious course over the nucleus of the abducent nerve (Fig. 243), where they form an elevation on the floor of the 4th ventricle known as the *facial colliculus*, then downwards and forwards to emerge from the lateral aspect of the pons together with VIII in the cerebellopontine angle.

The sensory and motor fibres pass together into the internal auditory meatus, at the bottom of which they leave the 8th nerve and enter the facial canal. Here they run laterally over the vestibule before bending sharply backwards over the promontory of the middle ear. This bend, or *genu of the facial nerve*, as it is called, marks the site of the facial ganglion and the point at which the secretomotor fibres for the lacrimal gland leave to form the greater superficial petrosal nerve. The facial nerve then passes downwards, medial to the middle ear, to reach the stylomastoid foramen (Fig. 263).

Just before entering this foramen it gives off the branch, known as the *chorda tympani*, which runs back through the middle ear to the infratemporal fossa where it joins the lingual nerve. Hence its taste fibres reach the anterior two-thirds of the tongue and its secretomotor fibres are conveyed to the submandibular ganglion.

On emerging from the stylomastoid foramen, the nerve supplies the stylohyoid and the posterior belly of digastric muscle. It then enters the parotid gland where it divides into five divisions for the supply of the facial muscles: the temporal, zygomatic, buccal, mandibular, and cervical branches (see Figs 208, 209, 264).

Pars intermedia (sensory root)

The genu and the geniculate ganglion

Nerve to stapedius

Stylomastoid foramen

Main trunk facial nerve

Internal auditory meatus

Greater superficial petrosal nerve

Chorda tympani

Fenestra vestibuli

Styloid process

Fig. 263 Distribution of the facial nerve within the temporal bone.

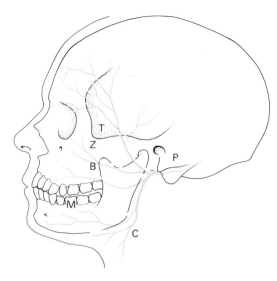

Fig. 264 Distribution of the facial nerve: T, temporal; Z, zygomatic; B, buccal; M, mandibular; C, cervical; and P, posterior auricular branch.

Clinical features

1 It is important to distinguish between 'nuclear' and 'infranuclear' facial palsies on the one hand and 'supranuclear' palsies on the other. Both nuclear and infranuclear palsies result in a facial paralysis which is complete and which affects *all* the muscles on one side of the face. In supranuclear palsies there is no involvement of the muscles above the palpebral fissure since the portion of the facial nucleus supplying these muscles receives fibres from both cerebral hemispheres. Furthermore, in such cases the patient may involuntarily use the facial muscles but will be unable to do so on request.

2 Supranuclear facial palsies most frequently result from vascular involvement of the corticobulbar pathways, e.g. in cerebral haemorrhage. Nuclear palsies may occur in poliomyelitis or other forms of bulbar paralysis, while infranuclear palsies may result from a variety of causes including compression in the cerebellopontine angle (as by an acoustic neuroma), fractures of the temporal bone and invasion by a malignant parotid tumour. However, by far the commonest cause of infranuclear facial paralysis is *Bell's palsy*, which is of unknown aetiology.

    When the intracranial part of the nerve is affected or when it is involved in fractures of the base of the skull there is usually loss of taste over the anterior two-thirds of the tongue and an associated loss of hearing (8th nerve damage).

The auditory (vestibulocochlear) nerve
(Fig. 265)

The 8th nerve consists of two sets of fibres: cochlear and vestibular. The *cochlear fibres* (concerned with hearing) represent the central processes of the bipolar spiral ganglion cells of the cochlea which traverse the internal auditory meatus to reach the lateral aspect of the medulla, at the cerebello-pontine angle (together with VII), where they terminate in the *dorsal* and *ventral cochlear nuclei*. The majority of the projection fibres from these nuclei cross to the opposite side, those from the dorsal nucleus forming the auditory striae in the floor of the 4th ventricle, those from the ventral nucleus forming the trape-

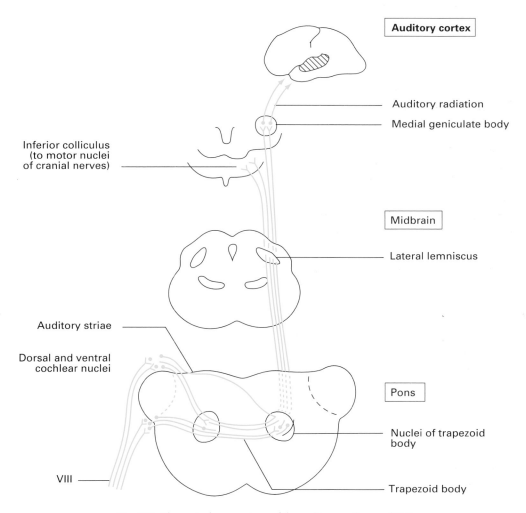

Fig. 265 The central connections of the auditory pathway of VIII.

zoid body in the ventral part of the pons. Most of these efferent fibres terminate in nuclei associated with the trapezoid body, either on the same or the opposite side, and then ascend in the lateral lemniscus to the *inferior colliculus* and the *medial geniculate body*; from the former, fibres reach the motor nuclei of the cranial nerves and form the pathway of auditory reflexes; from the latter, fibres sweep laterally in the *auditory radiation to the auditory cortex* in the superior temporal gyrus (Fig. 247).

The *vestibular fibres* (concerned with equilibrium) enter the medulla just medial to the cochlear division and terminate in the *vestibular nuclei*. Many of the efferent fibres from these nuclei pass to the cerebellum in the inferior cerebellar peduncle together with fibres bypassing the vestibular nuclei and passing directly to the cerebellum.

Other vestibular connections are to the nuclei of III, IV, VI and XI and to the upper cervical cord (via the vestibulospinal tract). These connections bring the eye and neck muscles under reflex vestibular control.

Clinical features

1   Lesions of the cochlear division result in deafness which may, or may not, be accompanied by tinnitus.

The differential diagnosis between middle ear deafness and cochlear (inner ear) or auditory nerve lesions can be made clinically by the use of a tuning fork. Air conduction (the fork being held beside the ear) is normally louder than bone conduction (the fork being held against the mastoid process). If the middle ear is damaged, the reverse will hold true.

2   Apart from injury to the cochlear nerve itself, unilateral lesions of the auditory pathway do not greatly affect auditory acuity because of the bilaterality of the auditory projections.

3   Temporal lobe tumours may give rise to auditory hallucinations if they encroach upon the auditory (superior temporal) gyrus.

4   Lesions of the vestibular division of the labyrinth or of the vestibulocerebellar pathway result in vertigo—a subjective feeling of rotation—nausea, ataxia and nystagmus.

The glossopharyngeal nerve (Fig. 210)

The glossopharyngeal nerve contains sensory fibres for the pharynx and the posterior one-third of the tongue (including the taste buds), motor fibres for the stylopharyngeus muscle and secretomotor fibres for the parotid gland. It is attached to the upper part of the medulla by four or five rootlets along the groove between the olive and the inferior cerebellar peduncle and leaves the skull by way of the jugular foramen in which it gives off its tympanic branch.

Below the jugular foramen the nerve courses downwards and forwards between the internal carotid artery and the internal jugular vein to reach the styloid process. From here it passes along the stylopharyngeus muscle to enter the pharnyx between the superior and middle constrictors. Here it breaks up into its terminal branches which supply the posterior one-third of the tongue and the mucous membrane of the pharynx (including the tonsil).

The *tympanic branch*, which is continued as the *lesser superficial petrosal nerve*, conveys the preganglionic parasympathetic fibres to the otic ganglion (parotid secretomotor fibres).

The only other branch of significance is the *carotid nerve* which arises just below the skull and runs down on the internal carotid artery to supply both the carotid body and carotid sinus. This small twig serves as the afferent limb of the baroreceptor and chemoreceptor reflexes from the carotid sinus and body respectively.

Clinical features

Complete section of the glossopharyngeal nerve results in sensory loss in the pharynx, loss of taste and common sensation over the posterior one-third of the tongue, some pharyngeal weakness and loss of salivation from the parotid gland. However, such lesions are frequently difficult to detect and rarely occur as isolated phenomena since there is so often associated involvement of the vagus or its nuclei.

The vagus nerve

The vagus has the most extensive distribution of all the cranial nerves, innervating the heart and the major part of the respiratory and alimentary tracts.

Central connections

The *dorsal nucleus* of the vagus (Fig. 242) is a mixed visceral afferent and efferent nucleus. It receives sensory fibres from the heart, the lower respiratory tract and the alimentary tract down to the transverse colon; in addition it gives rise to preganglionic parasympathetic motor fibres to the heart and the smooth muscles of the bronchi and gut.

From the *nucleus ambiguus* efferent fibres pass to the striped muscles of the pharynx and larynx.

Distribution

The nerve is connected to the side of the medulla by about ten filaments which lie in series with the glossopharyngeal nerve along the
groove between the olive and the inferior cerebellar peduncle. These
filaments unite to form a single bundle which passes beneath the cerebellum to the jugular foramen and is ensheathed in the same dural
covering as the accessory nerve. Two sensory ganglia are associated
with this part of the nerve: a superior, within the jugular foramen, and
an inferior, immediately beneath the skull.

The vagus then passes vertically downwards to the root of the
neck, lying in the posterior part of the carotid sheath between the
internal jugular vein and the internal and then common carotid arteries (Fig. 213). There are a number of important branches in the neck:
pharyngeal to the pharyngeal and palatal musculature by way of the
pharyngeal plexus; *superior laryngeal*, supplying the interior of the
larynx above the vocal folds and the cricothyroid and inferior constrictor muscles; and the superior and inferior *cardiac branches* which
are inhibitory to the heart.

Below the level of the subclavian arteries the course and relations
of the nerve on the two sides differ.

On the *right side* the *recurrent laryngeal branch* is given off as it
crosses the subclavian artery; beyond this the nerve descends through
the superior mediastinum in close association with the great veins.
Behind the root of the lung it takes part in the formation of the *pulmonary plexus* and then passes on to the oesophagus to form, with its
fellow, the *oesophageal plexus*.

The *left vagus* enters the thorax in close association with the great
arteries, lying at first lateral to the common carotid and then crossing
the arch of the aorta (Fig. 38). The *left recurrent laryngeal branch*,
which is given off as the vagus crosses the aortic arch, passes below
the ligamentum arteriosum, behind the arch and then ascends in the
groove between the trachea and the oesophagus (Fig. 34a). The vagus
then passes behind the root of the lung, enters into the formation of
the pulmonary plexus and passes on to the oesophagus.

The two vagi then enter the abdomen through the oesophageal
opening in the diaphragm, the left vagus passing on to the anterior
surface and the right passing to the posterior aspect of the stomach
(Fig. 55). Beyond this it is difficult to trace the course of the nerves,
but branches are given to the *coeliac, hepatic* and *renal plexuses* and,
by way of these plexuses, are distributed to the fore- and midgut and
to the kidneys.

Clinical features

1   Isolated lesions of the vagus nerve are uncommon but it may be involved in injuries or disease of related structures.
2   Vagotomy—see page 78.
3   Injuries to the recurrent laryngeal nerve—see 'The larynx', page 309.

The accessory nerve (Fig. 210)

The accessory nerve comprises a small cranial root, which is distributed by way of the vagus to the musculature of the palate, pharynx and larynx, and a larger spinal root which supplies the sternocleidomastoid and trapezius muscles.

The fibres of the *cranial root* are derived from the lower part of the nucleus ambiguus and leave the medulla, below the vagus, in four or five rootlets. Just before leaving the skull through the jugular foramen it is joined by the *spinal root*; this is formed by the union of fibres from an elongated nucleus in the anterior horn of the upper five cervical segments which leave the cord midway between the anterior and posterior roots and pass upwards through the foramen magnum. The two roots are united for only a short distance, the cranial root joining the inferior ganglion of the vagus immediately below the skull. The fibres of this root may be regarded as a caudal component of the vagus and are probably the source of the motor fibres to the palatal, pharyngeal and laryngeal musculature which leave the main trunk of the vagus in its pharyngeal and recurrent laryngeal branches. The spinal root passes backwards over the internal jugular vein to the sternocleidomastoid muscle which it pierces (and supplies) and then crosses the posterior triangle of the neck to enter and supply the deep surface of the trapezius.

Clinical features

1   Isolated lesions of the cranial root of the accessory nerve are infrequent; more commonly, it is involved concomitantly with the vagus when it gives rise to paresis of the laryngeal and pharyngeal muscles, resulting in dysphonia and dysphagia.
2   Division of the fibres of the spinal root (or lesions affecting their cells of origin) results in paresis of the sternocleidomastoid and trapezius muscles. This follows, for example, most block dissections of the lymph nodes of the neck, the nerve being sacrificed in clearing the posterior triangle.

The hypoglossal nerve

The hypoglossal nerve is entirely motor and supplies all the intrinsic and extrinsic muscles of the tongue (with the exception of the palatoglossus). From its nucleus, which lies in the floor of the 4th ventricle (Fig. 242), a series of about a dozen rootlets leave the side of the medulla in the groove between the pyramid and the olive. These rootlets unite to leave the skull by way of the anterior condylar, or hypoglossal, canal. Lying at first deep to the internal carotid artery and the jugular vein, the nerve passes downwards between these two vessels to just above the level of the angle of the mandible. Here it passes forwards over the internal and external carotid arteries, and gives off its descending and thyrohyoid branches. It then crosses the hyoglossus and genioglossus muscles to enter the tongue (Fig. 266). Its descending branch (*descendens hypoglossi*) actually derives from a twig of the 1st cervical nerve and therefore transmits C1 fibres. It passes more or less vertically downwards upon the internal carotid artery to join the descending cervical nerve (C2 and 3) to form a loop known as the *ansa cervicalis* (or *ansa hypoglossi*) just above the omohyoid muscle. From this loop branches are given to three infrahyoid muscles—sternothyroid, sternohyoid and omohyoid.

Clinical features

1 Division of the hypoglossal nerve, or lesions involving its nucleus, result in an ipsilateral paralysis and wasting of the muscles of the tongue. This is detected clinically by deviation of the protruded tongue to the affected side.

2 Supranuclear paralysis (due to an upper motor neuron lesion involving the corticobulbar pathways) leads to paresis but not atrophy of the muscles of the contralateral side.

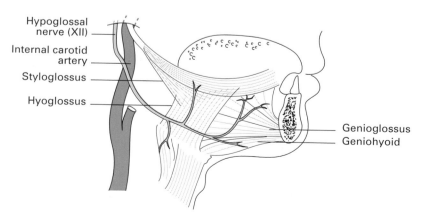

Hypoglossal nerve (XII)

Internal carotid artery

Styloglossus

Hyoglossus

Genioglossus
Geniohyoid

Fig. 266 The distal course of the hypoglossal nerve.

The special senses

The nose (see also 'The accessory nasal sinuses', page 342)

The *external nose* consists of a bony and cartilaginous framework closely overlaid by skin and fibrofatty tissues. The bones are the two nasal bones and the frontal processes of the maxilla. The *ala* is composed solely of fatty tissue at its lower free edge.

The *nasal cavity* is divided into right and left halves by a median *nasal septum* formed by the *perpendicular plate* of the *ethmoid bone*, the *septal cartilage* and the *vomer* (Fig. 267). Each cavity extends from the nostril (or anterior nares) in front to the posterior nasal aperture behind, communicating through the latter with the nasopharynx. The *lateral wall* is very irregular, due to the projection of the three *conchae* (superior, middle and inferior) and the underlying meatuses (Figs 225, 226).

The superior meatus receives the opening of the posterior ethmoidal air cells. Opening into the middle meatus are (from before backwards) the frontal and maxillary sinuses and the anterior and middle ethmoidal air cells. Only the nasolacrimal duct opens into the inferior meatus. The roof of the cavity is horizontal in its central portion, where it is formed by the cribriform plate of the ethmoid, but slopes downwards both anteriorly (the frontal and nasal bones) and posteriorly (the sphenoid). The *floor* corresponds to the roof of the mouth; it comprises the palatine process of the maxilla, the horizontal process of the palatine bone and the soft palate.

Frontal sinus

Nasal bone

Cartilage of septum

Nasal vestibule

Palatine process of maxilla

Perpendicular plate of ethmoid

Sphenoidal air sinus

Vomer

Horizontal plate of palatine bone

Fig. 267 The septum of the nose.

Mucous membrane

The *olfactory portion*, which is confined to the superior concha and the adjacent upper part of the septum, is thin and dull yellow in colour; it contains the olfactory receptors and supporting cells. The remaining *respiratory portion* is thick, vascular and moist with secretions of mucous glands; its epithelium is ciliated.

The upper part of the nasal cavity receives its arterial supply from the ethmoidal branches of the ophthalmic artery, a branch of the internal carotid. The sphenopalatine branch of the maxillary artery, a terminal of the external carotid, supplies the lower part of the cavity. Just within the vestibule of the nose, on the anteroinferior part of the septum, it links with a septal branch of the facial artery and it is from this zone, *Little's area*, that 90% of nose-bleeds occur. The veins drain downwards into the facial vein and upwards to the ethmoidal tributaries of the ophthalmic veins.

Clinical features

1   The skin of the external nose and its surrounds contains many sebaceous glands and hair follicles which may become blocked and infected. The significance of this fact is that the facial veins, which may become secondarily infected, communicate directly with the ophthalmic veins and hence with the cavernous sinus. For this reason, this zone is often known as the 'danger area of the face'.
2   The extensive relations of the nasal cavity are important in the spread of infection. Observe that it is in direct continuity with (i) the anterior cranial fossa (via the cribriform plate of the ethmoid bone); (ii) the nasopharynx and, through the pharyngotympanic tube, the middle ear; (iii) the paranasal air sinuses; (iv) the lacrimal apparatus and conjunctiva.
3   The septum is frequently deviated to one or other side, interfering both with inspiration and with drainage of the nose and accessory sinuses.

The ear

The external ear (Fig. 268)

This comprises the auricle and external auditory meatus. The *auricle*, for the most part, consists of a cartilaginous framework to which the skin is closely applied. The intrinsic and extrinsic muscles described for the ear are of no significance in man.

The *external auditory meatus* extends inwards to the tympanic

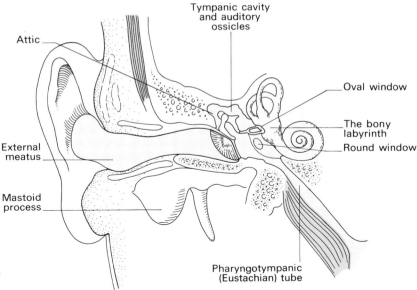

Attic

Tympanic cavity
and auditory
ossicles

Oval window

The bony
labyrinth

Round window

External
meatus

Mastoid
process

Pharyngotympanic
(Eustachian) tube

Fig. 268 General view of
the ear.

membrane. It is about 1.5 in (37 mm) long, and has a peculiar S-
shaped course, being directed first medially upwards and forwards,
then medially and backwards and, finally, medially forwards and
downwards. The outer third of the canal is cartilaginous and somewhat wider than the medial osseous portion. The whole canal is lined
by skin, which is closely adherent to the osseous portion but is separated from the cartilaginous part by the ceruminous glands in the subcutaneous tissue.

The *tympanic membrane*, or *ear drum* (Fig. 268; see Fig. 270),
separates the middle ear from the external auditory meatus. It is made
up of an outer cutaneous layer, continuous with the skin of the external auditory meatus, a middle fibrous layer and an inner mucous layer
continuous with the mucoperiosteum of the rest of the tympanic
cavity. It is oval in outline, a little less than 0.5 in (12 mm) in its greatest (vertical) diameter, and faces laterally, downwards and forwards;
it is slightly concave outwards. Since it is translucent (except at its
margin where it is attached to the medial aspect of the external auditory meatus), it is possible on examination to see the underlying
malleus and part of the incus. The greater part of the membrane is
taut and is known as the *pars tensa*, but above the lateral process of
the malleus there is a small triangular area where the membrane is
thin and lax—the *pars flaccida*. This area is bounded by two distinct
malleolar folds which reach down to the lateral process of the
malleus. The point of greatest concavity of the membrane is known as

the *umbo*; this marks the attachment of the handle of the malleus to the membrane.

The middle ear

The *middle ear*, or *tympanic cavity*, is the narrow slit-like cavity in the petrous part of the temporal bone containing the three auditory ossicles (Fig. 268).

The walls of the cavity and its important relations are as follows.

The *lateral wall* is formed mainly by the tympanic membrane, which divides it from the external auditory meatus, and above this by the squamous part of the temporal bone; the part of the cavity above the tympanic membrane is known as the *epitympanic recess* or *attic*; this part of the cavity contains the incus and the head of the malleus.

The *medial wall*, which separates the cavity from the internal ear, presents the *fenestra cochleae* (round window), closed by the secondary tympanic membrane; the *fenestra vestibuli* (oval window), occupied by the base of the stapes; the *promontory*, formed by the first turn of the cochlea; and the *prominence* caused by the underlying canal for the facial nerve (Fig. 263).

The *floor* is a thin plate of bone separating the cavity from the bulb of the jugular vein.

The *roof* is formed by the thin sheet of bone known as the *tegmen tympani*, which separates it from the middle cranial fossa and the temporal lobe of the brain.

Anteriorly, the cavity communicates with the pharynx by way of the pharyngotympanic or Eustachian tube.

Posteriorly, it communicates with the mastoid or tympanic antrum and the mastoid air cells.

The *mastoid antrum* is a small cavity in the posterior part of the petrous temporal bone connected to the epitympanic recess of the middle ear by way of the narrow *aditus*. Its importance is twofold: it is in communication with the mastoid air cells (hence the portal through which infection may spread to these spaces from the middle ear) and it is intimately related posteriorly to the sigmoid sinus and the cerebellum, both of which may be involved from a middle ear infection.

The *mastoid air cells* arise postnatally as diverticula from the tympanic antrum. They may invade not only the mastoid process but also the squamous part of the temporal bone. They are lined by a mucoperiosteum continuous anteriorly with that of the tympanic cavity.

The *pharyngotympanic (Eustachian) tube* reaches downwards, forwards and medially from the anterior part of the tympanic cavity to the lateral walls of the nasopharynx. In all it is about 1.5 in (37 mm) long, the first 0.5 in (12 mm) being bony while the rest is cartilaginous. It is lined by a ciliated columnar epithelium. The mucous membrane is thin in its bony part but the cartilaginous segment contains numerous mucous glands and, near its pharyngeal orifice, a considerable collection of lymphoid tissue termed the *tubal tonsil*. This may become swollen in infection, producing blockage of the tube. The tube is widest at its pharyngeal end and narrowest at the junction of the bony and cartilaginous portions.

Conduction of sound through the middle ear is by way of the malleus, incus and stapes. The *malleus* is the largest of the three and is described as having a handle, attached to the tympanic membrane, a rounded head, which articulates with the incus, and a lateral process, which can be seen through the tympanic membrane and from which the malleolar folds radiate. The *incus* comprises a body, which articulates with the malleus, and two processes, a short process attached to the posterior wall of the middle ear and a long process for articulation with the stapes. The shadow of the long process can often be seen through an auroscope running downwards behind the handle of the malleus. The *stapes* has a head for articulation with the incus, a neck and a base, which is firmly fixed in the fenestra vestibuli (the oval window).

Two small muscles are associated with these ossicles: the *stapedius*, which is attached to the neck of the stapes and is supplied by the facial nerve, and the *tensor tympani*, which is inserted into the handle of the malleus and is supplied by the mandibular division of V. Both serve to damp high-frequency vibrations.

The internal ear (Fig. 269)

The internal ear consists essentially of a complicated bony labyrinth made up of a central *vestibule*, which communicates posteriorly with three semicircular ducts and anteriorly with the spiral cochlea. This cavity contains a fluid known as *perilymph* and encloses the *membranous labyrinth*, comprising the *utricle* and *saccule*, which communicate respectively with the semicircular canals and the cochlear canal. The duct system is filled with *endolymph*.

In each component of the membranous labyrinth there are specialized sensory receptor areas known as the *maculae* of the utricle and saccule, the *ampullary crests* of the semicircular canals and the *spiral organ of Corti* in the cochlea.

The disposition of the semicircular canals in three planes at right angles to each other renders this part of the labyrinth particularly well suited to signal changes in position of the head. The organ of Corti is

Fig. 269 Detail of the membranous labyrinth.

adapted to record the sound vibrations transmitted by the stapes at the oval window.

The eye and associated structures (For optic nerve and visual pathway see pages 393–5.)

The eyeball (Fig. 271)

The eyeball, which is just under 25 mm in all diameters, is formed by segments of two spheres of different size: a prominent anterior segment, which is transparent and forms about one-sixth of the eyeball, and a larger posterior segment, which is opaque and comprises five-sixths of a sphere. A line joining the points of greatest curvature of these two segments (the anterior and posterior poles of the eye) is known as the *optical axis* of the eye. The optic nerve enters the eye about 0.125 in (3 mm) to the nasal (medial) side of the posterior pole.

The eyeball is formed by three coats: a fibrous outer coat, a vascular middle coat and an inner neural coat—the retina.

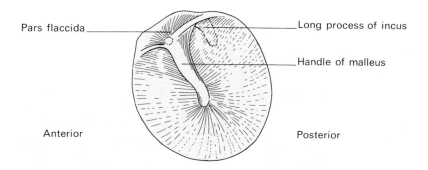

Fig. 270 The tympanic membrane as seen through an auroscope.

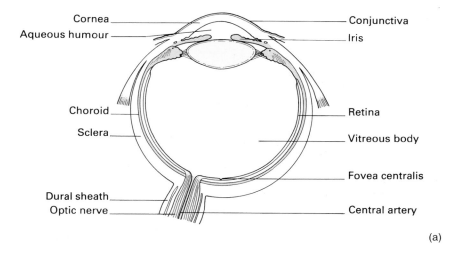

Cornea
Aqueous humour
Conjunctiva
Iris
Choroid
Sclera
Retina
Vitreous body
Fovea centralis
Dural sheath
Optic nerve
Central artery

(a)

Conjunctiva
Sinus venosus sclerae
Circular and radiating ciliary muscles
Sclera
Choroid
Cornea
Anterior chamber
Iris
Posterior chamber
Suspensory ligament
Lens

Fig. 271 (a) The eyeball in section. (b) Detail of the ciliary region.

(b)

The fibrous coat

The fibrous coat comprises a transparent anterior part, the cornea, and an opaque posterior portion, the sclera. Peripherally, the *cornea* is continuous with the sclera at the *sclerocorneal junction*. The *sclera* is a tough, fibrous membrane which is responsible for the maintenance of the shape of the eyeball and which receives the insertion of the extraocular muscles. Posteriorly, it is pierced by the optic nerve, with whose dural sheath it is continuous.

The vascular coat

This is made up of the choroid, the ciliary body and the iris.

The *choroid* is a thin but highly vascular membrane lining the inner surface of the sclera. Posteriorly it is pierced by the optic nerve and anteriorly it is connected to the iris by the ciliary body.

The *ciliary body* includes the *ciliary ring*, a fibrous ring continuous with the choroid, the *ciliary processes*, a group of sixty to eighty

folds arranged radially between the ciliary ring and the iris and connected posteriorly to the suspensory ligament of the lens, and the
ciliary muscles, an outer radial and inner circular layer of smooth
muscle responsible for the changes in convexity of the lens in
accommodation.

The *iris* is the contractile disc surrounding the pupil. It consists of
four layers:

1   an anterior mesothelial lining;
2   a connective tissue stroma containing pigment cells;
3   a group of radially arranged smooth muscle fibres—the dilator of
the pupil (supplied by the sympathetic system) and a circular group,
the pupillary sphincter (supplied by the parasympathetic fibres in the
oculomotor nerve);
4   a posterior layer of pigmented cells which is continuous with the
ciliary part of the retina.

The neural coat

The *retina* is formed by an outer pigmented and an inner nervous
layer, and is interposed between the choroid and the hyaloid membrane of the vitreous. Anteriorly, it presents an irregular edge, the *ora
serrata*, while posteriorly the nerve fibres on its surface collect to form
the optic nerve. Its appearance as seen through an ophthalmoscope is
shown in Fig. 272. Near its posterior pole there is a pale yellowish
area, the *macula lutea*, the site of central vision, and just medial to
this is the pale *optic disc* formed by the passage of nerve fibres
through the retina, corresponding to the 'blind spot'. The *central
artery of the retina* emerges from the disc and then divides into upper
and lower branches; each of these in turn divides into a nasal and temporal branch. Histologically, the retina consists of a number of layers
but from a functional point of view only three need be considered: an
inner receptor cell layer—the layer of rods and cones—an intermediate layer of bipolar neurones, and the layer of ganglion cells, whose
axons form the superfical layer of optic nerve fibres (Fig. 255).

Contents of the eyeball

Within the eyeball are found: the lens, the aqueous humour and the
vitreous body. The *lens* is biconvex and is placed between the vitreous
and the aqueous humour, just behind the iris. The *aqueous humour* is
a filtrate of plasma secreted by the vessels of the iris and ciliary body
into the *posterior chamber* of the eye (i.e. the space between the lens
and the iris). From here it passes through the pupillary aperture into
the *anterior chamber* (between the cornea and the iris) and is re-

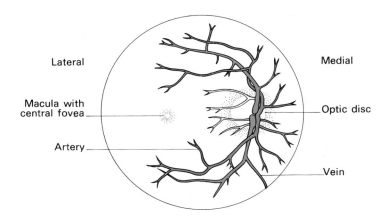

Lateral

Medial

Macula with
central fovea

Optic disc

Artery

Vein

Fig. 272 The right fundus
oculi as seen through an
ophthalmoscope.

absorbed into the ciliary veins by way of the *sinus venosus sclerae* (or *canal of Schlemm*). The *vitreous body*, which occupies the posterior four-fifths of the eyeball, is a thin transparent gel contained within a delicate membrane — the *hyaloid membrane* — and pierced by the lymph-filled *hyaloid canal*. The anterior part of the hyaloid membrane is thickened, receives attachments from the ciliary processes and gives rise to the *suspensory ligament of the lens*. This ligament is attached to the capsule of the lens in front of its equator and serves to retain it in position. It is relaxed by contraction of the radial fibres of the ciliary muscle and so allows the lens to assume a more convex form.

The orbital muscles (Fig. 262)

These are the levator palpebrae superioris and the extra-ocular muscles; the medial, lateral, superior and inferior recti and the superior and inferior obliques. The four *recti* arise from a tendinous ring around the optic foramen and the medial part of the superior orbital fissure and are inserted into the sclera anterior to the equator of the eyeball. The lateral rectus is supplied by the 6th nerve, the others by the 3rd. The *superior oblique* arises just above the tendinous ring and is inserted by means of a long tendon which loops around a fibrous pulley on the medial part of the roof of the orbit into the sclera just lateral to the insertion of the superior rectus. It is supplied by the 6th nerve. The *inferior oblique* passes like a sling from its origin on the medial side of the orbit around the undersurface of the eye to insert into the sclera between the superior and lateral recti; it is supplied by III.

Both the oblique muscles insert behind the equator of the eyeball.

The eyeball is capable of elevation, depression, adduction, abduction and rotation. The medial and lateral recti move the eyeball in one axis only.

The other four muscles move it on all three axes:

- rectus superior—elevation, adduction and medial rotation.
- rectus inferior—depression, adduction and lateral rotation.
- superior oblique—depression, abduction and medial rotation.
- inferior oblique—elevation, abduction and lateral rotation.

Pure elevation and depression of the eyeball is produced by one rectus acting with its opposite oblique—rectus superior with inferior oblique producing pure elevation and rectus inferior with the superior oblique producing pure depression.

A useful mnemonic is that the superior oblique is 'the tramp's muscle'—it moves the eye 'down and out'!

The actions of these muscles are shown in Fig. 273.

The *fascial sheath of the eye* (*Tenon's capsule*) is the membrane enclosing the eyeball from the optic nerve behind to the sclerocorneal junction in front. It is pierced by the vessels and nerves of the eye and by the tendons of the extra-ocular muscles. It is thickened inferiorly, where it forms the suspensory ligament.

The eyelids and conjunctiva

Of the two eyelids, the upper is the larger and more mobile, but apart from the presence of the levator palpebrae superioris in this lid, the structure of the eyelids is essentially the same. Each consists of the following layers, from without inwards: skin, loose connective tissue, fibres of the orbicularis oculi muscle, the tarsal plates, of very dense fibrous tissue, tarsal glands and conjunctiva. The eyelashes arise along the mucocutaneous junction and immediately behind the lashes there are the openings of the *tarsal* (*Meibomian*) *glands*. These are

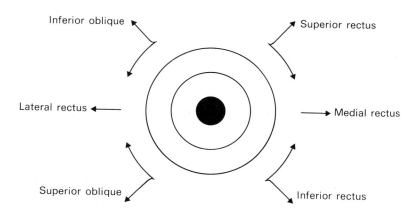

Fig. 273 The direction of action of the muscles acting on the eyeball from the primary position (i.e. looking directly forwards).

large sebaceous glands whose secretion helps to seal the palpebral fissure when the eyelids are closed and forms a thin layer over the exposed surface of the open eye; if blocked, they distend into Meibomian cysts.

The *conjunctiva* is the delicate mucous membrane lining the inner surface of the lids from which it is reflected over the anterior part of the sclera to the cornea. Over the lids it is thick and highly vascular, but over the sclera it is much thinner and over the cornea it is reduced to a single layer of epithelium. The line of reflection from the lid to the sclera is known as the conjunctival fornix; the superior fornix receives the openings of the lacrimal glands.

Movements of the eyelids are brought about by the contraction of the orbicularis oculi and levator palpebrae superioris muscles. The width of the palpebral fissure at any one time depends on the tone of these muscles and the degree of protrusion of the eyeball.

The lacrimal apparatus

The *lacrimal gland* is situated in the upper, lateral part of the orbit in what is known as the lacrimal fossa. The main part of the gland is about the size and shape of an almond, but it is connected to a small terminal process which extends into the posterior part of the upper lid. The gland is drained by a series of 8–12 small ducts which open into the lateral part of the superior conjunctival fornix whence its secretion is spread over the surface of the eye by the action of the lids.

The tears are drained by way of the *lacrimal canaliculi* whose openings, the *lacrimal puncta*, can be seen on the small elevation near the medial margin of each eyelid known as the *lacrimal papilla*. The two canaliculi, superior and inferior, open into the *lacrimal sac*, which is situated in a small depression on the medial surface of the orbit. This in turn drains through the nasolacrimal duct into the anterior part of the inferior meatus of the nose. The nasolacrimal duct, which not uncommonly becomes obstructed, is about 0.5 in (12 mm) in length and lies in its own bony canal in the medial wall of the orbit.

The autonomic nervous system

The nervous system is divided into two great subgroups: the *cerebrospinal system*, made up of the brain, spinal cord and the peripheral cranial and spinal nerves, and the *autonomic system* (also termed

the *vegetative, visceral* or *involuntary system*), comprising the autonomic ganglia and nerves. Broadly speaking, the cerebrospinal system
is concerned with the responses of the body to the external environment. In contrast, the autonomic system is concerned with the control
of the internal environment, exercised through the innervation of the
non-skeletal muscle of the heart, blood vessels, bronchial tree, gut
and the pupils and the secretomotor supply of many glands, including
those of the alimentary tract and its outgrowths, the sweat glands,
and, as a rather special example, the adrenal medulla.

The two systems should not be regarded as being independent of
each other, for they are linked anatomically and functionally.
Anatomically, autonomic nerve fibres are transmitted in all of the
peripheral and some of the cranial nerves; moreover, the higher connections of the autonomic system are situated within the spinal cord
and brain. Functionally, the two systems are closely linked within the
brain and cord.

The characteristic feature of the autonomic system is that its efferent nerves emerge as medullated fibres from the brain and spinal cord,
are interrupted in their course by a synapse in a peripheral ganglion
and are then relayed for distribution as fine non-medullated fibres. In
this respect they differ from the cerebrospinal efferent nerves, which
pass without interruption to their terminations (Fig. 274).

The autonomic system is subdivided into the sympathetic and
parasympathetic systems on anatomical, functional, and to a considerable extent, pharmacological grounds.

Anatomically, the sympathetic nervous system has its motor cell
stations in the lateral grey column of the thoracic and upper two
lumbar segments of the spinal cord. The parasympathetic system is
less neatly defined anatomically since it is divided into a cranial
outflow, which passes along the cranial nerves III, VII, IX and X, and
a sacral outflow, with cell stations in the 2nd, 3rd and sometimes 4th
sacral segments of the cord.

Functionally, the sympathetic system is concerned principally
with stress reactions of the body. When this system is stimulated, the
pupils dilate, peripheral blood vessels constrict, the force, rate and
oxygen consumption of the heart increase, the bronchial tree dilates,
visceral activity is diminished by inhibition of peristalsis and increase
of sphincter tone, glycogenolysis takes place in the liver, the adrenal
medulla is stimulated to secrete, and there is cutaneous sweating and
pilo-erection. The sympathetic pelvic nerves inhibit bladder contraction and are motor to the internal vesical sphincter.

Coronary blood flow is increased, partly by a direct sympathetic
effect and partly produced by indirect factors, which include more
vigorous cardiac contraction, reduced systole, relatively increased

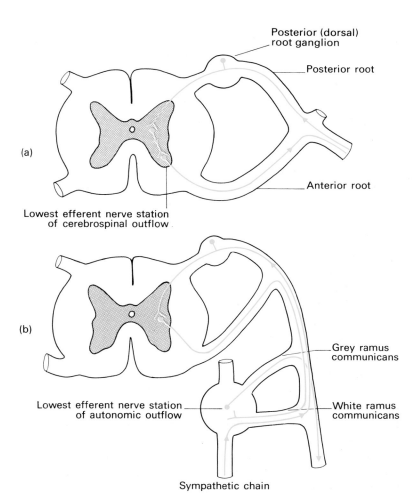

Posterior (dorsal) root ganglion

Posterior root

(a)

Anterior root

Lowest efferent nerve station of cerebrospinal outflow

(b)

Grey ramus communicans

Lowest efferent nerve station of autonomic outflow

White ramus communicans

Sympathetic chain

Fig. 274 The essential difference between the cerebrospinal and autonomic outflows: (a) the cerebrospinal system has its lowest efferent nerve cell stations within the c.n.s.; (b) the autonomic system has its lowest efferent cell stations in a peripheral ganglion (here illustrated by a typical sympathetic nerve ganglion).

diastole and an increased concentration of vasodilator metabolites.

The parasympathetic system tends to be antagonistic to the sympathetic system (Table 6). Its stimulation results in constriction of the pupils, diminution in the rate, conduction and excitability of the heart, an increase in gut peristalisis with sphincter relaxation and enhanced alimentary glandular secretion. In addition, the pelvic parasympathetic nerves inhibit the vesical internal sphincter and are motor to the detrusor muscle of the bladder.

The sympathetic system tends to have a 'mass action' effect; stimulation of any part of it results in a widespread response. In contrast, parasympathetic activity is usually discrete and localized. This difference can be explained, at least in part, by differences in anatomical peripheral connections of the two systems, as will be shown below.

Table 6 Summary of effects of sympathetic and parasympathetic stimulation.

	Sympathetic stimulation	Parasympathetic stimulation
Eye	Pupil dilates	Pupil constricts; accommodation of lens
Lacrimal gland	Vasoconstrictor	Secretomotor
Heart	Increase in force, rate, conduction and excitability	Decrease in force, rate, conduction and excitability
Lung	Bronchi dilate	Bronchi constrict; secretomotor to mucous glands
Skin	Vasoconstrictor Pilo-erection Secretomotor to sweat glands	—
Salivary glands	Vasoconstrictor	Secretomotor
Musculature of alimentary canal	Peristalsis inhibited	Peristalsis activated; sphincters relax
Acid secretion of stomach	—	Secretomotor
Pancreas	—	Secretomotor
Liver	Glycogenolysis	—
Suprarenal	Secretomotor	—
Bladder	Detrusor inhibited Sphincter stimulated	Detrusor stimulated Sphincter inhibited
Uterus	Uterine contraction Vasoconstriction	Vasodilatation

It is useful to think of the two systems as acting synergistically. For example, reflex slowing of the heart is effected partly from increased vagal and partly from decreased sympathetic stimulation. In addition, some organs receive their autonomic innervation from one system only; for example, the suprarenal medulla and the cutaneous arterioles receive only sympathetic fibres, whereas neurogenic gastric secretion is entirely under parasympathetic control via the vagus nerve.

Pharmacologically, the sympathetic postganglionic terminals release adrenaline and noradrenaline, with the single exception of the terminals to the sweat glands which, in common with all the parasympathetic postganglionic terminations, release acetylcholine.

Visceral afferents

As well as the efferent system, there are afferent visceral fibres which are concerned with the afferent arc of autonomic reflexes and with the conduction of visceral pain stimuli. These nerves have their cell stations in the dorsal root ganglia of the spinal nerves or of the ganglia of the cranial nerves concerned with the autonomic system. The fibres from the viscera ascend in the autonomic plexuses; those from the body wall are conveyed in the peripheral spinal nerves. The afferent course from any structure is therefore along the same pathway as the efferent autonomic fibres which supply the part.

The afferent fibres ascend centrally to the hypothalamus and thence to the orbital and frontal gyri of the cerebral cortex along as yet indeterminate pathways. Normally, we are unaware of the afferent impulses from the viscera unless they become sufficiently great to exceed the pain threshold when they are perceived as visceral pain, e.g. the pain of coronary ischaemia or intestinal colic.

The sympathetic system

The efferent fibres of the sympathetic system arise in the lateral grey column of the spinal cord (see Fig. 274) from segments T1 to L2. From each of these segments small medullated axons emerge into the corresponding anterior primary ramus and pass via a white ramus communicans into the sympathetic trunk.

The spinal segments responsible for the sympathetic innervation of the various parts of the body are approximately as follows:

- head and neck, T1–2;
- upper limb, T2–5;
- thoracic viscera, T1–4;
- abdominal viscera, T4–L2;
- pelvic viscera, T10–L2;
- lower limb, T11–L2.

Stimulation of a single white ramus communicans would thus obviously have widespread effects—the anatomical basis of the 'mass action' response of sympathetic stimulation.

The sympathetic trunk

The sympathetic trunk on each side is a ganglionated nerve chain which extends from the base of the skull to the coccyx in close relationship to the vertebral column, maintaining a distance of about 1 inch (2.5 cm) from the midline throughout its course. Commencing in the superior cervical ganglion beneath the skull base, the chain descends closely behind the posterior wall of the carotid sheath, enters the thorax anterior to the neck of the first rib, descends over the heads of the upper ribs and then on the sides of the bodies of the last three or four thoracic vertebrae. The chain then passes into the abdomen behind the medial arcuate ligament of the diaphragm and descends in a groove between psoas major and the sides of the lumbar vertebral bodies, overlapped by the abdominal aorta on the left and the inferior vena cava on the right. The chain then passes behind the common iliac vessels to enter the pelvis anterior to the ala of the sacrum and then descends medial to the anterior sacral foramina. The sympathetic trunks end below by meeting each other at the *ganglion impar* on the anterior face of the coccyx.

The details of the cervical, thoracic and lumbar portions of the trunk are given on pages 331, 50 and 166 respectively.

The sympathetic trunk bears a series of ganglia along its course which contain motor cells with which preganglionic medullated fibres enter into synapse and from which non-medullated postganglionic axons originate. Developmentally, there was originally one ganglion for each peripheral nerve, but by a process of fusion these have been reduced in man to three cervical, twelve or less thoracic, two to four lumbar and four sacral ganglia. Only the ganglia of T1 to L2 receive white rami directly; the higher and lower ganglia must receive their preganglionic supply from medullated nerves which travel through their corresponding ganglia without relay and which then ascend or descend in the sympathetic chain. Still other preganglionic fibres pass intact through the ganglia to peripheral visceral ganglia for relay.

There are thus three fates which may befall white rami (Fig. 275).
1   To enter into synapse from the corresponding sympathetic ganglion (this applies only to the T1 to L2 segments).
2   To ascend or descend in the sympathetic chain with relay in higher or lower ganglia.
3   To traverse the ganglia intact and relay in peripheral ganglia.

Pharmacologically, the sympathetic postganglionic terminals release adrenaline and noradrenaline, with a single exception of the sweat glands, which, in common with all the parasympathetic postganglionic terminations, release acetylcholine.

Distribution

The branches of the sympathetic ganglionic chain have somatic and visceral distribution.

Somatic distribution

Each spinal nerve receives one or more grey rami from a sympathetic ganglion which distributes postganglionic non-medullated sympa-

Fig. 275 The three fates of sympathetic white rami. These may (A) relay in their corresponding ganglion and pass to their corresponding spinal nerve for distribution, (B) ascend or descend in the sympathetic chain and relay in higher or lower ganglia, or (C) pass without synapse to a peripheral ganglion for relay.

thetic fibres to the segmental skin area supplied by the spinal nerve. These fibres are vasoconstrictor to the skin arterioles, sudomotor to sweat glands and pilomotor to the cutaneous hairs.

Visceral distribution

Postganglionic fibres to the head and neck and to the thoracic viscera arise from the ganglion cells of the sympathetic chain. Those to the head ascend along the internal carotid and vertebral arteries, whereas those to the thoracic organs are distributed by the cardiac, pulmonary and oesophageal plexuses.

The abdominal and pelvic viscera, however, are supplied by post-ganglionic fibres which have their cell stations in more peripherally placed prevertebral ganglia — the coeliac, hypogastric and pelvic plexuses — which receive their preganglionic fibres from the splanchnic nerves (Fig. 276). These nerves are detailed on page 50.

The suprarenal medulla has a unique nerve supply comprising a rich plexus of preganglionic fibres which pass without relay from the coeliac ganglion to the gland. These fibres end in direct contact with

Fig. 276 The coeliac plexus.

the chromaffin medullary cells, and liberate acetylcholine (as in all autonomic ganglia) which stimulates the secretion of adrenaline and noradrenaline by the adrenal medulla.

The chromaffin cells of the adrenal medulla may thus be regarded as sympathetic cells which have not developed postganglionic fibres; indeed, embryologically both the medulla and the sympathetic nerves have a common origin from the neural crest.

Clinical features

Cervical sympathectomy—see page 333.
Lumbar sympathectomy—see page 167.

The parasympathetic system

As already stated, this system has a cranial and a sacral component. Its medullated preganglionic fibres synapse with ganglion cells which lie close to, or actually in the walls of, the viscera supplied. Postganglionic fibres therefore have only a short and direct course to their

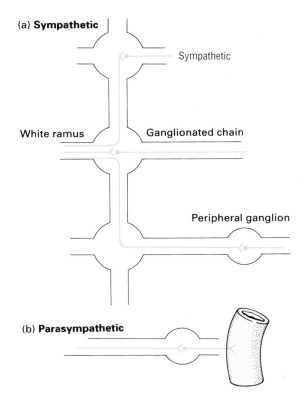

(a) Sympathetic

Sympathetic

White ramus

Ganglionated chain

Peripheral ganglion

Fig. 277 The anatomical basis of widespread sympathetic and local parasympathetic response. (a) The widespread distribution of postganglionic fibres from a single sympathetic white ramus. (b) The localized distribution of postganglionic parasympathetic fibres.

(b) Parasympathetic

effector cells and there is thus the anatomical pathway of a local discrete response to parasympathetic stimulation (Fig. 277).

Cranial outflow

The cranial component of the parasympathetic system is conveyed in cranial nerves III, VII, IX and X, of which X (the vagus) is the most important and the most widely distributed. The functions of this group of nerves can be summarized as follows:

1 pupils—constrictor to pupil, motor to ciliary muscle (accommodation);
2 salivary glands—secretomotor;
3 lacrimal glands—secretomotor;
4 heart—inhibitor of cardiac conduction, contraction, excitability and impulse formation (with consequent slowing of the heart and diminution of its contraction force);
5 lungs—bronchoconstrictor, secretomotor to mucous glands;
6 alimentary canal—motor to gut muscles as far as the region of the ascending colon; inhibitor to the pyloric sphincter; secretomotor to the glands and adnexae of the stomach and intestine.

The parasympathetic distribution of III, VII and IX is carried out via four ganglia from which postganglionic fibres relay. These ganglia also transmit (without synapse and therefore without functional connection) sympathetic and sensory fibres which have similar peripheral distribution. These ganglia are the ciliary (see page 396), pterygopalatine (see page 400), submandibular (see page 401) and otic (see page 401).

The 10th (vagal) distribution conveys by far the most important and largest contributions of the parasympathetic system. It is responsible for all the functions of the parasympathetic cranial outflow enumerated above, apart from the innervation of the eye and the secretomotor supply to the salivary and lacrimal glands. The efferent fibres are derived from the dorsal nucleus of X and are distributed widely in the cardiac, pulmonary and alimentary plexuses. Postganglionic fibres are relayed from tiny ganglia which lie in the walls of the viscera concerned; in the gut these constitute the submucosal *plexus of Meissner* and the myenteric *plexus of Auerbach*.

The sacral outflow

The anterior primary rami of S2, 3 and occasionally 4 give off nerve fibres termed the *pelvic splanchnic nerves* or *nervi erigentes*, which join the sympathetic pelvic plexus for distribution to the pelvic organs. Tiny ganglia in the walls of the viscera then relay postganglionic fibres.

The sacral parasympathetic system has been termed by Cannon 'the mechanism for emptying'. It supplies visceromotor fibres to the muscles of the rectum and inhibitor fibres to the internal anal sphincter, motor fibres to the bladder wall and inhibitor fibres to the internal vesical sphincter. In addition, vasodilator fibres supply the erectile cavernous sinuses of the penis and the clitoris.

Afferent parasympathetic fibres

Visceral afferent fibres from the heart, lung and the alimentary tract are conveyed in the vagus nerve. Sacral afferents are conveyed in the pelvic splanchnic nerves and are responsible for visceral pain experienced in the bladder, prostate, rectum and uterus. The reference of pain from these structures to the sacral area, buttocks and posterior aspect of the thighs is explained by the similar segmental supply of the sacral dermatomes.

Note that although afferent fibres are conveyed in both sympathetic and parasympathetic nerves, they are completely independent of the autonomic system. They do not relay in the autonomic ganglia and have their cell stations, just like somatic sensory fibres, in the dorsal ganglia of the spinal and cranial nerves. They simply use the autonomic nerves as a convenient anatomical conveyor system from the periphery to the brain.

Glossary of eponyms

In spite of their being 'unscientific', eponymous terms are still commonly used—more so among clinicians, it is true, than among professional anatomists. This glossary gives brief biographical details of the names mentioned in the text. Entries appear alphabetically according to the person whose name is used adjectivally in the term: *valves of Ball* under Ball, *column of Burdach* under Burdach, and so on.

Alcock's canal Fascial tunnel on the lateral wall of the ischiorectal fossa which conveys the pudendal vessels and nerve.
Benjamin Alcock (1801–?), Professor of Anatomy, first in Dublin and then Cork. He emigrated to America and disappeared from the scene.

Argyll Robertson pupil The pupil does not respond to light but reacts to accommodation. It is classically seen in neurosyphilis.
Douglas Argyll Robertson (1837–1909), ophthalmic surgeon, Royal Infirmary, Edinburgh.

Auerbach's plexus Nerve plexus between the circular and longitudinal muscle coats of the intestine.
Leopold Auerbach (1828–1897), Professor of Pathology, Breslau.

valves of Ball Valve-like folds which connect the distal extremities of the columns of Morgagni (q.v.) in the upper half of the anal canal.
Sir Charles Ball (1851–1916), Regius Professor of Surgery in Dublin and an early pioneer of rectal surgery.

Bartholin's gland The greater vestibular gland. Mucus-secreting gland in the posterior labium majus.
Caspar Bartholin (1655–1738), Professor of Anatomy, Copenhagen.

Batson's venous plexus The 'valveless vertebral veins' communicate with the prostatic plexus of veins and explain the readiness with which carcinoma of the prostate spreads to the pelvis and lumbar vertebrae.
Oscar Batson (1894–1979), Professor of Anatomy, University of Pennsylvania.

Blalock's operation The right subclavian artery is anastomosed end-to-side into the right pulmonary artery in order to overcome the pulmonary stenosis of the tetralogy of Fallot (q.v.).
Alfred Blalock (1899–1964), Professor of Surgery, Johns Hopkins Hospital, Baltimore.

Broca's area The anterior portion of the inferior frontal gyrus; on the dominant side it is the motor area for speech.
Pierre Broca (1824–1880), Professor of Clinical Surgery in Paris.

Brown-Séquard syndrome Produced by hemisection of the spinal cord.
Charles Edouard Brown-Séquard (1817–1894), born in Mauritius, practised as a neurologist in Paris, at Harvard and at the National Hospital, Queen's Square, London.

Brunner's glands The characteristic submucosal acinar glands of the duodenum.
Johann Brunner (1653–1727), Swiss anatomist who became Professor of Anatomy at Heidelberg and later at Strasbourg. It is said that his father-in-law, J. J. Wepfer, actually discovered these glands!

Bryant's triangle Used in measurement of the hip.
Thomas Bryant (1828–1914), surgeon at Guy's Hospital; President of the Royal College of Surgeons of England.

column of Burdach The lateral tract (fasiculus cuneatus) of the posterior column of the spinal cord.
Charles Burdach (1776–1847), Professor of Anatomy and Medicine, Konigsberg.

Caldwell–Luc operation For drainage of the maxillary sinus.
George Walter Caldwell (1866–1946), ENT surgeon, New York.

Henri Luc (1855–1925), ENT surgeon, Paris.

Camper's fascia The superficial fatty layer of the superficial fascia of the lower abdominal wall. Peter Camper (1722–1789), Professor of Medicine in Amsterdam and then Professor of Medicine, Surgery, Anatomy and Botany in Groningen.

Cloquet's gland Lymph node situated in the femoral canal. Jules Cloquet (1790–1883), Professor of Anatomy and Surgery, Paris.

Colles' fascia The perineal fascia.

Colles' fracture Fracture of the lower end of the radius with dorsal displacement. Abraham Colles (1773–1843), Professor of Anatomy and Surgery at the Royal College of Surgeons in Ireland.

the ligaments of Cooper Fibrous septa in the breast.

the ligament of Cooper The iliopectineal fascia. Sir Astley Paston Cooper (1768–1841), surgeon at Guy's Hospital; President of the Royal College of Surgeons.

organ of Corti The sound receptor organ in the cochlea. Alfonso Corti (1822–1888), Italian histologist who worked mainly on the retina and the ear.

Cowper's glands Two glands situated in the deep perineal pouch which drain into the bulbous urethra. William Cowper (1666–1698), surgeon in London.

Cushing's syndrome Endocrine abnormality associated with hyperplasia or tumour of the adrenal cortex or of a basophil adenoma of the anterior pituitary. Harvey Cushing (1869–1939), pioneer neurosurgeon at the Peter Bent Brigham Hospital, Boston, USA.

fascia of Denonvilliers Fascia which separates the prostate from the rectum. Charles Pierre Denonvilliers (1808–1872), Professor of Anatomy in Paris.

the arcuate line of Douglas The lower margin of the posterior rectus sheath.

pouch of Douglas The recto-uterine peritoneal pouch.

James Douglas (1675–1742); anatomist and obstetrician in London.

Dupuytren's contracture Contraction and fibrosis of the palmar (and occasional plantar) fascia. Baron Guillaume Dupuytren (1777–1835), surgeon to the Hôtel Dieu, Paris.

Edinger–Westphal nucleus Supplies the parasympathetic fibres of the oculomotor nerve. Ludwig Edinger (1855–1918), German anatomist and neurologist. Karl Westphal (1833–1890), neurologist in Berlin. Described the nucleus in the adult two years after this had been demonstrated in the fetus by Edinger.

Erb–Duchenne paralysis Results from injury to the C5 and C6 roots of the brachial plexus. Wilhelm Erb (1840–1921), Professor of Neurology in Heidelberg. G. B. A. Duchenne (1806–1875), neurologist in Paris; also described Duchenne's muscular dystrophy.

Eustachian tube The pharyngotympanic tube. Bartolomeo Eustachi (1524–1574), Professor of Anatomy, Rome.

Fallopian tube The uterine tube. Gabrielle Fallopio (1523–1562), Professor of Anatomy in Padua and a pupil of Vesalius.

Fallot's tetralogy Congenital heart disease comprising pulmonary stenosis, right ventricular hypertrophy, ventricular septal defect and overriding of the aorta. Etienne Fallot (1850–1911), Professor of Medicine, Marseilles.

duct of Gartner Mesonephric duct remnant in the female. Hermann Gartner (1785–1827), Danish surgeon.

Gimbernat's ligament The lacunar portion of the inguinal ligament. Manuel Gimbernat (1734–1816), Professor of Anatomy in Barcelona.

column of Goll The medial tract of the posterior column of the spinal cord (fasiculus gracilis). Friedrich Goll (1829–1903), both a neurologist and anatomist; Professor of Anatomy in Zurich.

Hartmann's pouch Dilatation of the gall bladder proximal to its neck.
Henri Hartmann (1860–1952), Professor of Surgery in the Faculty of Medicine, Paris.

the antrum of Highmore The maxillary sinus.
Nathaniel Highmore (1613–1685), physician in Sherborne, Dorset. His claim to fame is tenuous since the maxillary sinus was well known to previous anatomists and was illustrated by Leonardo da Vinci.

Hilton's law Nerves crossing a joint supply the muscles acting on the joint and the joint itself.
John Hilton (1805–1878), surgeon at Guy's Hospital, London.

Horner's syndrome Ptosis and constriction of the pupil resulting from interruption of the sympathetic innervation to the eyelids and pupil.
Johann Horner (1831–1886), Professor of Ophthalmology, Zurich.

Houston's valves The three lateral folds of the rectum.
John Houston (1802–1845), lecturer in Surgery, Dublin.

Hunter's canal The subsartorial canal.
John Hunter (1728–1793), surgeon at St George's Hospital, London. He described ligation of the femoral artery in the subsartorial canal for popliteal aneurysm.

Killian's dehiscence The potential gap between the two parts of the interior constrictor muscle through which a pharyngeal pouch protrudes.
Gustav Killian (1860–1921), ENT surgeon in Berlin.

Klumpke's paralysis Injury to the lowest root of the brachial plexus.
Augusta Dejerine-Klumpke (1859–1927), neurologist in Paris. Married another famous neurologist, Joseph Dejerine.

Kocher's incision The subcostal approach to the gall bladder.

Kocher's manoeuvre Mobilization of the duodenum by incising its lateral peritoneal attachment.

Kocher's method For reduction of dislocation of the shoulder.
Theodore Kocher (1841–1917), Professor of Surgery in Berne, Switzerland.

islands of Langerhans Clumps of insulin-secreting cells in the pancreas.
Paul Langerhans (1847–1888), Professor of Pathology, Freiburg. He described the islet cells in his doctorate studies in 1869, at the age of 22.

nerve of Latarjet The fibres of the vagus nerve which supply the gastric antrum.
André Latarjet (1876–1947), Professor of Anatomy, Lyons.

Ludwig's angina Infection of the submandibular region.
Wilhelm Ludwig (1790–1865), Professor of Surgery in Tubingen, Germany.

McBurney's point Two-thirds of the way laterally along the line from the umbilicus to the anterior superior iliac spine; the usual site of maximum of tenderness in acute appendicitis and the centre of the skin incision for appendicectomy.
Charles McBurney (1845–1913), Professor of Surgery, New York. A pioneer of early surgery in acute appendicitis.

Mackenrodt's ligaments The transverse cervical (or cardinal) ligaments of the female pelvis.
Alwin Mackenrodt (1859–1925), Professor of Gynaecology, Berlin.

the vein of Mayo A constant vein which crosses the junction of the pylorus with the duodenum.
Charles Mayo (1865–1939); with his father and his brother, William, founded the Mayo Clinic, Rochester, Minnesota. The vein of Mayo was described earlier by Latarjet (q.v.).

Meckel's cartilage The cartilage of the first branchial arch.

Meckel's diverticulum The remains of the embryonic vitellointestinal duct and present in approximately 2 per cent of subjects.
Johann Meckel (1781–1833), Professor of Anatomy in Halle, Germany. His grandfather was Professor of Anatomy in Berlin who described the pterygopalatine ganglion and the dural space with contains the ganglion of the trigeminal nerve. Johann's father was also Professor of Anatomy in Halle.

Meibomian glands The tarsal glands of the eyelid. If blocked, they distend into Meibomian cysts.

Heinrich Meibom (1638–1700), Professor of Medicine in Helmstadt.

Meissner's plexus The nerve plexus in the submucosal layer of the intestine.
George Meissner (1829–1905), anatomist and physiologist, Professor successively in Basle, Freiburg and Gottingen. He also described Meissner's corpuscles, the cutaneous sensory end-organs.

columns of Morgagni Vertical columns of mucosa in the anal canal.

hydatid of Morgagni The appendix epididymis, the remnant of the mesonephros.
Giovanni Morgagni (1682–1771), Professor of Anatomy in Padua—a post he held for 59 years!

Morison's pouch The right subhepatic space.
James Rutherford Morison (1853–1939), Professor of Surgery, University of Durham. He had served as a surgical dresser to Joseph Lister.

glands of Montgomery The modified sebaceous glands of the areola of the nipple.
William Montgomery (1797–1859), practised obstetrics in Dublin.

Mullerian duct The paramesonephric duct.
Johannes Muller (1801–1858), Professor of Anatomy, Berlin.

Nelaton's line A line joining the anterior superior iliac spine to the ischial tuberosity. The greater trochanter normally lies distal to this line.
Auguste Nelaton (1807–1873), Professor of Surgery in Paris and surgeon to Napoleon III.

sphincter of Oddi The sphincter around the termination of the common bile duct.
Ruggero Oddi (1845–1906), surgeon in Rome. The sphincter had already been described by Glisson in the 17th century!

Pancoast's syndrome Invasion of the brachial plexus by an apical tumour of the lung.
H. K. Pancoast (1875–1939), Professor of Radiology in the University of Pennsylvania, the first such appointment in the USA.

Parkinson's disease Tremor and rigidity due to lesions of the substantia nigra.
James Parkinson (1755–1824), medical practitioner, Shoreditch, London. His small book *An Essay on the Shaking Palsy* in 1817 was based mainly on close observation of one case.

Passavant's ridge Produced by contraction of the superior pharyngeal constrictor in deglutition.
Philipp Passavant (1815–1893), surgeon in Frankfurt.

Poupart's ligament The inguinal ligament.
Francois Poupart (1661–1708), surgeon at the Hôtel Dieu, Paris.

Pringle's manoeuvre Compression of the hepatic artery at the foramen of Winslow in the control of haemorrhage.
James Hogarth Pringle (1863–1941), surgeon at the Royal Infirmary, Glasgow.

Queckenstedt's test Compression of the internal jugular vein during lumbar puncture produces a rise in C.S.F. pressure.
H. H. G. Queckenstedt (1876–1918), neurologist in Rostock. He describes his test in 1916 while serving in the German army in World War I. He was killed in a road accident two days before the armistice.

cave of Retzius The retropubic space.
Andreas Retzius (1797–1860), Professor of Anatomy at the Karolinska Institute, Stockholm.

Romberg's sign Ataxia when the eyes are closed because of loss of position sense. Characteristic of posterior column lesions.
Moritz Romberg (1795–1873), Director of the University Hospital in Berlin and author of the first systematic book on neurology.

duct of Santorini The accessory pancreatic duct.
Giovanni Santorini (1681–1737), Professor of Anatomy in Venice.

Sappey's plexus The plexus of lymphatics below the nipple.
Marie Sappey (1810–1896), Professor of Anatomy, Paris.

Scarpa's fascia The fibrous layer of superficial fascia of the lower abdomen.
Antonio Scarpa (1747–1832), Professor of Anatomy and Surgery at Pavia, Italy. His name is also attached to the femoral triangle.

canal of Schlemm The sinus venosus sclerae, draining the aqueous humour from the anterior chamber of the eye.
F. S. Schlemm (1795–1858), Professor of Anatomy, Berlin.

Semon's law In partial damage of the recurrent

laryngeal nerve, the abductors of the vocal cords are affected more than the adductors.
Felix Semon (1849–1921), graduated in Berlin but emigrated to England and became laryngologist at St Thomas's Hospital, London.

Tenon's capsule The fascial sheath of the eye.
Jacques Tenon (1724–1816), surgeon at the Salpetrière, Paris.

the suspensory ligament of Treitz Peritoneal fold from the right crus of the diaphragm to the duodenal termination.
Wenzel Treitz (1819–1872), Professor of Pathology in Prague.

Trendelenburg's test A clinical test for hip stability.
Friedrich Trendelenburg (1844–1924), Professor of Surgery, Leipzig. He attempted a pulmonary embolectomy (unsuccessfully) in 1908.

the bloodless fold of Treves The ileocaecal fold.
Sir Frederick Treves (1853–1923), surgeon at the London Hospital. Drained the appendix abscess of King Edward VII in 1902.

ampulla of Vater The ampulla of the common bile duct.
Abraham Vater (1684–1751), Professor of Anatomy, Wittenberg.

Volkmann's contracture Produced by ischaemic fibrosis of the forearm muscles.
Richard von Volkmann (1830–1889), Professor of Surgery in Halle and one of the pioneers of Lister's antiseptic surgical technique.

Waldeyer's ring The ring of lymphoid tissue comprising the nasopharyngeal tonsil, the palatine tonsils and the lymphoid nodules on the dorsum of the tongue.
Heinrich Waldeyer (1836–1921), Professor of Anatomy, first in Strasbourg and then Berlin. He also described the plasma cell in 1875.

Wharton's duct The duct of the submandibular salivary gland.
Thomas Wharton (?1616–1673), physician at St Thomas's Hospital, London. His name is also given to the mucoid substance of the umbilical cord (Wharton's jelly).

circle of Willis The arterial anastomosis at the base of the brain.
Thomas Willis (1621–1675), practised medicine in Oxford and then London. Buried in Westminster Abbey.

foramen of Winslow The opening into the lesser sac (epiploic foramen).
Jacob Winslow (1669–1760), born in Denmark and became Professor of Anatomy and Surgery in Paris.

Wirsung's duct The main pancreatic duct.
Johann Wirsung (1600–1643), Professor of Anatomy in Padua, where he was assassinated!

Wolffian body and ducts The mesonephros and its ducts.
Caspar Wolff (1733–1794), born in Berlin, Professor of Anatomy in St Petersburg. One of the pioneers of embryology.

Wormian bones Occasional accessory bones between the parietal and occipital bones.
Ole Worm (1588–1654), Professor of Anatomy, Copenhagen.

Index